BIODIVERSITY HOTSPOT OF THE IRANO-ANATOLIAN REGION

Cover photo credits:

Top left: Iranian Cheetah. Tasnim News Agency. https://en.wikipedia.org/wiki/Asiatic_cheetah#/media/File:Iranian_Cheetah_roars.jpg. https://creativecommons.org/licenses/by/4.0/

Top middle: Morel (Morchella esculenta agg.). Photo by Andreas Kunze. https://commons.wikimedia.org/wiki/File:2011-04-30_Morchella_esculenta.jpg. https://creativecommons.org/licenses/by-sa/3.0/deed.en

Top right: Azolla filiculoides in the Jardim Botânico do Porto, Portugal. Photo by Krzysztof Ziarnek, Kenraiz. https://commons.wikimedia.org/wiki/File:Azolla_filiculoides_kz01.jpg. https://creativecommons.org/licenses/by-sa/4.0/deed.en

Bottom: Turkey: aerial view of the breathtaking landscape around the town of Dogubayazıt, the easternmost district of Turkey, bordering Iran, seen from the top of the mountain of the famous Ishak Pasha Palace.

Biodiversity Hotspots of the World

BIODIVERSITY HOTSPOT OF THE IRANO-ANATOLIAN REGION

T. Pullaiah, PhD

Editor

APPLE ACADEMIC PRESS

First edition published 2027

Apple Academic Press Inc.
1265 Goldenrod Circle, NE,
Palm Bay, FL 32905 USA

760 Laurentian Drive, Unit 19,
Burlington, ON L7N 0A4, Canada

CRC Press
2385 NW Executive Center Drive,
Suite 320, Boca Raton FL 33431

4 Park Square, Milton Park,
Abingdon, Oxon, OX14 4RN, UK

Apple Academic Press exclusively co-publishes with CRC Press, an imprint of Taylor & Francis Group, LLC

Library and Archives Canada Cataloguing in Publication

CIP data on file with Canada Library and Archives

Library of Congress Cataloging-in-Publication Data

CIP data on file with US Library of Congress

ISBN: 978-1-77964-575-3 (hbk)
ISBN: 978-1-77964-576-0 (ebk)
DOI: 10.1201/9781779645760

ABOUT THE BOOK SERIES: BIODIVERSITY HOTSPOTS OF THE WORLD

Series Editor: **T. Pullaiah, PhD**
Former Professor, Department of Botany, Sri Krishnadevaraya University, Andhra Pradesh, India

The term 'biodiversity' came into common usage after the 1986 National Forum on Biodiversity, held in Washington, DC, and the publication of selected papers from that event, titled *Biodiversity,* edited by Wilson (1988). Wilson credits Walter G. Rosen for coining the term. In 1988, British ecologist Norman Myers published a seminal paper identifying 10 tropical forest "hotspots." Myers described ten tropical forest "hotspots" on the basis of extraordinary plant endemism and high levels of habitat loss, albeit without quantitative criteria for the designation of "hotspot" status. A subsequent analysis added eight additional hotspots, including four from Mediterranean-type ecosystems (Myers, 1990).

A biodiversity hotspot is a biogeographic region that is both a significant reservoir of biodiversity and is threatened with destruction. Conservation International introduced two strict quantitative criteria to qualify as a hotspot: a region had to contain at least 1,500 vascular plants as endemics (>0.5% of the world's total), and it had to have 30% or less of its original vegetation (extent of historical habitat cover) remaining. These efforts culminated in an extensive global review (Mittermeier et al., 1999) and scientific publication (Myers et al., 2000) that introduced seven new hotspots on the basis of both the better-defined criteria and new data. In 2005, an additional analysis brought the total number of biodiversity hotspots to 34, based on the work of nearly 400 specialists. In 2011, the Forests of East Australia was identified as the 35th hotspot by a team of researchers from the Commonwealth Scientific and Industrial Research Organisation (CSIRO) working with Conservation International. In February 2016, the North American Coastal Plain was recognized as meeting the criteria and became the Earth's 36th hotspot.

Very few books have been written/edited to document the biodiversity of these hotspots. Hence, these new volumes which aim to be a reference source for all foresters, wildlife biologists, conservationists, botanists, zoologists, policymakers, and others concerned with the conservation of the biodiversity of our planet.

REFERENCES

Mittermeier, R. A.; Myers, N.; Robles-Gil, P.; Mittermeier, C. G. *Hotspots: Earth's Biologically Richest and Most Endangered Terrestrial Ecoregions*; CEMEX/Agrupación Sierra Madre: Mexico City, 1999.

Myers, N. The Biodiversity Challenge: Expanded Hot-Spots Analysis. *Environmentalist* **1990,** *10,* 243–256.

Myers, N.; Mittermeier, R. A.; Mittermeier, C. G.; da Fonseca, G. A. B.; Kent, J. Biodiversity Hotspots for Conservation Priorities. *Nature* **2000,** *403,* 853–858. https://doi.org/10.1038/35002501

Wilson, E. O.; Peter, F. M., Eds. Biodiversity; The National Academies Press: Washington, DC, 1988. https://doi.org/10.17226/989

BIODIVERSITY HOTSPOTS OF THE WORLD

North and Central America

- California Floristic Province
- Madren Pine-Oak woodlands
- Mesoamerica
- North American Coastal Plain

The Carribbean

- Caribbean Islands

South America

- Atlantic Forest
- Cerrado
- Chilean Winter Rainfall-Valdivian Forests
- Tumbes-Choco-Magdalena
- Tropical Andes

Europe

- Mediterranean Basin

Africa

- Cape Floristic Region
- Coastal Forests of Eastern Africa
- Eastern Afromontane
- Guinean Forests of West Africa
- Horn of Africa
- Madagascar and Indian Ocean Islands
- Maputaland-Pondoland-Albany
- Succulent Kaoo

Central Asia

- Mountains of Central Asia

South Asia

- The Himalaya/Eastern Himalaya
- Indo-Burma
- Western Ghats–Sri Lanka

Southeast Asia and Asia-Pacific

- East Melanesian Islands
- New Caledonia
- New Zealand
- Philippines
- Polynesia–Micronesia
- Eastern Australian Temperate Forests

- Southwest Australia
- Sundaland and Nicobar Islands
- Wallacea

East Asia

- Japan
- Mountains of Southwest China

West Asia

- Caucasus
- Irano-Anatolian

BOOKS IN THE SERIES TO DATE:

- **Biodiversity Hotspot of the Western Ghats and Sri Lanka**
  Editor: T. Pullaiah
- **Biodiversity Hotspot of the Himalaya**
  Editor: T. Pullaiah
- **Biodiversity Hotspot of the Indo-Burma Region**
  Editor: T. Pullaiah
- **Biodiversity Hotspot of the Coastal Forests of Eastern Africa**
  Editors: T. Pullaiah and Ally J. K. Nkawabi
- **Biodiversity Hotspot of the Mediterranean Basin**
  Editors: T. Pullaiah and Loris Galli
- **Biodiversity Hotspot of the Eastern Afromontane**
  Editors: T. Pullaiah and Teshome Soromessa
- **Biodiversity Hotspot of the Sundaland**
  Editor: T. Pullaiah
- **Biodiversity Hotspot of the Caucasus**
  Editors: T. Pullaiah and Levan Mumladze
- **Biodiversity Hotspot of the Irano-Anatolian Region**
  Editor: T. Pullaiah

ALSO FORTHCOMING FROM THE BOOK SERIES:

- **Biodiversity Hotspot of the Caribbean Islands**
  Editors: T. Pullaiah and Chinthapalli Bhaskar Rao
- **Biodiversity Hotspot of the North American Coastal Plain**
  Editors: T. Pullaiah and Edwin Bridges
- **Biodiversity Hotspot of the Eastern Australian Temperate Forests**
  Editors: T. Pullaiah and Chris Dickman

- **Biodiversity Hotspot of the of Philippines**
  Editors: T. Pullaiah, Crusty E. Tinio, and Lerma Sj. Maldia
- **Biodiversity Hotspot of Madagascar and Indian Ocean Islands**
  Editors: T. Pullaiah and Zo Lalaina Rakotobe
- **Biodiversity Hotspot of the Guinean Forest of West Africa**
  Editors: T. Pullaiah and Kouami Kokou
- **Biodiversity Hotspot of the Tropical Andes**
  Editors: T. Pullaiah and Monica Moraes
- **Biodiversity Hotspot of the Atlantic Forest**
  Editors: T. Pullaiah and Filipe Nascimento
- **Wallacea: An Enigmatic Biodiversity Hotspot**
  Editors: B. Michaux and T. Pullaiah
- **Biodiversity Hotspot of the Cerrado**
  Editors: T. Pullaiah and Silvia Cristina de Jesus
- **Biodiversity Hotspot of the Chilean Winter Rainfall Valdivian Forest**
  T. Pullaiah and Cristian Echeverria
- **Biodiversity Hotspot of Mesoamerica**
  Editors: T. Pullaiah and Irene Pisanty
- **Biodiversity Hotspot of Tumbes-Choco-Magdalena**
  Editors: T. Pullaiah, Diego Tirira, and Pablo Lozano
- **Biodiversity Hotspot of the Horn of Africa**
  Editors: Teshome Soromessa and T. Pullaiah

ABOUT THE EDITOR

T. Pullaiah, PhD
Department of Botany, Sri Krishnadevaraya University, Anantapur, Andhra Pradesh, India

T. Pullaiah, PhD, is a former Professor at the Department of Botany, Sri Krishnadevaraya University, Andhra Pradesh, India, where he taught for more than 35 years. He has held several positions at the university, including Dean, Faculty of Biosciences; Head, Department of Botany; Head, Department of Biotechnology; and Member, Academic Senate. He served as President of the Indian Botanical Society (2014), President of the Indian Association for Angiosperm Taxonomy (2013), and was elected Fellow of the Andhra Pradesh Akademi of Sciences. He was awarded the Panchanan Maheshwari Gold Medal, the Prof. P. C. Trivedi Medal, the Dr. G. Panigrahi Memorial Lecture Award of the Indian Botanical Society, the Prof. Y. D. Tyagi Gold Medal of the Indian Association for Angiosperm Taxonomy, and a Best Teacher Award from the Government of Andhra Pradesh. Under his guidance, 54 students obtained their doctoral degrees. He has authored 80 books, edited 50 books, and published over 340 research papers. His books include *Mangroves: Biodiversity, Livelihoods and Conservation* (Springer); *Ethnobotany of India* (five volumes, Apple Academic Press); *Global Biodiversity* (four volumes, Apple Academic Press); *Monograph on Brachystelma and Ceropegia in India* (CRC Press); *Apomixis in Angiosperms* (CRC Press); *Invasive Alien Species* (four volumes, Wiley International); *Advances in Cell and Molecular Diagnostics* (Elsevier); *Camptothecin and Camptothecin Producing Plants* (Elsevier); *Paclitaxel* (Elsevier); *Red Sanders: Silviculture and Conservation* (Springer); *Genetically Modified Crops* (two volumes, Springer); *Flora of Andhra Pradesh* (five volumes); *Flora of Eastern Ghats* (seven volumes); *Flora of Telangana* (three volumes); *Encyclopedia of World Medicinal Plants* (seven volumes, 2nd edition); and *Encyclopedia of Herbal Antioxidants* (three volumes). Dr. Pullaiah is editor of the *Journal Plant Biosystems*

(Springer) and *Encyclopedia of Life Sciences* (Wiley). Dr. Pullaiah is also the editor of the book series *Biodiversity Hotspots of the World.* He was a member of the Species Survival Commission of the International Union for Conservation of Nature (IUCN). Professor Pullaiah received his PhD from Andhra University, India, attended Moscow State University, Russia, and worked as a Postdoctoral Fellow during 1976–1978.

CONTENTS

CONTRIBUTORS

Meral Avcı
Department of Geography, Istanbul University, İstanbul, Türkiye

Sedat Avcı
Department of Geography, Istanbul University, İstanbul, Türkiye

Meenam Bhatia
Department of Botany, Daulat Ram College, University of Delhi, Delhi, India

Meriç Çakır
Department of Soil Science and Ecology, Faculty of Forestry, Çankırı Karatekin University, Çankırı, Türkiye

Tolga Coşkun
Technology Transfer Office, Ömer Halisdemir University, Niğde, Türkiye; Department of Biology, Faculty of Arts and Sciences, Middle East Technical University (METU), Ankara, Türkiye

Anshul Dhyani
Department of Botany, Hindu College, University of Delhi, Delhi, India

Ibrahim Tamer Emecan
Vocational School of Technical Sciences, Istanbul University–Cerrahpaşa, Istanbul, Türkiye

Anıl Esmeyar
School of Foreign Languages, Cappadocia University, Nevşehir, Türkiye

Priti Giri
Department of Botany, University of Delhi, Delhi, India

Halil Güngör
Department of Biology, Faculty of Science, Muğla Sıtkı Koçman University, Muğla, Türkiye

Mihrican Yaratanakul Güngör
Health Services Vocational School, Kocaeli University, Kocaeli, Türkiye

Pınar Gürbüzer
Department of Hydrobiology, Faculty of Fisheries, Sinop University, Sinop, Türkiye; Department of Astronautical Engineering, University of Turkish Aeronautical Association, Ankara, Türkiye

Ali Thorabi Haghighi
Water, Energy and Environmental Engineering Research Unit, Faculty of Technology, University of Oulu, Oulu, Finland

Hossein Mostafavi
Department of Biodiversity and Ecosystem Management, Environmental Sciences Research Institute, Shahid Beheshti University, Tehran, Iran

Milad Nouri
Soil and Water Research Institute, Agricultural Research, Education and Extension Organization (AREEO), Karaj, Iran

Najmeh Tabasinezhad
Department of Biology, Faculty of Sciences, Shahid Bahonar University of Kerman, Kerman, Iran

Azad Teimori
Department of Biology, Faculty of Sciences, Shahid Bahonar University of Kerman, Iran

İbrahim Türkekul
Department of Biology, Faculty of Arts and Sciences, Tokat Gaziosmanpaşa University, Tokat, Türkiye

Prem L. Uniyal
Department of Botany, University of Delhi, Delhi, India

Himani Yadav
Department of Botany, University of Delhi, Delhi, India

Hasan Yaşayacak
Department of Biodiversity, General Directorate of Nature Conservation and National Parks, Ankara, Türkiye

Nese Yilmaz
Department of Freshwater Resources and Management, Faculty of Aquatic Sciences, Istanbul University, Fatih, Istanbul, Türkiye

ABBREVIATIONS

CEPF	Critical Ecosystem Partnership Fund
EDW	elevation-dependent warming
GLWD	Global Lakes and Wetlands Database
IoE	Institution of Eminence
ITCZ	Intertropical Convergence Zone
LDR	longwave downwelling radiation
NFMC	National Facility for Marine Cyanobacteria
NPK	nitrogen, phosphorus, and potassium
PE	potential evapotranspiration
PSA	public service announcement
UGC	University Grants Commission
WRB	World Reference Base

PREFACE

The concept of biodiversity hotspots—regions marked by exceptional endemism and severe habitat loss—was introduced to highlight areas most urgently in need of conservation attention. Forests and vegetation are being degraded at an alarming rate, resulting in further loss of biodiversity. There is growing concern over the loss of this precious gift of nature. Hence, there is an urgent need to take stock of biodiversity in these hotspots and implement appropriate conservation measures. The present series is a humble attempt to summarize the biodiversity of these regions. Very few books have been written or edited to document the biodiversity of these hotspots, and none comprehensively covers all recognized biodiversity hotspots.

Some aspects could not be covered due to two reasons. First, expertise on certain topics, such as the entire Irano-Anatolian region, is not readily available. Second, some experts were too busy with their professional commitments.

I thank all the contributors for their critical review of the biodiversity of the Irano-Anatolian region. I sincerely thank all the chapter authors for their contributions to this volume. I also thank Rakesh Kumar and Sandra Sickels for their support and encouragement. Their patience and perseverance are greatly appreciated.

—***Editor***

CHAPTER 1

GEOGRAPHY, GEOLOGY, AND CLIMATE OF THE IRANO-ANATOLIAN HOTSPOT

MERAL AVCI* and SEDAT AVCI*

Department of Geography, Istanbul University, İstanbul, Türkiye

*Corresponding authors

ABSTRACT

The Irano-Anatolian Hotspot extends in a northwest-southeast direction, spanning 27°34'–41°44' N latitude and 30°59'–61°31' E longitude, covering approximately 1.39 million km^2. This hotspot forms part of the Iran-Alp-Himalayan orogenic belt, encompassing portions of Anatolia and Iran. The western section of the Irano-Anatolian Hotspot has an average elevation of approximately 1500 meters, while the elevation in the eastern section gradually increases to 2000 meters. Aridity represents a major climatic determinant influencing the floristic composition and distribution of natural vegetation in the Irano-Anatolian Hotspot. Despite its location in an arid and semi-arid zone, the Irano-Anatolian Hotspot contains essential rivers and lakes, serving as critical water resources for both ecosystems and economic activities. The dominant soil types in the region include Leptosols and Calcisols. The Irano-Anatolian Hotspot belongs to the Irano-Turanian Floristic Region, which is part of the Holarctic Floristic Realm of Eurasia and serves as a key center of endemism. The Irano-Turanian Floristic Region supports exceptional biological diversity, driven by complex topography, distinct elevational gradients associated with extensive mountain ranges, and a unique paleogeographic history. Given the potential impacts of climate change and the increasing anthropogenic

Biodiversity Hotspot of the Irano-Anatolian Region. T. Pullaiah (Ed.)

DOI: 10.1201/9781779645760-1

pressures, the Irano-Anatolian Hotspot represents a high-priority region for conservation efforts.

1.1 INTRODUCTION

The Irano-Anatolian hotspot, part of the Holarctic Floristic Realm, is one of the 36 globally recognized biodiversity hotspots (Hoffman et al., 2016). It is bordered by the Caucasus Hotspot to the north and the Mediterranean Basin Hotspot to the west (Figure 1.1). Geographically, the Irano-Anatolian Hotspot extends approximately in a northwest-southeast direction, spanning between 27°34' and 41°44' N latitude and 30°59' and 61°31' E longitude, covering an area of approximately 1.39 million km². The largest portion of this hotspot lies within Türkiye and Iran, while smaller portions extend into Georgia, Armenia, Azerbaijan, and Turkmenistan to the north, as well as Iraq to the south.

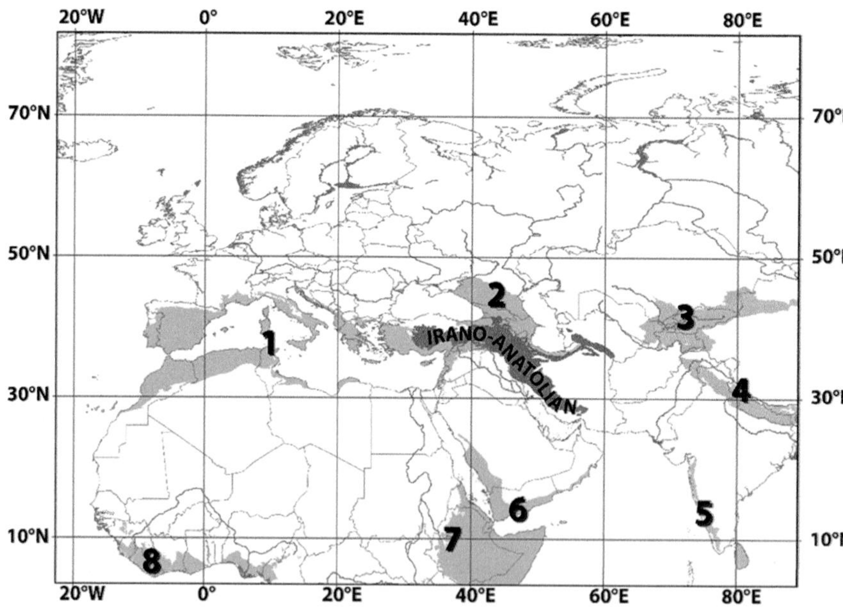

FIGURE 1.1 Locations of Irano-Anatolian hotspot areas and nearby hotspots (1: Mediterranean Basin; 2: Caucasus; 3: Mountains of Central Asia; 4: Himalaya; 5: Western Ghats and Sri Lanka; 6:Horn of Africa; 7: Eastern Afromontane; 8: Guinea Forest of West Africa).

The northern boundary of the Irano-Anatolian Hotspot follows the southern slopes of the North Anatolian Mountains and the Caucasus Mountains. Further east, the high elevations of the Alborz Mountains, located south of the Caspian Sea, are included within the hotspot. This mountainous region extends eastward, ending at the Kopet Mountains south of Ashgabat. The southern boundary begins in the west, north of the Taurus Mountains, and forms a broad arc as it extends eastward, encompassing the Southeastern Taurus range. In the eastern extension of this arc, the Zagros Mountains, which generally trend in a northwest-southeast direction, continue toward the southeast. In Anatolia, the area between the northern and southern mountain ranges consists of plains and plateaus, which are located within the Irano-Anatolian Hotspot. In Iran, the region between the Alborz and Zagros Mountains also takes the form of a plateau. Further east, two vast plains (Dasht-i Kavir and Dasht-i Lut), along with the Central Iranian Range, cause the Irano-Anatolian Hotspot to split into two distinct branches (Figure 1.2).

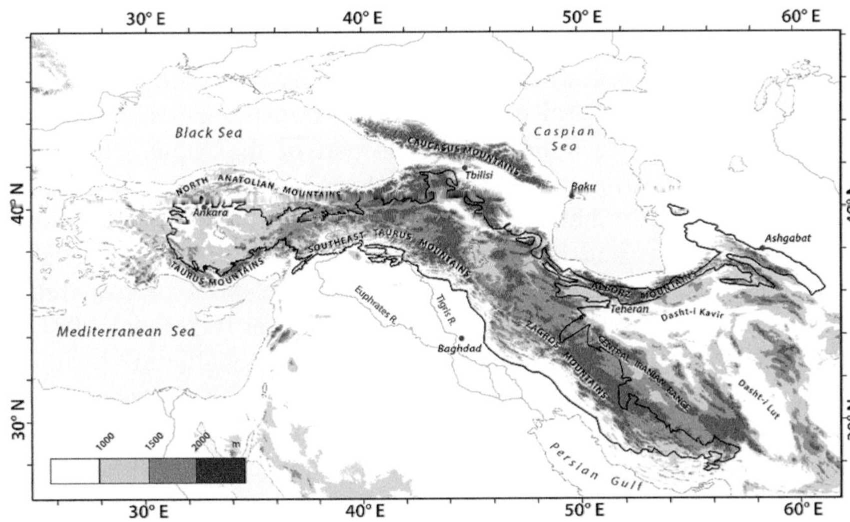

FIGURE 1.2 Physiography of Irano-Anatolian hotspot.

1.2 GEOLOGY AND GEOMORPHOLOGY

According to the theory of plate tectonics, the Pangaea supercontinent was divided into two parts by the east-west oriented Tethyan Ocean.

The northern part is known as Laurasia, while the southern part is called Gondwanaland. There are multiple theories regarding the opening and closing of the Tethyan Ocean between the Late Proterozoic and Neogene periods. The closure of the Tethyan Ocean led to the formation of the Alpine-Himalayan orogenic belt (Ghorbani, 2021; Stampfli and Borel, 2002; Stampfli et al., 2001; Şengör and Yilmaz, 1981). The Irano-Anatolian Hotspot is spread across the section of the orogenic belt that includes parts of Anatolia and Iran. The western part of the Irano-Anatolian Hotspot is located in the Anatolia area of Türkiye, which is part of the Asian continent. In this region, the area between the North Anatolian Mountains, which surround northern Anatolia, and the Taurus Mountains to the south, consists of depressions separated by low ridges.

The basement of this relatively lower area is formed by the Paleozoic metamorphic massif. During different periods, this formation was occupied by the Tethys Sea, and its surface has been covered by younger formations from the Paleogene and Neogene periods. Where this massif has fractured, magma has risen to the surface, and volcanism has occurred repeatedly during different geological periods (Kuzucuoğlu et al., 2019; Yılmaz et al., 1987, 1998). As a result of this, volcanic mountains, volcanic cones, and features like craters have formed, while the existing ones have become younger.

A similar formation is also observed in the Iranian section of the Irano-Anatolian Hotspot. Tectonically, Iran is part of the Alpine-Himalayan orogenic belt. The process related to the closure of the multi-phase Tethys Ocean explains the formation of Iran's Paleozoic, Mesozoic, and Cenozoic sedimentary basins and magmatic zones. The characteristics of Paleo-Tethys represent the Paleozoic conditions, while those of Neo-Tethys correspond to the Mesozoic and Cenozoic conditions (Ghorbani, 2021).

The western part of the Irano-Anatolian Hotspot within the Anatolian region has an average elevation of approximately 1500 meters. Moving eastward, the average elevation increases, reaching up to 2000 meters. The high central plateau, situated between the North Anatolian Mountains to the north and the Taurus Mountains to the south, is divided into eastern and western sections by a topographic threshold known as the "Anatolian Diagonal," which serves as a biogeographical boundary. In the westernmost part of the Irano-Anatolian Hotspot, the lowest area in Central Anatolia is the Tuz Lake Basin and its surroundings, which lack external drainage. Low-elevation thresholds separate this basin from other lowland areas such as the Konya-Ereğli Plain and the Cihanbeyli Plateau, both with elevations below 1000 meters, as well as the Develi Plain, which stands at 1100 meters. The highlands of Central Anatolia predominantly consist of volcanic terrain, including notable volcanic mountains such as Melendiz Mountain (3268 m), Hasan Mountain (2963 m), and Erciyes Mountain (3917 m). In Eastern Anatolia, which is topographically higher than Central Anatolia, the Taurus Mountains split into two branches. The southern branch, forming a broad arc, is referred to as the Southeast Taurus Mountains and lies within the Irano-Anatolian Hotspot. Some of its major peaks include Akdağ (2510 m), Maden Mountain (2230 m), Genç Mountain (2940 m), and Hakkâri Mountain (4135 m). The northern branch extends toward the northeast and comprises mountain ranges such as the Munzur Mountains (3366 m) and the Karasu-Aras Mountains, which include Palandöken Mountain (3167 m), Çakmak Mountain (3063 m), and Perili Mountain (3231 m). The eastern half of the Irano-Anatolian Hotspot within Türkiye is deeply incised by the Euphrates and Tigris rivers and their tributaries. In this region, flatlands are generally confined by depressions. The elevations of these plains vary significantly, with some lying below 1000 meters and others reaching up to 2000 meters (e.g., Iğdır Plain at 850 m, Malatya Plain at 900 m, Uluova at 1000 m, Erzincan Plain at 1250 m, Erzurum Plain at 1850 m, Muş Plain at 1250 m, and Yüksekova at 2000 m). Lake Van, the largest lake in Türkiye (3713 km^2), is located in a tectonic basin. Lava flows from Mount Nemrut blocked its outflow, leading to the lake's formation. The surface elevation of Lake Van is 1646 meters above sea level. West of the lake, a chain of volcanic mountains extends in a roughly southwest-northeast direction. Along this alignment are several major volcanic peaks, including Mount Nemrut (2935 m), Mount Süphan (4058 m), Aladağ (3358 m), Mount Tendürek (3533 m), and Mount Ağrı (Ararat) (5137 m), the latter being the highest peak in Türkiye.

The eastern portion of the Irano-Anatolian Hotspot is predominantly located in Iran. This elevated region, also referred to as the Iranian Plateau, is bounded to the south by the Zagros Mountains, which extend in a roughly northwest-southeast direction. The Zagros Mountains represent the eastward continuation of the Southeastern Taurus Mountains in Anatolia. When considering their Anatolian segment, these mountain ranges extend over a distance of more than 1500 km, reaching as far south as the Gulf of Oman. The Zagros Mountains are among the relatively young morphological features shaped by the north-south collision between the Arabian Plate and the Laurasian Plate (Ghorbani, 2021).

The formation of the Zagros Mountains belt is the result of a prolonged and intricate orogenic process that commenced in the Paleozoic era and continues to the present day. Before subduction commenced in the Late Cretaceous, the Neo-Tethys Ocean was a narrow but spreading oceanic basin positioned between the Arabian and Eurasian Plates. At that time, the northeastern margin of the Arabian Plate formed a passive continental setting, which later evolved into the Zagros Mountains. From the Paleozoic to the Cenozoic, this continental shelf experienced the deposition of thick sequences of shallow-marine limestones, dolomites, siltstones, shales, and evaporitic sediments (Kent, 1978; Noroozi et al., 2020b; Sembroni et al., 2024). The Zagros Mountains have an average elevation of approximately 2000 meters. Notable peaks within the range include Dena (4409 m) in the south, as well as Zard-Kuh (4220 m) and Oshtorankuh (4050 m) in the north (Noroozi et al., 2020b). The Zagros Mountains are separated from the more easterly Central Iran Range by the Isfahan-Babak Valley. Both the Central Iran Range and this valley lie outside the Irano-Anatolian Hotspot. The Iranian Plateau has been shaped by volcanic activity, seismic events, and fluvial inundation and flooding, particularly during the period spanning from the late Pliocene to the early Quaternary (Ghorbani, 2013).

To the north of the Zagros Mountains, the Lake Urmia Depression is located. This lake, situated at an elevation of 1250 meters above sea level, has a surface area that fluctuates between 4750 and 6100 km^2 (Eimanifar and Mohebbi, 2007). Southeast of Lake Urmia, the mountainous region that includes Alam-Kuh (4850 m) gradually loses elevation eastward, eventually dropping below 1000 meters as it transitions into the Dasht-e Kavir in the southeast. North of the Irano-Anatolian Hotspot, the Alborz Mountains and the Kopet Dagh range are located. Mount Damavand, the highest peak in the Alborz Mountains, stands at 5671 meters, making it the tallest peak in both Iran and Southwest Asia. The highest point in the Kopet Dagh range is Kuh-e Quchan, with an elevation of 3191 meters (Noroozi et al., 2020a).

1.3 CLIMATE

The Irano-Anatolian Hotspot is not influenced by a single air mass. Instead, both Polar air masses from the north and Tropical air masses from the south intermittently affect the region. There are transitional seasons between warm and cold periods. During summer, the Intertropical Convergence Zone (ITCZ) shifts further north, allowing the continental

Tropical (cT) air mass, which has distinct continental characteristics, to dominate the Irano-Anatolian Hotspot region (Figure 1.3). Tropical air masses are generally characterized by high temperatures, and the cT air mass is notably dry. The maritime Tropical (mT) air mass remains too far west to influence the Irano-Anatolian Hotspot during the summer. Although Polar air masses retreat further north during this season, they occasionally penetrate the region, causing temperature drops. Meanwhile, the intrusion of maritime Tropical (mT) air masses into the southern part of the Irano-Anatolian Hotspot contributes to summer precipitation (Avcı and Avcı, 2022; Erinç, 1969; Heydarizad et al., 2022; Mohammadzadeh et al., 2020; Türkeş, 2010).

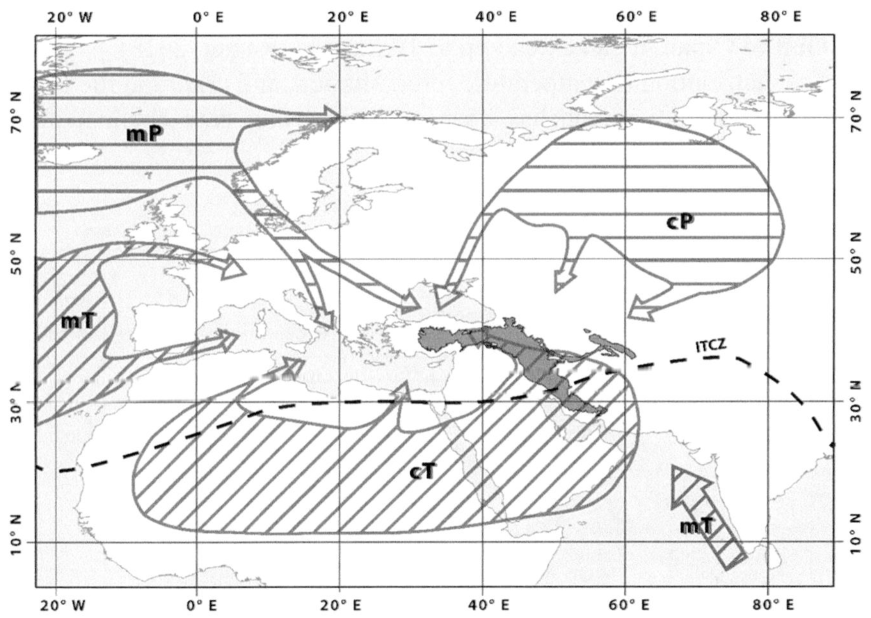

FIGURE 1.3 Air masses influencing the "Irano-Anatolian hotspot" during summer (mP: maritime Polar; cP: continental Polar; mT: maritime Tropical; cT: continental Tropical; ITCZ: Intertropical Convergence Zone; mE: maritime Equatorial).

During the winter season, as Tropical air masses retreat southward, Polar air masses become dominant over the Irano-Anatolian Hotspot region (Figure 1.4). The maritime Tropical (mT) air mass, which is located to the west of the Mediterranean Basin, does not significantly influence the region during winter. In contrast, the interaction between the continental Tropical (cT) air mass and the maritime Polar (mP) air mass over the Mediterranean generates a cyclonic system (Med). The eastward movement of this system, influenced by topography, leads to precipitation, particularly along coastal areas, while precipitation amounts decrease in the interior regions. However, due to the intensification of frontal activity over Anatolia, winter remains the wettest season. This is followed by the transitional seasons of spring and autumn (Erinç, 1969). A similar seasonal distribution of precipitation is observed in the Iranian section of the Irano-Anatolian Hotspot. In general, winter is the wettest season, while summer receives the least precipitation. The southern regions of Iran are notably more arid. Along the Caspian Sea coast, the number of dry months is fewer than four, whereas in the south, particularly around the Persian Gulf and the Gulf of Oman, it can extend up to 10 months (Raziei, 2018).

The distribution of temperature, precipitation, and aridity in the Irano-Anatolian Hotspot region has been assessed using data obtained from CHELSA (Brun et al., 2022; Karger et al., 2017).

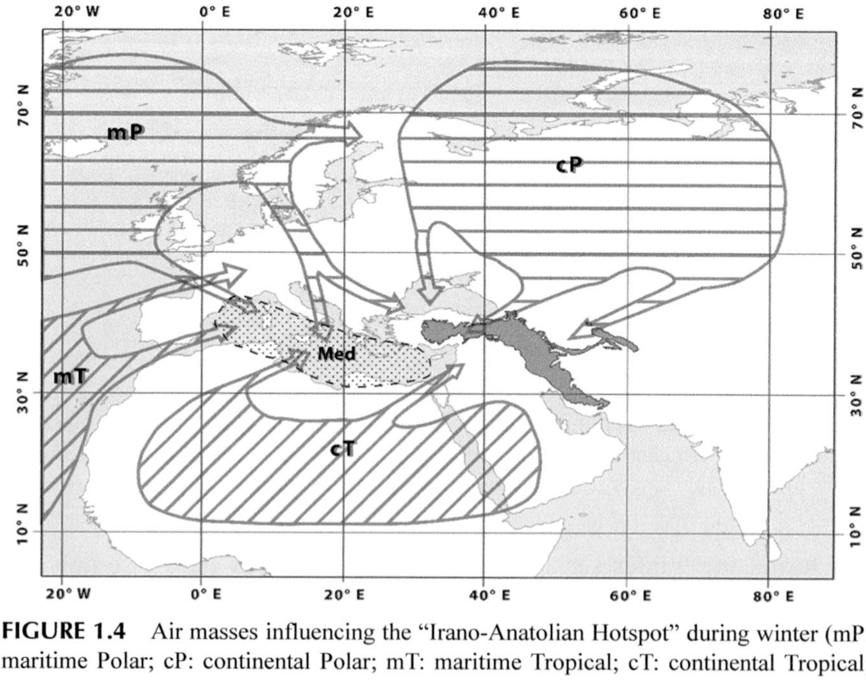

FIGURE 1.4 Air masses influencing the "Irano-Anatolian Hotspot" during winter (mP: maritime Polar; cP: continental Polar; mT: maritime Tropical; cT: continental Tropical; Med: Mediterranean cyclonic center).

1.3.1 TEMPERATURE

The distribution of temperature in the Irano-Anatolian Hotspot clearly reflects the influence of topography. Lowland areas located to the south and southwest of the Zagros Mountains are among the hottest regions. In the part of the Irano-Anatolian Hotspot that lies within Anatolia, the Central Anatolian plains, and the Malatya and Elazığ plains are warmer compared to the mountainous areas and plateaus with higher elevations. In Iran, the area around Lake Urmia and the depression between the Central Iranian Range and Zagros Mountain are warmer than the plateau areas and mountainous areas with higher elevations. The northern part of Eastern Anatolia, the eastern parts of the Southeast Taurus Mountains, and the Alborz Mountains are the coldest areas in the region. Annual average temperatures in the Irano-Anatolian Hotspot typically range from 5 to 15°C (Figure 1.5). In areas with relatively low altitudes, annual average temperatures range from 10 to 15°C, while in the southern and southwestern parts of the Zagros Mountains, temperatures can exceed 15°C due to lower altitudes. In mountainous areas, annual average temperatures are below 5°C. This is particularly evident in the mountainous regions of eastern Anatolia. The hottest month is August, with temperatures generally ranging between 15 and 25°C (Figure 1.6). While temperatures in lowland areas exceed 25°C, in the higher parts of the mountain's temperatures are below 10°C. The coldest month is January, with temperatures generally ranging between 0 and 10°C (Figure 1.7). In the southern and southwestern parts of the Zagros Mountains, temperatures exceed 10°C, while in the higher parts of the mountain's temperatures are below 0°C.

1.3.2 PRECIPITATION

In the Irano-Anatolian Hotspot region, the annual average precipitation is generally between 250 and 500 mm, except in the higher elevations (Figure 1.8). Around the Tuz Lake, in the Iğdır Plain north of Mount Ararat, and partly to the east and north of Lake Urmia, the average rainfall is below 250 mm. The direction of the mountains is an important factor in the distribution of precipitation. In mountainous areas, the average precipitation ranges from 500 to 750 mm. The high mountain areas that receive the most precipitation are the Munzur Mountains and the Southeastern Taurus Mountains in the Anatolian Peninsula, and the Zagros Mountains further east. In the peaks of these mountains, the annual precipitation can exceed 1000 mm.

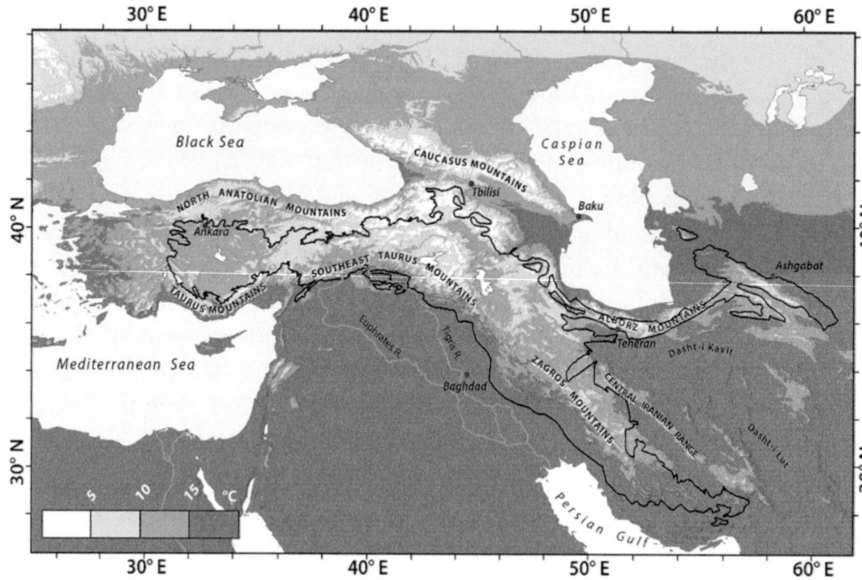

FIGURE 1.5 Distribution of annual mean temperatures in the Irano-Anatolian hotspot.

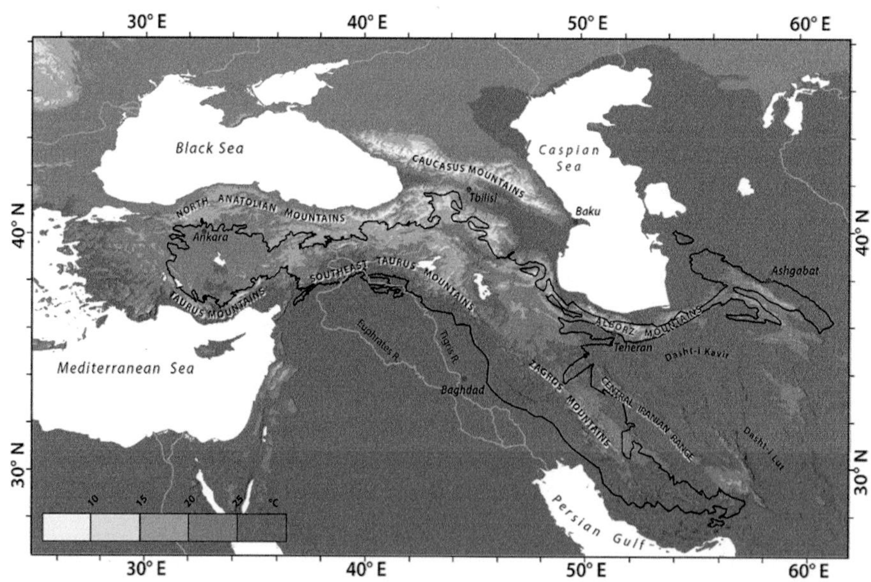

FIGURE 1.6 Distribution of August mean temperatures in the Irano-Anatolian hotspot.

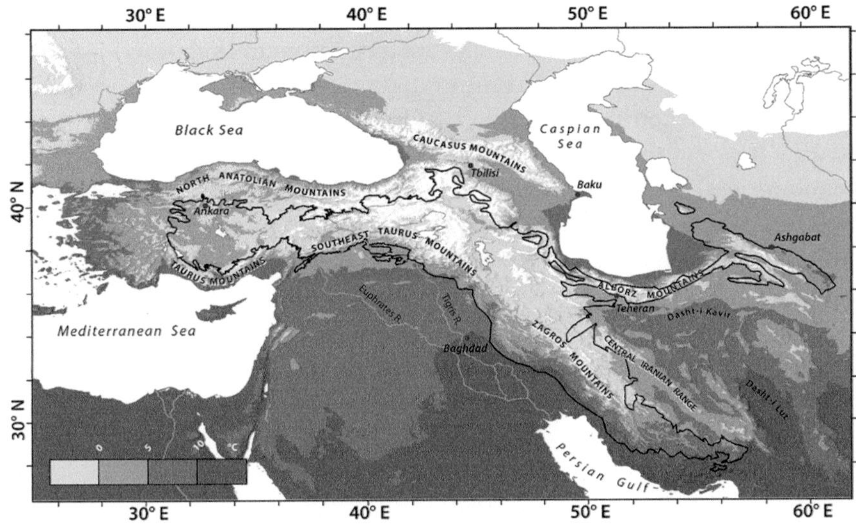

FIGURE 1.7 Distribution of January mean temperatures in the Irano-Anatolian hotspot.

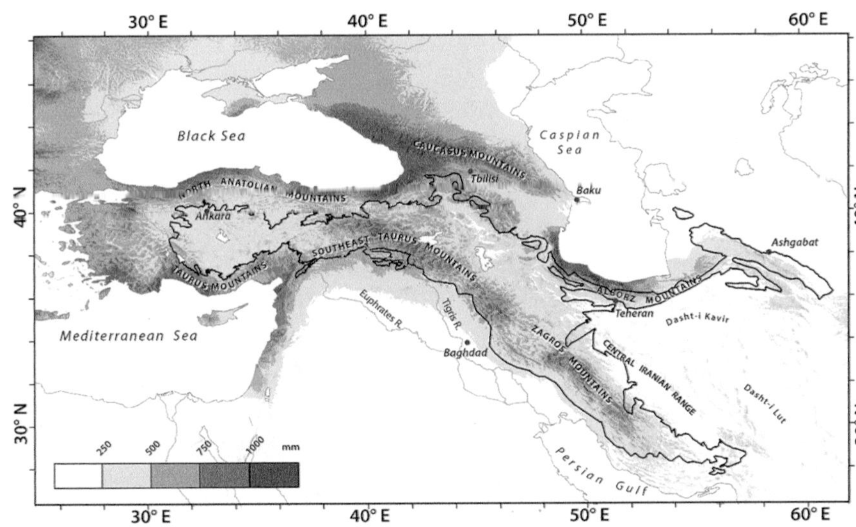

FIGURE 1.8 Distribution of annual precipitation in the Irano-Anatolian hotspot.

1.3.3 ARIDITY

In the Irano-Anatolian Hotspot region, drought is a key factor that limits both life and economic activities. Water, which is essential for both plant and animal life, is equally important for human activities. The deficiency of water is a notable feature of this region. Drought is defined based on whether precipitation is sufficient to match the potential evapotranspiration (PE). When comparing maps showing the distribution of precipitation and PE, it is evident that only the eastern half of the Southeast Taurus Mountains and part of the Upper Euphrates Basin have precipitation levels that meet the PE (Figure 1.9). In other areas, precipitation is less than PE, resulting in varying degrees of drought.

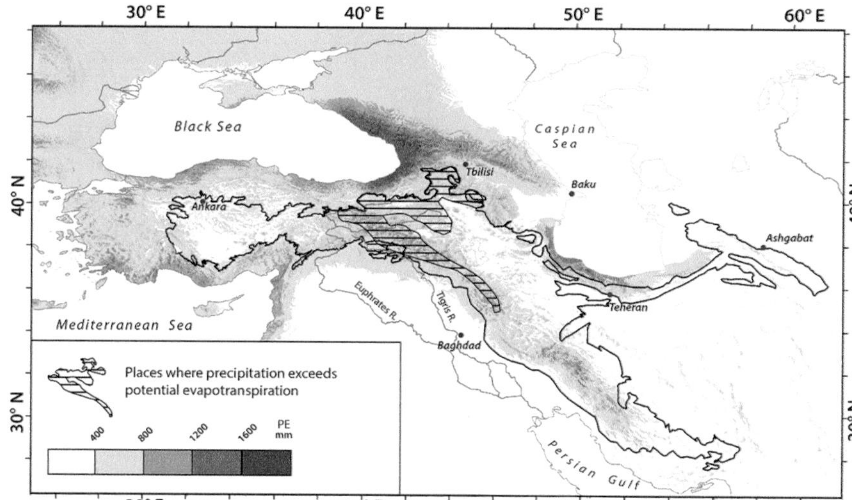

FIGURE 1.9 Distribution of precipitation and areas where potential evapotranspiration is exceeded. Precipitation exceeds potential evapotranspiration in limited areas, while water deficits occur in other regions.

Drought is a significant climatic factor influencing the floristic composition and distribution of natural vegetation in the Irano-Anatolian Hotspot region. It also limits traditional dry farming practices and necessitates irrigation in agricultural production. Irrigated agriculture is only possible in limited areas. The predominant form of livestock farming in the Irano-Anatolian Hotspot region is pasture-based, with some areas also exhibiting pastoral nomadism. Due to the preference of animals for certain plant species for grazing, changes have occurred in the composition of natural vegetation. One example of this is found in the steppes around Tuz Lake in central Anatolia, where *Artemisia* species dominate the flora (Avcı, 2019).

1.3.4 CLIMATE TYPES

The Köppen-Geiger method was used to determine the climatic characteristics of the Irano-Anatolian Hotspot region, and a climate type map was created using CHELSA data (Beck et al., 2023; Chen and Chen, 2013; Kottek et al., 2006; Taşoğlu et al., 2024) (Figure 1.10). Accordingly, in the Irano-Anatolian Hotspot region, a dry climate (B) is observed around Tuz Lake and Lake Urmia, a temperate climate (C) is found in the depressions and lowland areas, and a cold climate (D) prevails in the mountainous regions.

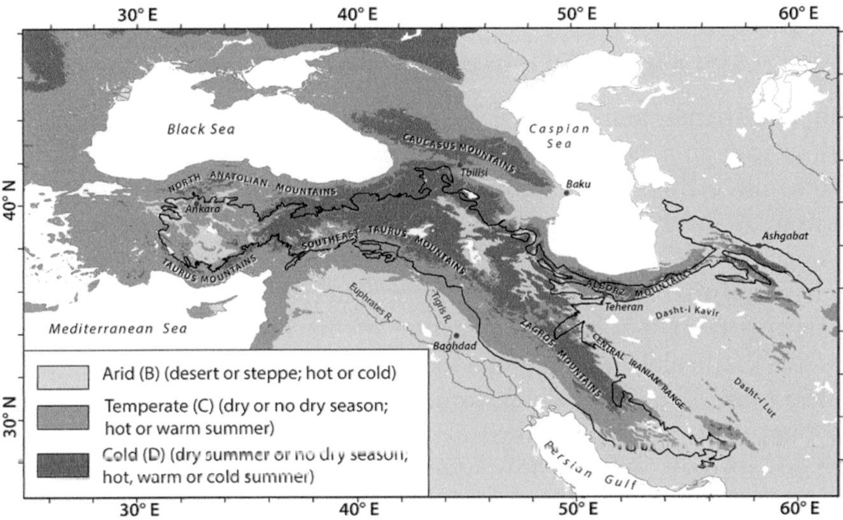

FIGURE 1.10 Main climate types in the Iran-Anatolia Hotspot according to the Köppen-Geiger.

According to the Köppen-Geiger classification, the Irano-Anatolian Hotspot region exhibits cold steppe (cold semiarid climate) BSk, cold desert (cold arid climate) BWk, and hot desert (hot dry climate) BWh climate types. All these climate types, denoted by B, are categorized under the Arid climate group. BSk occurs to the south of Tuz Lake in Central Anatolia, as well as around Lake Urmia. BWk is found in the depressions west of the Central Iranian Range, while BSh appears to the west of the Zagros Mountains.

Temperate climates (C) in the Irano-Anatolian Hotspot region are characterized by dry and hot summers (Csa) and dry and warm summers (Csb). The Csa climate is typical of the Mediterranean climate and is found widely in the western and southern parts of the Irano-Anatolian Hotspot. This climate type occurs on the southern slopes of the Southeast Taurus Mountains and Zagros Mountains, as well as in the lowlands of Central Anatolia and around Malatya and Elazığ in Eastern Anatolia. In the Irano-Anatolian Hotspot region, continental and cold climate types (D) dominate the higher elevations. In all mountainous areas, except for the Zagros Mountains, the climate is characterized by dry and warm summers (Dsb). The climate in the Zagros Mountains is distinct from other areas due to its hot summers (Dsa). The peaks of the mountains, on the other hand, have a dry and cool summer climate (Dsc).

1.4 RIVERS AND WATER BODIES

The Irano-Anatolian Hotspot, despite being located in an arid and semi-arid region, contains significant rivers and lakes that serve as vital water sources for sustaining life and economic activities. The Anatolian portion of the Irano-Anatolian Hotspot is richer in rivers and lakes compared to its eastern half in Iran, where the climate and topography limit the availability of such water bodies (Maghsoudi, 2021). In the western part of the Anatolian section of the Irano-Anatolian Hotspot lies the Kızılırmak, which is the longest river in Türkiye, with a length of 1355 km. The river, which forms a broad arc, flows northward and eventually discharges into the Black Sea outside the Hotspot boundaries. The Kızılırmak, which flows through sedimentary, magmatic, and metamorphic formations, has a reddish hue due to its high clay content, which is the reason for its name "Red River" (Akbulut et al., 2009). The Aras River, located to the north of the Irano-Anatolian Hotspot, is situated within a tectonic basin. It has a length of 1072 km. Forming part of the borders between Türkiye, Armenia, Azerbaijan, and Iran, the Aras River originates in Türkiye. Outside the boundaries of the Hotspot, it merges with the Kura River and ultimately discharges its waters into the Caspian Sea. Most of the rivers located to the north in the eastern half of the Hotspot (those on the northern slopes of the Alborz Mountains) drain into the Caspian Sea basin. The largest of these rivers is the Sefid, which originates southeast of Tabriz. The eastern portion of the Irano-Anatolian Hotspot in Anatolia is drained by the Euphrates and Tigris rivers. The Euphrates has a length of 2780 km, with approximately 1000 km of its course lying within the Hotspot.

The Tigris, which stretches for 1750 km, has a much smaller portion (about 200 km) within the Irano-Anatolian Hotspot. The Karkheh River, originating from the Zagros Mountains, merges with the Tigris from the east. The Euphrates and Tigris rivers converge north of the Persian Gulf to form the Shatt al-Arab. To the south, the Karun River also joins the Shatt al-Arab and flows into the Persian Gulf. There are no major rivers located east of the Zagros Mountains; the rivers in this region are predominantly temporary streams.

The majority of rivers in the Irano-Anatolian Hotspot have various-sized dams constructed on them. The Kızılırmak River, with the Hirfanlı Dam (263 km^2), and the Euphrates River, with the Keban Dam (675 km^2), Karakaya Dam (268 km^2), and Atatürk Dam (817 km^2), are utilized for irrigation and energy production (TUIK, 2014). Another significant dam is the Aras Dam (145 km^2) on the Aras River, which is located at the border between Iran and Azerbaijan. This dam is one of the largest with a reservoir area exceeding 100 km^2.

The tectonic characteristics of the Irano-Anatolian Hotspot have facilitated the formation of numerous lakes. Tuz Lake, Lake Van, Lake Urmia, and Lake Bakhtegan are examples of endorheic lakes. The tectonic activity shaping the geological features of the Irano-Anatolian Hotspot has led to the formation of multiple depressions. While these depressions have been filled with sediments over time, some have evolved into lakes. Among the freshwater lakes of varying sizes in the Irano-Anatolian Hotspot, notable examples include Lake Beyşehir, Lake Hazar, and Lake Sevan.

Lake Beyşehir, located in the western part of the Irano-Anatolian Hotspot, has a surface area of 656 km^2 and a depth of 7 meters. Situated at an elevation of 1121 meters above sea level, this lake formed in one of the depressions that developed northward in response to the uplift of the Taurus Mountains during the Miocene (Gürbüz et al., 2021). Lake Hazar, positioned along the East Anatolian Fault Zone, covers an area of 81 km^2. It occupies a pull-apart depression and is situated at an altitude of 1248 meters above sea level (Eriş et al., 2018). Similarly, Lake Sevan in Armenia is another lake that has formed within a pull-apart depression resulting from tectonic activity. Volcanic activity occurred in the surrounding area throughout the Holocene. Lake Sevan, located at approximately 1900 meters above sea level, has a surface area of 1242 km^2 and a depth of about 95 meters (Karakhanian et al., 2001; Vardanian, 2009). Both Lake Hazar and Lake Sevan are utilized for hydroelectric power generation and irrigation, leading to significant declines in their water levels. In the 1930s, Lake Sevan covered an area of 3475 km^2; however, by the 2000s, its surface area had decreased to 1416 km^2 due to extensive water extraction for agricultural use. The lake is surrounded by some of Armenia's most important agricultural lands (Vardanian, 2009).

In addition to freshwater lakes, some lakes contain saline or alkaline waters. Tuz Lake in Central Anatolia and Lake Urmia in western Iran are among the highly saline lakes, with salt concentrations exceeding 300 grams per liter (Avcı, 2003). These lakes are also utilized for salt production.

In the central part of the steppe landscape, within the Tuz Lake plain, three fault-controlled lakes stand out: Tuz Lake, Tersakan, and Bolluk. Tuz Lake is the second largest lake in Türkiye, covering an area of 1665 km^2. The two adjacent saltwater bodies, Tersakan and Bolluk lakes, also contain significant amounts of sodium sulfate (Özsayın et al., 2019). The Quaternary evolution of the Tuz Lake depression exhibits similarities to the evolution of other lake depressions in Central Anatolia (Erol, 1969, 1978; Lahn, 1948). The geological structure of Lake Urmia and its surrounding area comprises rocks of various ages, ranging from the Precambrian to the Quaternary, as well as relatively recent lacustrine deposits (Eimanifar and Mohebbi, 2007). Lake Bakhtegan in Iran is another saline lake. The diversion of its inflowing rivers for irrigation and drinking water supply led to its complete desiccation in 2007, and no water accumulated in the lake until the early 2010s. As a result, dust and salt particles from the dried lakebed were dispersed into the surrounding environment, causing damage to nearby forest vegetation (Amirkhani et al., 2022). Lake Van, the largest lake in Türkiye, contains alkaline waters. It has been suggested that the lake's depression was once connected to an external drainage system from the southwest, but lava flows from the Nemrut Volcano blocked the outlet, leading to the formation of the present lake (Yılmaz et al., 1998). The lake has terraces that indicate Quaternary lake-level fluctuations and are dated to various periods (Görür et al., 2015; Özalp, 2020).

1.5 SOIL TYPES

The classification of soil types within the Irano-Anatolian Hotspot is based on the World Reference Base (WRB) system (FAO, 2025; ISRIC, 2025; IUSS Working Group WRB, 2022). The most widespread soil types in the hotspot are Leptosols and Calcisols (Figure 1.11).

Leptosols are among the most extensive soil groups worldwide. They are predominantly found in mountainous regions and areas where extensive erosion has exposed hard rock near the surface (ISRIC, 2025). Lithosols, which are azonal soils with an incomplete solum, correspond to Leptosols in various international classification systems (e.g., USA, FAO). Additionally, Leptosols align with the Lithic subgroups of other soil classifications. Depending on the parent rock, Leptosols on calcareous substrates are referred to as Rendzinas, whereas those on acidic rocks are called Rankers. A mollic horizon (a dark-colored surface horizon with moderate to high organic matter content) ranging from 10 to 25 cm in thickness can develop in Leptosols on bedrock with a calcium carbonate content exceeding 40% (Spaargaren, 2008g). In the Irano-Anatolian Hotspot, Leptosols cover approximately 41% of the region, predominantly occurring in the Southeastern Taurus Mountains, Zagros Mountains, Karasu-Aras Mountains, Alborz Mountains, and the Central Kopet Mountains.

In the Irano-Anatolian Hotspot, Calcisols are prevalent in relatively low-lying areas, covering approximately 31.7% of the region. These soils develop in arid regions dominated by carbonate-rich parent material and exhibit significant secondary carbonate accumulation. The distribution of Calcisols includes Central Anatolia, the area east of Lake Urmia, the eastern Zagros Mountains, and both the western and eastern Kopet Mountains. The natural vegetation is sparse, primarily consisting of xerophytic shrubs, trees, and/or ephemeral grasses (Spaargaren, 2008a).

Another widely distributed soil type in the Irano-Anatolian Hotspot is Regosols, which cover approximately 9% of the area. Regosols are characterized as very weakly developed mineral soils in unconsolidated materials, featuring only an ochric surface horizon. These soils extend in a belt around the northern and southeastern regions of Lake Urmia in Iran, as well as in the southern parts of the Zagros Mountains. Predominantly found in arid, semi-arid, and mountainous regions, Regosols typically develop on sand deposits, loess, and young glacial tills. The primary factors influencing the morphology of Regosols are parent material and climate (Spaargaren, 2008e).

Kastanozems are primarily found in the dry regions of the world's steppe zones. These soils are distinguished by the presence of a mollic horizon, which is a deep, dark-colored surface horizon enriched with a significant accumulation of organic matter and high base saturation. Additionally, Kastanozems exhibit the accumulation of secondary calcium carbonate within 100 cm from the soil surface (ISRIC, 2025). Kastanozems are also identified as one of the typical soil types of loess deposits. Another soil type that develops in loess regions, Phaeozems, forms in more humid climates and is generally dark red in color, whereas Kastanozems are found in drier areas. Kastanozem soils are also referred to as chestnut soils due to their brown solum, which includes the A and B horizons (Chesworth, 2008). Kastanozems soils cover approximately 4.0% of the Irano-Anatolian Hotspot. The northern and eastern regions of Lake Van and the northern parts of the Karasu-Aras Mountains constitute the main distribution areas of these soils.

Luvisols constitute approximately 3.9% of the region and are primarily found in the southern parts of the Southeastern Taurus Mountains. Luvisols typically form various unconsolidated materials, including glacial till, as well as eolian, alluvial, and colluvial deposits, and are characterized by an argic B horizon. In Soil Taxonomy, these soils are classified as Alfisols (Spaargaren, 2008d).

Cambisols typically feature a cambic horizon situated between the A horizon and the relatively unaltered C horizon. This horizon exhibits soil-specific properties rather than those of the parent material and is distinguishable from the C horizon by its color. In Soil Taxonomy, Cambisols are classified as Inceptisols. They are characterized by the slight to moderate weathering of the parent material and the absence of significant illuvial clay, organic matter, aluminum, and/or iron compounds. Cambisols can develop under various types of vegetation. These soils are found in central and eastern Anatolia, Georgia, and Armenia, as well as in the northern regions of the Kopet Mountains, covering approximately 3.6% of the soils within the Irano-Anatolian Hotspot (Spaargaren, 2008b).

In the Irano-Anatolian hotspot, the proportions of Solonchaks and Chernozems are less than 2%, while the proportions of *Vertisols*, *Arenosols*, *Anthrosols*, *Gleysols*, and *Fluvisols* are less than 1%. Solonchaks in the east of the Iranian Anatolian hotspot area, particularly in the desert regions between the Alborz and Zagros mountains, are notable for their profiles, which exhibit high concentrations of soluble salts. These soils feature a salic horizon, which refers to a surface or subsurface horizon enriched with soluble salts (Spaargaren, 2008f). The vegetation is predominantly composed of halophytic species. Chernozems are found in narrow zones in the northernmost part of the Iranian Anatolian hotspot region. These soils are characterized by a relatively thick, dark-colored surface horizon enriched with organic matter (Spaargaren, 2008c). Features such as rock outcrops, salt flats, and dunes each constitute less than 1%.

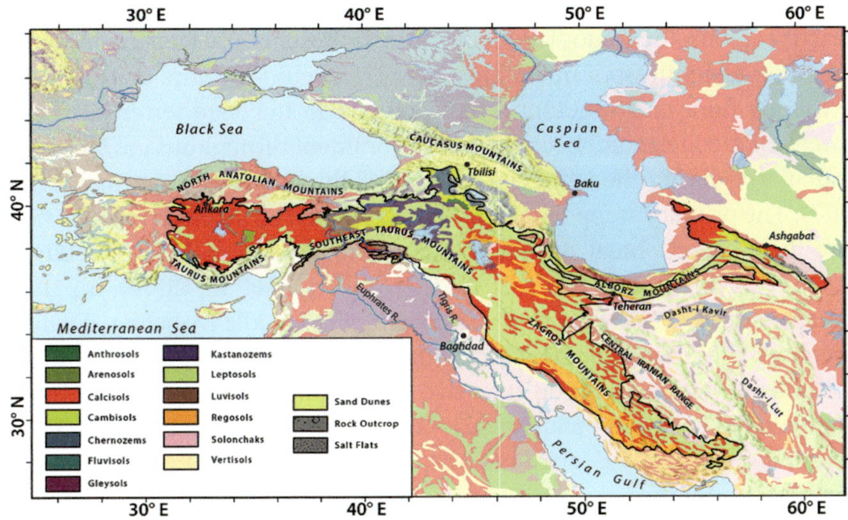

FIGURE 1.11 Soil types in the Iran-Anatolia hotspot.

1.6 BIOGEOGRAPHY OF IRANO-ANATOLIAN HOTSPOT

The region defined as the Irano-Anatolian Hotspot is part of the Irano-Turanian Floristic Region, which lies within the Holarctic Floristic Realm of Eurasia and serves as a significant center of endemism. This area is bordered to the south by the ecologically and biogeographically distinct Mediterranean and Saharo-Sindian regions, while to the north, it neighbors the Euro-Siberian regions. Continentality is the most crucial bioclimatic factor distinguishing the Irano-Turanian Floristic region from the Mediterranean and Saharo-Sindian regions. This characteristic also contributes to the distinct floristic differences among the subregions within the Irano-Turanian region (Djamali et al., 2012). Divided into four subregions, the western-central part of the Irano-Turanian Floristic region, identified as the Irano-Anatolian Hotspot, is a key center of speciation and endemism.

In the second half of the 19th century, Boissier described the eastern part of this region in Flora Orientalis, while Grisebach referred to it as a steppe region. However, it was formally designated as the Irano-Turanian

Region following Alexander Eig's (1894–1938) proposal in 1931 (Eig, 1946). Since the 19th century, the predominantly arid and semi-arid lands of what is now known as the Irano-Turanian Floristic region have been recognized as an essential part of Eurasia. As one of the world's largest floristic regions, the Irano-Turanian region represents a crucial evolutionary and biodiversity hotspot in the biosphere (Herrando-Moraira et al., 2023).

Steppe flora and vegetation have existed in the Mediterranean Basin since the Paleogene and became more prominent throughout the Neogene, particularly during the Miocene (Avcı, 2022b; Jimenez-Moreno and Suc, 2007). The primary steppe lineages that colonized the Mediterranean Basin have their geographical origins in Central and Western Asia (Bozkurt et al., 2022; Manafzadeh et al., 2013, 2016; Roquet et al., 2009), a pattern that also applies to *Rhaponticoides*. During the Middle Miocene, Anatolia served as the ancestral area for *Rhaponticoides*. By the Late Miocene (8 Ma), the genus diverged into two geographical lineages—the Irano-Turanian and Mediterranean clades—leading to diversification both in the eastern regions and westward into the Mediterranean and Europe (Bozkurt et al., 2022). Studies on *Astragalus* sect. *Caprini*, which is represented by numerous species, further emphasize the importance of the Irano-Turanian Floristic Region as a source area for the colonization of the Mediterranean Basin (Maylandt et al., 2025). Similarly, this region has also been identified as a source of xerophytic taxa for neighboring regions such as the Mediterranean and Saharo-Sindian zones (Manafzadeh et al., 2013). The mountainous areas of the Irano-Turanian Floristic region likely served as refugia for xerophytic elements during Quaternary interglacials, while during glacial maxima, these areas may have facilitated the expansion of cold-adapted species into other regions (Djamali et al., 2012).

The Irano-Turanian Floristic Region harbors exceptional biodiversity, driven by its complex topography, pronounced elevation gradients shaped by extensive mountain ranges, and unique paleogeographic history. As both a genetic reservoir and a hotspot for plant diversification, this region supports a remarkable array of endemic genera (Ansell et al., 2011; Herrando-Moraira et al., 2023; Manafzadeh et al., 2016; Peterson et al., 2019; Roquet et al., 2009; Shamsabad et al., 2021; Zohary, 1973).

The western-central part of the Irano-Turanian Floristic Region, known as the Irano-Anatolian Hotspot, encompasses a large portion of Türkiye. Within this region, aside from high mountain areas, two main vegetation zones can be distinguished. The first consists of forest communities, which have been largely degraded, with *Quercus* being the dominant genus (Avcı, 2004; Kırsal Çevre, 2019). The second vegetation zone comprises natural steppe areas. Natural steppe habitats in this region include the surroundings of Tuz Lake and the Konya-Ereğli basin in Central Anatolia, the Akçakale-Ceylanpınar area in Southeastern Anatolia, and the Iğdır Plain in Eastern Anatolia (Avcı, 2019). These areas receive notably low precipitation and host several remarkable species, such as *Halothamnus glaucus*, a characteristic element of true semi-desert communities, and *Halothamnus hierochunticus*, whose distribution in Türkiye represents the northernmost extension of the western Mesopotamian plains near the Syrian border (Avcı, 1996; Freitag et al., 2009). The steppes that have emerged as a result of deforestation in Eastern Anatolia and Central Anatolia are not natural but rather secondary/anthropogenic steppe areas, covering vast expanses. In the regions of Central Anatolia, Eastern Anatolia, and Southeastern Anatolia, anthropogenic steppes and natural steppes are often intermingled. As a result, distinguishing the boundaries between natural and anthropogenic steppes is complicated (Avcı, 1996, 2019). The Irano-Anatolian hotspot is characterized by dominant genera such as *Astragalus*, *Acantholimon*, *Artemisia*, *Bromus*, *Stipa*, *Cousinia*, *Centaurea*, and *Verbascum* (Ketenoğlu et al., 2022). Among woody plants, species of *Quercus* (e.g., *Q. pubescens*, *Q. robur* subsp. *robur*, *Q. libani*, *Q. macranthera* subsp. *syspirensis*, and *Q. brantii*) are common (Avcı, 1997), alongside members of the Rosaceae family, particularly the Pomoideae and Prunoideae subfamilies. While many genera typical of the Iran-Turan region's eastern part, such as *Acanthophyllum*, *Calligonum*, *Cousinia*, *Ferula*, and *Eremostachys*, are represented by fewer species in Anatolia, certain plants that have their centers of development in the Iran-Turan region, such as *Centaurea*, *Verbascum*, *Achillea*, *Aethionema*, *Alyssum*, and *Isatis*, are characteristic of this area.

In the western part of the Irano-Anatolian hotspot in Anatolia, within the steppe of Central Anatolia, cushion-shaped spiny chamaephytes such as *Astragalus microcephalus*, *A. angustifolius*, *A. plumosus*, *A. baibutensis*, *Acantholimon acerosum*, *A. ulicinum*, *Onobrychis cornuta*, and *Minuartia juniperina* are commonly found. Alongside these, perennial grasses from the Poaceae family, including *Festuca valesiaca*, *Bromus tomentellus*, *Rostraria cristata* var. *cristata*, *Stipa lessingiana*, and *S. holosericea*, are widespread (Ketenoğlu et al., 2022). The Tuz Lake area within this section of the Irano-Anatolian hotspot in Anatolia has particular significance.

During climatic changes in the Pleistocene, the area covered by the lake underwent substantial modifications. Currently, the soils surrounding the lake, with high salt content, serve as the habitat for halophytic plants. Among these halophytes, species of *Salicornia*, *Limonium*, and *Atriplex* are prevalent. In the most saline areas along the lake's shore, *Salicornia europaea*, a species of sea blite, forms communities with coverages of up to 95% (Avcı, 2019). As one moves away from the lake, the species composition also changes. Within the Irano-Anatolian hotspot area remaining in Anatolia, plants growing on soils formed in areas with gypsum-rich bedrock, such as around Çankırı and further east near Sivas, are subjected to stress and must adapt to harsh conditions. The soils resulting from the weathering of gypsum are low in plant nutrients due to the exchange of calcium ions. Conversely, chemical properties such as the high concentration of sulfate ions, which are toxic to plants, lead to the development of gypsophilous plant communities. Species such as *Scrophularia gypsicola, Thymus spathulifolius, Gypsophila heteropoda* subsp. *minutiflora, Reaumuria sivasica, Campanula sivasica, Allium sivasicum,* and *Centaurea yildizii* are examples of plants found in gypsum-rich soils around Sivas (Avcı, 2005, 2022a).

In the eastern part of the Irano-Anatolian hotspot within Anatolia, *Pinus sylvestris* and *Quercus* sp. forests have been restricted to certain areas due to anthropogenic factors. The Eastern Anatolian steppes are highly significant for livestock grazing. In regions where grazing pressure is intense, species such as *Elymus hispidus* subsp. *hispidus*, *Agropyron repens*, *Alopecurus pratensis*, *Alchemilla caucasica*, *Bromus tomentellus*, *Centaurea depressa*, *Galium verum*, *Lotus corniculatus*, *Medicago x varia*, *Onobrychis cornuta*, *Trifolium hybridum*, and *Vicia sativa* are commonly found (Ketenoğlu et al., 2022). Within the Irano-Anatolian hotspot, the Fertile Crescent represents a key region where numerous plant species were domesticated (Adıgüzel and Aytaç, 2010).

The relatively elevated threshold known as the Anatolian Diagonal separates the western and eastern parts of the Irano-Anatolian hotspot in Anatolia. During climatic changes in the Pleistocene, this threshold functioned not only as a migration route for plants and animals but also as a barrier between eastward and westward dispersal (Avcı, 1996; Davis, 1971; Ekim and Güner, 1986). From a neotectonic perspective, the Anatolian Diagonal corresponds to a left-lateral shear zone, extending in a northeast-southwest direction between the North Anatolian Fault Zone and the Cyprus Arc. It is bounded by the Central Anatolian Fault Zone and the Eastern Anatolian Fault Zone, spanning approximately 70 km in width and 850 km in length (Seyitoğlu et al., 2022).

East of this threshold, the landscape transitions into a higher plateau with more continental climatic conditions (Avcı, 1996). Along the Diagonal and on the Central Anatolian volcanoes, Irano-Turanian elements become dominant. Generally, the proportion of Irano-Turanian taxa increases eastward and with elevation, whereas Mediterranean taxa become less prevalent. The southeasternmost section of the Irano-Anatolian hotspot in Anatolia, encompassing the Hakkari region, shares significant floristic similarities with the Eastern Anatolian plateaus and hosts numerous endemic taxa. Additionally, *Nigella unguicularis, Ranunculus diversifolius*, and *R. myosuroides*, which extend northward from more arid regions, exemplify xerophytic Mesopotamian plant species (Parolly, 2020).

This threshold coincides with the distributional limits of many populations and taxa. Furthermore, the geographic distribution of biodiversity hotspots for Nitidulidae beetles and butterflies is predominantly concentrated east of the Anatolian Diagonal (Avcı, 1996; Gür, 2016). It also serves as an important barrier affecting the distribution of various animal species, including *Salamandra infraimmaculata orientalis*, *Triturus vittatus cilicensis*, *Anatololacerta danfordi*, *Eirenis barani*, and *Hyla savignyi* (Gül, 2013). The potential impact of the Anatolian Diagonal on aphid diversity in Türkiye has also been investigated. Of the 219 species identified in the western part of the Irano-Turanian floristic region, 91 were not recorded in the eastern and southeastern regions of Anatolia. Conversely, of the 226 species identified in the eastern and southeastern parts, 111 were absent in the western region. These initial findings suggest that the Anatolian Diagonal has a strong influence on aphid diversity in Türkiye (Şenol et al., 2019). A morphometric study on the wing morphology of *Platycnemis dealbata* further supports the idea that the Anatolian Diagonal acts as a barrier to population connectivity (Salur et al., 2024). Similarly, the last surviving species of the genus *Zonocypris*, most of whose species are now extinct, are geographically restricted to the southern regions of Türkiye, while fossil forms exhibit a much wider distribution in the northern regions. This suggests that the Anatolian Diagonal has played a crucial role as a physical barrier, separating the fossil forms (eastern and northern regions) from the extant forms (southeastern region) of the genus in Türkiye (Külköylüoğlu et al., 2021). Notably, the distribution of genetic variation in oak gall wasp (Cynipidae) species across the Anatolian Diagonal is striking. The observed genetic differentiation among oak gall wasp species suggests a significant disruption in the population genetic structure, further indicating that the Diagonal may act as a barrier to gene flow in these species (Mutun, 2016).

In the eastern section of the Irano-Anatolian hotspot, beyond the Anatolian Peninsula, the Zagros Mountains represent one of the most significant mountain ranges. This mountain range serves as a barrier between the Mesopotamian Plain and the Central Iranian Plateau. The Zagros Mountains were formed by folding processes resulting from the collision of the Arabian Plate with the Eurasian continent during the Cretaceous–Paleocene period, a process that continues to this day. The oldest rocks in the Zagros Mountains date back to the Lower Cambrian and primarily consist of sandstone and dolomites (Noroozi et al., 2020b).

An examination of the climate and vegetation changes in the Zagros Mountains over time reveals periods dominated by both warm and cold conditions. Various palynological studies conducted on lakes such as Lake Mirabad, Lake Zeribar, and Lake Maharlou have provided valuable insights into the climatic and vegetation shifts in the surrounding Zagros region (Djamali et al., 2009; El-Moslimany, 1987; van Zeist and Bottema, 1977; van Zeist and Wright, 1963). Particularly in the Late Holocene, it is evident that *Quercus brantii* forests and *Pistacia-Amgdalus* shrub communities were widespread, and human activities became notably prominent after 3700 cal BP (Djamali et al., 2009). The Zagros Mountains, with over 3500 plant taxa recorded in its flora, lie entirely within both the Irano-Turanian floristic region and the Irano-Anatolian hotspot. However, in addition to elements typical of the Mediterranean Flora Region, it also hosts species from the Euro-Siberian Flora, such as *Pterocarya fraxinifolia* and *Zelkova carpinifolia*. Some isolated species, like *Hypericum dogonbadanicum* (closely related to relatives in the Somali-Masai region) and *Clinopodium kallaricum* (found in tropical regions of Africa), are also found in the Zagros Mountains. This indicates the complex phytogeographical history of this mountainous area (Noroozi et al., 2020b).

The Zagros Mountains are also noteworthy for their endemic plant species. Approximately 21% of the recorded taxa are endemic, with six genera showing endemism within this area. Most of these endemic species are found in the higher mountainous regions (Noroozi et al., 2019). The oak forests of the Zagros are predominantly located in the western part of the range, at elevations between 1000 and 2500 meters above sea level.

In the northern part of the Zagros, *Quercus brantii* is the dominant oak species, while in the southern slopes, alongside *Q. brantii, Q. infectoria,* and *Q. libani* also occur (Noroozi et al., 2020b).

The Alborz Mountains, located between the Talish Mountains in the west and the Kopet Dagh in the east, are composed of formations that have been uplifted since the Late Cretaceous as a result of the collision between the Eurasian and Gondwanan plates. This mountainous area primarily consists of carbonate rocks and classics, with volcanic activity during the Quaternary period giving rise to prominent elevations such as Damavand and Alam Kuh (Noroozi et al., 2020a). Glaciation occurred in the higher elevations of the Alborz Mountains during the Pleistocene (Ferrigno, 1991). Distinct climate types are observed on the northern and southern slopes of the Alborz Mountains. The northern slope experiences a more humid temperate climate, while the region within the Irano-Anatolian Hotspot reflects the arid characteristics of the Mediterranean climate (Noroozi et al., 2020a). According to the Köppen-Geiger climate classification, the section of the Alborz Mountains within the Irano-Anatolian Hotspot is predominantly characterized by cold climates (Dsb), marked by mild and dry summers. As elevation increases, precipitation levels rise, and the area undergoes cold climates with either mild (Dfb) or cool (Dfc) summers. The Alborz Mountains lie at the boundary between the Euro-Siberian Floristic Region and the Irano-Turanian Floristic Region. The Irano-Anatolian Hotspot region is primarily composed of steppes in the higher elevations of mountainous areas (Akhani et al., 2013). In total, 3617 native vascular plant species from 876 genera and 146 families have been recorded in the Alborz. Dominant species in this region include *Acanthophyllum acerosum, Aethionema grandiflorum, Amygdalus communis, Arabis aucheri, Astragalus chrysostachys, Astragalus microcephalus, Chaerophyllum macropodum, Corydalis verticillaris, Elymus tauri, Iris acutiloba, Isatis kotschyana, Physoptychis gnaphalodes, Salvia atropatana, Silene aucheriana, Thymus pubescens*, and *Verbascum cheiranthifolium* (Noroozi, 2020a).

The Kopet Dagh forms the eastern extension of the Alborz Mountains. The Kopet Dagh and the surrounding Khorassan region to the southeast were shaped by a sedimentary basin during the Middle Triassic orogeny. This basin has been continuously accumulating sediment from the Jurassic to the Neogene. Sandstone, limestone, marl, and dolomites have formed layers of varying thickness (Memariani, 2020). According to the Köppen-Geiger climate classification, the Kopet Dagh region exhibits BSk, Csa,

Dsb, and Dsc climate types. The cold steppe climate (BSk) occupies the largest area in the Kopet Dagh. As elevation increases, the climate transitions to a dry and warm summer climate (Dsb) and a dry and cold summer climate (Dsc). These are cold climate types (D type). To the west of the Kopet Dagh, in narrow areas, the Mediterranean climate type with hot, dry summers (Csa) is observed. In these regions, along with herbaceous plants, drought-tolerant woody plant communities consisting of trees and shrubs are also found.

The Khorassan-Kopet Dagh region lies at the intersection of various areas of the Irano-Turanian flora region and the Hyrcanian section of the Euro-Siberian region. The number of recorded taxa in the flora of this mountainous area is close to 2600. The geographical location and other characteristics of the region have led to the development of a variety of habitats, which has increased species richness. Some of the largest plant families represented in this mountainous area include Asteraceae, Fabaceae, Poaceae, Brassicaceae, Apiaceae, Caryophyllaceae, Chenopodiaceae, and Lamiaceae (Memariani, 2020).

The typical large taxonomic groups of the Irano-Turan region inhabit a variety of habitats created by the heterogeneous topography of the Middle East and Central Asia, where elevation changes occur vertically. Many species of these taxa (such as those in the genera *Acantholimon* and *Astragalus*) are primarily endemics, adapted to narrow habitats. The relatively low geographical latitudes of the Irano-Turan region, along with the opportunity for species to respond to climate change through altitudinal migration, suggest that the floristic elements of the Irano-Turan flora were less affected by Pleistocene glaciations compared to the Mediterranean and Euro-Siberian regions (Djamali et al., 2012). This is also evident for the areas within the Anatolian section of the Irano-Anatolian hotspot. The maximum extent of glaciers in the Anatolian Mountains during the Quaternary has been studied. Traces of glaciation from the Pleistocene are primarily limited to the higher elevations of the Taurus Mountains in the south and the mountains in the eastern part of Anatolia (Akçar et al., 2015; Bayrakdar et al., 2015; Erinç, 1952).

The biogeographic evidence suggests that the boundaries of the Irano-Turanian floristic region shifted primarily laterally during the climatic changes of the Pleistocene (Djamali et al., 2012). As a result of these shifts, the relatively homogenous and extensive area of the Irano-Anatolian hotspot is exceptionally rich in taxonomic and genetic diversity, with high beta diversity among its major vegetation types. Most of the Irano-Turanian endemic species are found in the higher elevations of the mountains (Noroozi et al., 2018). The Eastern Anatolian highlands are a center of diversity for *Gypsophila*, which also hosts many endemic species. In the mountain-steppe zone, there are over 20 taxa, including species that grow at higher altitudes. These include *G. adenophylla, G. bicolor, G. briquetiana, G. bitlisensis, G. elegans, G. graminifolia, G. nabelbekii, G. peshmenii, G. pulvinaris*, and *G. ruscifoli* (Parolly, 2020). These high plateaus host a rich geophyte flora, which is also notable for its endemism.

Considering the potential impacts of future climate change, even though species adapted to cold conditions may shift to higher elevations, the limited extent of these areas, coupled with the increasing human influence, suggests that the survival of these species will be challenging (Noroozi et al., 2023). Moreover, the donor character of the Irano-Turanian floristic region, where new populations are established in neighboring areas after biodiversity loss events (Ansell et al., 2011; Manafzadeh et al., 2013), highlights the particular importance of the Irano-Anatolian hotspot region for conservation efforts.

KEYWORDS

- **Anatolian diagonal**
- **Anatolian peninsula**
- **exceptional biodiversity**
- **Iranian plateau**
- **Irano-Turanian floristic region**
- **steppe flora**

REFERENCES

Adıgüzel, N.; Aytaç, Z. Ceylanpınar Steppes. In *Important Plant Areas in Turkey: 122 Key Turkish Botanical Sites*; Byfield, A.; Atay, S.; Özhatay, N., Eds.; WWF Turkey: İstanbul, 2010; pp 363–264.

Akbulut, N. E.; Bayarı, S.; Akbulut, A.; Şahin, Y. Rivers of Turkey. In *Rivers of Europe*; 2009; pp 643–672. https://doi.org/10.1016/b978-0-12-369449-2.00017-5.

Akçar, N.; Yavuz, V.; Yeşilyurt, S.; Ivy-Ochs, S.; Reber, R.; Bayrakdar, C.; et al. Synchronous Last Glacial Maximum across the Anatolian Peninsula. *Geological Society* **2015**, *433* (1), 251–269. https://doi.org/10.1144/sp433.7.

Akhani, H.; Mahdavi, P.; Noroozi, J.; Zarrinpour, V. Vegetation Patterns of the Irano-Turanian Steppe along a 3,000 m Altitudinal Gradient in the Alborz Mountains of Northern Iran. *Folia Geobotanica* **2013**, *48* (3), 229–255. https://doi.org/10.1007/s12224-012-9147-8.

Amirkhani, M.; Zarei, H.; Radmanesh, F.; Pande, S. An Operational Sociohydrological Model to Understand the Feedbacks between Community Sensitivity and Environmental Flows for an Endorheic Lake Basin, Lake Bakhtegan, Iran. *J. Hydrol.* **2022**, *605*, 127375. https://doi.org/10.1016/j.jhydrol.2021.127375.

Ansell, S. W.; Stenøien, H. K.; Grundmann, M.; Russell, S. J.; Koch, M. A.; Schneider, H.; Vogel, J. C. The Importance of Anatolian Mountains as the Cradle of Global Diversity in *Arabis alpina*, a Key Arctic–Alpine Species. *Ann. Bot.* **2011**, *108* (2), 241–252. https://doi.org/10.1093/aob/mcr134.

Avcı, M. Anadolu Bozkırları: Özellikleri ve Dağılışları. In *Alıç Ağacının Gölgesinde Anadolu Bozkırları*; Ekim, T.; Gür, M. K., Eds.; İş Bankası Kültür Yayınları: İstanbul, 2019; pp 59–90.

Avcı, M. Diversity and Endemism in Turkey's Vegetation. *Journal of Geography–Coğrafya Dergisi* **2005**, *13*, 27–55.

Avcı, M. Jeoloji. In *Resimli Türkiye Florası*, 2nd ed.; Güner, A.; Ekim, T.; Kandemir, A.; Menemen, Y.; Yıldırım, H.; Aslan, S.; et al., Eds.; ANG Vakfı Nezahat Gökyiğit Botanik Bahçesi Yayınları: İstanbul, 2022a; Vol. 1, pp 29–49.

Avcı, M. Karasu–Tuzla Çayı Arasındaki Sahada Orman Kalıntıları. *Journal of Geography–Coğrafya Dergisi* **1997**, *5*, 179–224. https://dergipark.org.tr/tr/download/article-file/231133 (accessed April 21, 2026).

Avcı, M. Paleocoğrafya. In *Resimli Türkiye Florası*, 2nd ed.; Güner, A.; Ekim, T.; Kandemir, A.; Menemen, Y.; Yıldırım, H.; Aslan, S.; et al., Eds.; ANG Vakfı Nezahat Gökyiğit Botanik Bahçesi Yayınları: İstanbul, 2022b; Vol. 1, pp 51–78.

Avcı, M. The Floristic Regions of Turkey and a Geographical Approach for Anatolia. *Review of the Department of Geography, University of Istanbul* **1996**, *3*, 59–91.

Avcı, M. *The Last Shelters for the Remaining Forests of the Central Anatolian Region: The Vegetation on Karacadağ and Karadağ Volcanoes*; Çantay: İstanbul, 2004.

Avcı, M.; Avcı, S. İklim. In *Resimli Türkiye Florası*, 2nd ed.; Güner, A.; Ekim, T.; Kandemir, A.; Menemen, Y.; Yıldırım, H.; Aslan, S.; et al., Eds.; ANG Vakfı Nezahat Gökyiğit Botanik Bahçesi Yayınları: İstanbul, 2022; Vol. 1, pp 107–119.

Avcı, S. The Important Mine for Economic Geography: Salt (History, Importance and Salt Economy of the World). *Coğrafya Dergisi* **2003**, *11*, 21–45. https://dergipark.org.tr/en/download/article-file/231185 (accessed April 20, 2026).

Bayrakdar, C.; Çılğın, Z.; Döker, M. F.; Canpolat, E. Evidence of an Active Glacier in the Munzur Mountains, Eastern Turkey. *Turkish J. Earth Sci.* **2015**, *24* (1), 56–71. https://doi.org/10.3906/yer-1403-7.

Beck, H. E.; McVicar, T. R.; Vergopolan, N.; Berg, A.; Lutsko, N. J.; Dufour, A.; et al. High-Resolution (1 km) Köppen–Geiger Maps for 1901–2099 Based on Constrained CMIP6 Projections. *Scientific Data* **2023**, *10*, 724. https://doi.org/10.1038/s41597-023-02549-6.

Bozkurt, M.; Calleja Alarcón, J. A.; Uysal, T.; Garcia-Jacas, N.; Ertuğrul, K.; Susanna, A. Biogeography of *Rhaponticoides*, an Irano-Turanian Element in the Mediterranean Flora. *Scientific Reports* **2022**, *12*, 22019. https://doi.org/10.1038/s41598-022-24947-3.

Brun, P.; Zimmermann, N. E.; Hari, C.; Pellissier, L.; Karger, D. CHELSA-BIOCLIM+ A Novel Set of Global Climate-Related Predictors at Kilometre-Resolution. *EnviDat* **2022**. https://doi.org/10.16904/envidat.332.

Chen, D.; Chen, H. W. Using the Köppen Classification to Quantify Climate Variation and Change: An Example for 1901–2010. *Environ. Dev.* **2013**, *6*, 69–79. https://doi.org/10.1016/j.envdev.2013.03.007.

Chesworth, W. Biome. In *Encyclopedia of Soil Science*; Chesworth, W., Ed.; Springer: Dordrecht, The Netherlands, 2008; pp 60–68.

Davis, P. H. Distribution Patterns in Anatolia with Particular Reference to Endemism. In *Plant Life of South-West Asia*; Davis, P. H.; Harper, P. C.; Hedge, I. C., Eds.; The Botanical Society of Edinburgh, University Press: Aberdeen, 1971; pp 15–27.

Djamali, M.; Brewer, S.; Breckle, S. W.; Jackson, S. T. Climatic Determinism in Phytogeographic Regionalization: A Test from the Irano-Turanian Region, SW and Central Asia. *Flora* **2012**, *207* (4), 237–249. https://doi.org/10.1016/j.flora.2012.01.009.

Djamali, M.; De Beaulieu, J.; Miller, N.; Andrieu-Ponel, W.; Ponel, P.; Lak, R.; et al. Vegetation History of the SE Section of the Zagros Mountains during the Last Five Millennia; A Pollen Record from the Maharlou Lake, Fars Province, Iran. *Veget. Hist. Archaeobot.* **2009**, *18*, 123–136. https://doi.org/10.1007/s00334-008-0178-2.

Eig, A. Sinopsis of the Phytosociological Units of Palestine. *Palestine J. Bot.* **1946**, *3* (4), 183–246.

Eimanifar, A.; Mohebbi, F. Urmia Lake (Northwest Iran): A Brief Review. *Saline Syst.* **2007**, *3*, 5. https://doi.org/10.1186/1746-1448-3-5.

Ekim, T.; Güner, A. The Anatolian Diagonal: Fact or Fiction? *Proc. R. Soc. Edinburgh, Sect. B, Biol. Sci.* **1986**, *89*. https://doi.org/10.1017/S0269727000008915.

El-Moslimany, A. The Late Pleistocene Climates of the Lake Zeribar Region (Kurdistan, Western Iran) Deduced from the Ecology and Pollen Production of Nonarboreal Vegetation. *Vegetatio* **1987**, *72*, 131–139. https://doi.org/10.1007/BF00039834.

Erinç, S. Glacial Evidences of the Climatic Variations in Turkey. *Geogr. Ann.* **1952**, *1*, 89–98.

Erinç, S. *Klimatoloji ve Metodları*; İstanbul University Geography Institute: İstanbul, 1969.

Eriş, K. K.; Ön, S. A.; Çağatay, M. N.; Ülgen, U. B.; Ön, Z. B.; Gürocak, Z.; et al. Late Pleistocene to Holocene Paleoenvironmental Evolution of Lake Hazar, Eastern Anatolia, Turkey. *Quat. Int.* **2018**, *486*, 4–16. https://doi.org/10.1016/j.quaint.2017.09.027.

Erol, O. *Geology and Geomorphology of Tuzgölü Basin*; TÜBİTAK-MTA Report No. 4220: Ankara, 1969.

Erol, O. The Quaternary History of the Lake Basins of Central and Southern Anatolia. In *The Environmental History of the Near and Middle East since the Last Ice Age*; Brice, W., Ed.; Academic Press: London, 1978; pp 111–139.

FAO. *World Reference Base*; FAO Soil Portal; 2025. https://www.fao.org/soils-portal/data-hub/soil-classification/world-reference-base/en/ (accessed April 21, 2026).

Ferrigno, J. G. Glaciers of the Middle East and Africa—Glaciers of Iran. In *Satellite Image Atlas of Glaciers of the World: Middle East and Africa*; Williams, R. S.; Ferrigno, J.

G., Eds.; U.S. Geological Survey Professional Paper 1386-G: Washington, 1991; pp G31–G47.

Freitag, H.; Atamov, V.; Çetin, E.; Aslan, M. The Genus *Halothamnus* Jaub. & Spach (Chenopodiaceae) in Turkey. *Turkish J. Bot.* **2009**, *33* (5). https://doi.org/10.3906/bot-0902-14.

Ghorbani, M. *The Economic Geology of Iran*; Springer: Dordrecht, Heidelberg, New York, and London, 2013.

Ghorbani, M. *The Geology of Iran: Tectonic, Magmatism and Metamorphism*; Springer Nature Switzerland AG: Cham, Switzerland, 2021. https://doi.org/10.1007/978-3-030-71109-2.

Görür, N.; Çağatay, M.; Zapçı, C.; Sakınç, M.; Akkök, R.; Şile, H.; Örçen, S. The Late Quaternary Tectono-Stratigraphic Evolution of Lake Van, Turkey. *Bull. Miner. Res. Explor.* **2015**, *151*, 1–46.

Gül, S. Ecological Divergence between Two Evolutionary Lineages of *Hyla savignyi* (Audouin, 1827) in Turkey: Effects of the Anatolian Diagonal. *Anim. Biol.* **2013**, *63* (3), 1–11. https://doi.org/10.1163/15707563-00002412.

Gür, H. The Anatolian Diagonal Revisited: Testing the Ecological Basis of a Biogeographic Boundary. *Zool. Middle East* **2016**, *62* (3), 189–199. https://doi.org/10.1080/09397140.2016.1226544.

Gürbüz, A.; Kazancı, N.; Hakyemez, H. Y.; Leroy, S. A.; Roberts, N.; Saraç, G.; et al. Geological Evolution of a Tectonic and Climatic Transition Zone: The Beyşehir-Suğla Basin, Lake District of Turkey. *Int. J. Earth Sci.* **2015**, *110* (3), 1077–1107. https://doi.org/10.1007/s00531-021-02007-x.

Herrando-Moraira, S.; Roquet, C.; Calleja, J. A.; Chen, Y. S.; Fujikawa, K.; Galbany-Casals, M.; et al. Impact of Climatic Changes in the Pliocene–Pleistocene Transition on Irano-Turanian Species: The Radiation of Genus *Jurinea* (Compositae). *Mol. Phylogenet. Evol.* **2023**, *189*, 107928. https://doi.org/10.1016/j.ympev.2023.107928.

Heydarizad, M.; Gimeno, L.; Amiri, S.; Minaei, M.; Mohammadabadi, H. G. A Comprehensive Overview of the Hydrochemical Characteristics of Precipitation across the Middle East. *Water* **2022**, *14*, 2657. https://doi.org/10.3390/w14172657.

Hoffman, M.; Koenig, K.; Bunting, G.; Costanza, J.; Kristen, J. W. Biodiversity Hotspots, version 2016.1; **2016**. https://doi.org/10.5281/zenodo.3261807.

ISRIC. *World Soil Information*. ISRIC World Soil Information; **2025**. https://www.isric.org/explore/world-soil-distribution (accessed April 21, 2026).

IUSS Working Group WRB. *World Reference Base for Soil Resources: International Soil Classification System for Naming Soils and Creating Legends for Soil Maps*, 4th ed.; International Union of Soil Sciences: Vienna, Austria, 2022.

Jimenez-Moreno, G.; Suc, J. P. Middle Miocene Latitudinal Climatic Gradient in Western Europe: Evidence from Pollen Records. *Palaeogeogr. Palaeoclimatol. Palaeoecol.* **2007**, *253* (1, 2), 208–225. https://doi.org/10.1016/j.palaeo.2007.03.040.

Karakhanian, A.; Tozalakyan, P.; Grillot, J. C.; Philip, H.; Melkonyan, D.; Paronyan, P.; Arakelyan, S. Tectonic Impact on the Lake Sevan Environment (Armenia). *Environ. Geol.* **2001**, *40* (3), 279–288.

Karger, D. K.; Conrad, O.; Böhner, J.; Kawohl, T.; Kreft, H.; Soria-Auza, R. W.; et al. Climatologies at High Resolution for the Earth's Land Surface Areas. *Sci. Data* **2017**, *4*, 170122. https://doi.org/10.1038/sdata.2017.122.

Kent, P. E. Middle East—The Geological Background. *Q. J. Eng. Geol. Hydrogeol.* **1978**, *11* (1), 1–7.

Ketenoğlu, O.; Vural, M.; Kurt, L.; Körüklü, T. Vejetasyon. In *Resimli Türkiye Florası*; Güner, A., Ed.; ANG Vakfı Nezahat Gökyiğit Botanik Bahçesi Yayınları: İstanbul, 2022; 2nd ed.; Vol. 1, pp 181–265.

Kırsal Çevre. *İç Anadolu'nun Kalıntı Ormanları*; Kırsal Çevre ve Ormancılık Sorunları Araştırma Derneği: Ankara, 2019.

Kottek, M.; Grieser, J.; Beck, C.; Rudolf, B.; Rubel, F. World Map of the Köppen–Geiger Climate Classification Updated. *Meteorol. Z.* **2006**, *6*, 259–263. https://doi.org/10.1127/0941-2948/2006/0130.

Külköylüoğlu, O.; Yavuzatmaca, M.; Yılmaz, O.; Tunoğlu, C.; Akdemir, D.; Nazik, A.; Tuncer, A. Zoogeographical and Stratigraphical Distribution of the Genus *Zonocypris*: Supportive Evidence for Anatolian Diagonal and Description of a New Species from Turkey. *BSGF Earth Sci. Bull.* **2021**, *192*, 1.

Kuzucuoğlu, C.; Şengör, A. C.; Çiner, A. The Tectonic Control on the Geomorphological Landscape of Turkey. In *Landscapes and Landforms of Turkey*; Kuzucuoğlu, C.; Kazancı, N.; Çiner, A., Eds.; Springer Nature Switzerland AG: Cham, Switzerland, 2019; pp 17–40.

Lahn, E. *Türkiye Göllerinin Jeolojisi ve Jeomorfolojisi Hakkında Bir Etüd*; MTA Enstitüsü: Ankara, 1948.

Maghsoudi, M. *Desert Landscapes and Landforms of Iran*; Springer Nature Switzerland AG: Cham, Switzerland, 2021.

Manafzadeh, S.; Salvo, G.; Conti, E. A Tale of Migrations from East to West: The Irano-Turanian Floristic Region as a Source of Mediterranean Xerophytes. *J. Biogeogr.* **2013**, *41* (2). https://doi.org/10.1111/jbi.12185.

Manafzadeh, S.; Staedler, Y. M.; Conti, E. Visions of the Past and Dreams of the Future in the Orient: The Irano-Turanian Region from Classical Botany to Evolutionary Studies. *Biol. Rev.* **2016**. https://doi.org/10.1111/brv.12287.

Maylandt, C.; Kirchner, P.; Pirkebner, D.; Frajman, B.; de Giles, J. P.; Schönswetter, P.; Carnicero, P. Evolution, Range Formation and a Revised Taxonomy of the Disjunctly Distributed European Members of *Astragalus* sect. *Caprini*, an Intricate Group Including Highly Endangered Species of Dry Grasslands. *Mol. Phylogenet. Evol.* **2025**, *204*, 108242. https://doi.org/10.1016/j.ympev.2024.108242.

Memariani, F. The Khorassan-Kopet Dagh Mountains. In *Plant Biogeography and Vegetation of High Mountains of Central and South-West Asia*; Noroozi, J., Ed.; Springer Nature Switzerland AG: Cham, Switzerland, 2020; pp 93–116.

Mohammadzadeh, H.; Eskandari, J.; Heydarizad, M. The Effects of Moisture Sources and Local Parameters on the 18O and 2H Contents of Precipitation in the West of Iran and the East of Iraq. *Tellus B: Chem. Phys. Meteorol.* **2020**, *72* (1), 25–39. https://doi.org/10.1080/16000889.2020.1721224.

Mutun, S. Review of Oak Gall Wasps Phylogeographic Patterns in Turkey Suggests a Main Role of the Anatolian Diagonal. *Turkish J. For.* **2016**, *17* (Special Issue), 1–6. https://doi.org/10.18182/tjf.65861.

Noroozi, J.; Minaei, M.; Khalvati, S.; Kaveh, A.; Nafisi, H.; Nazari, B.; et al. Hotspots of (Sub)Alpine Plants in the Irano-Anatolian Global Biodiversity Hotspot Are Insufficiently Protected. *Divers. Distrib.* **2023**, *29*, 244–253. https://doi.org/10.1111/ddi.13656.

Noroozi, J.; Talebi, A.; Doostmohammadi, M. The Alborz Mountain Range. In *Plant Biogeography and Vegetation of High Mountains of Central Asia and South-West Asia*; Noroozi, J., Ed.; Springer Nature Switzerland AG: Cham, Switzerland, 2020a; pp 117–149.

Noroozi, J.; Talebi, A.; Doostmohammadi, M.; Bagheri, A. The Zagros Mountain Range. In *The Zagros Mountain Range*; Noroozi, J., Ed.; Springer Nature Switzerland AG: Cham, Switzerland, 2020b.

Noroozi, J.; Talebi, A.; Doostmohammadi, M.; Manafzadeh, S.; Asgarpour, Z.; Schneeweis, G. M. Endemic Diversity and Distribution of the Iranian Vascular Flora across Phytogeographical Regions, Biodiversity Hotspots and Areas of Endemism. *Sci. Rep.* **2019**, *9*, 12991. https://doi.org/10.1038/s41598-019-49417-1.

Noroozi, J.; Talebi, A.; Dostmohammadi, M.; Rumpf, S. B.; Linder, P.; Schneeweiss, G. M. Hotspots within a Global Biodiversity Hotspot—Areas of Endemism Are Associated with High Mountain Ranges. *Sci. Rep.* **2018**, *8*, 10345. https://doi.org/10.1038/s41598-018-28504-9.

Özalp, S. Late Pleistocene–Holocene Lake Terraces, Water Level Change, and Active Tectonics: Eastern Coast of Lake Van, Eastern Anatolia, Turkey. *Quat. Int.* **2020**, *542*, 54–64. https://doi.org/10.1016/j.quaint.2020.02.037.

Özsayın, E.; Gürbüz, A.; Kuzucuoğlu, C.; Erdoğu, B. Salted Landscapes in the Tuz Gölü (Central Anatolia): The End Stage of a Tertiary Basin. In *Landscapes and Landforms of Turkey*; Kuzucuoğlu, C.; Kazancı, N.; Çiner, A., Eds.; Springer Nature Switzerland: Cham, 2019; pp 339–352.

Parolly, G. The Anatolian High-Mountain Ranges—Plant Diversity between Two Seas. In *Plant Biogeography and Vegetation of High Mountains of Central and South-West Asia*; Noroozi, J., Ed.; Springer Nature Switzerland: Cham, 2020; pp 215–286. https://doi.org/10.1007/978-3-030-45212-4_7.

Peterson, A.; Harpke, D.; Peterson, J.; Harpke, A.; Peruzzi, L. A Pre-Miocene Irano-Turanian Cradle: Origin and Diversification of the Species-Rich Monocot Genus *Gagea* (Liliaceae). *Ecol. Evol.* **2019**, *9* (10), 5870–5890. https://doi.org/10.5061/dryad.97np7bt.

Raziei, T. An Analysis of Daily and Monthly Precipitation Seasonality and Regimes in Iran and the Associated Changes in 1951–2014. *Theor. Appl. Climatol.* **2018**, *134*, 913–934. https://doi.org/10.1007/s00704-017-2317-0.

Roquet, C.; Sanmartin, I.; Garcia-Jacas, N.; Saez, L.; Susanna, A.; Wikström, N.; Aldasoro, J. J. Reconstructing the History of Campanulaceae with a Bayesian Approach to Molecular Dating and Dispersal–Vicariance Analyses. *Mol. Phylogenet. Evol.* **2009**, *52* (3), 575–587. https://doi.org/10.1016/j.ympev.2009.05.014.

Salur, A.; Miroğlu, A.; Akyıldız, G. K. The Impact of the Anatolian Diagonal on Biodiversity in Turkey: A Geometric Morphometric Study on the Wing Morphology of *Platycnemis dealbata* (Insecta: Odonata). *Biol. Bull. Russ. Acad. Sci.* **2024**, *51*, 1379–1389. https://doi.org/10.1134/S1062359023605827.

Sembroni, A.; Reitano, R.; Faccenna, C.; Callieri, P. The Geologic Configuration of the Zagros Fold and Thrust Belt: An Overview. *Mediterr. Geosci. Rev.* **2024**, *6* (2), 61–86. https://doi.org/10.1007/s42990-024-00118-6.

Şengör, A. C.; Yılmaz, Y. Tethyan Evolution of Turkey: A Plate Tectonic Approach. *Tectonophysics* **1981**, *75* (3, 4), 203–241.

Şenol, Ö.; Görür, G.; Beğen, H. A. Contributions of the Anatolian Diagonal Effect on Turkish Aphid Diversity. *Artvin Coruh Univ. J. For. Fac.* **2019**, *20* (1), 102–109.

Seyitoğlu, G.; Tunçel, E.; Kaypak, B.; Esat, K.; Gökkaya, E. The Anatolian Diagonal: A Broad Left-Lateral Shear Zone between the North Anatolian Fault Zone and the Aegean/Cyprus Arcs. *Türkiye Jeol. Bül.* **2022**, *65* (2), 93–116.

Shamsabad, M. M.; Moharrek, F.; Assadi, M.; Feliner, G. N. Biogeographic History and Diversification Patterns in the Irano-Turanian Genus *Acanthophyllum* s.l. (Caryophyllaceae). *Plant Biosyst.* **2021**, *155* (3), 425–435. https://doi.org/10.1080/11263504.2020.1756974.

Spaargaren, O. Calcisols. In *Encyclopedia of Soil Science*; Chesworth, W., Ed.; Springer: Dordrecht, The Netherlands, 2008a; pp 79, 80.

Spaargaren, O. Cambisols. In *Encyclopedia of Soil Science*; Chesworth, W., Ed.; Springer: Dordrecht, 2008b; pp 80, 81.

Spaargaren, O. Chernozems. In *Encyclopedia of Soil Science*; Chesworth, W., Ed.; Springer: Dordrecht, The Netherlands, 2008c; pp 108, 109.

Spaargaren, O. Leptosols. In *Encyclopedia of Soil Science*; Chesworth, W., Ed.; Springer: Dordrecht, The Netherlands, 2008g; pp 437, 438.

Spaargaren, O. Luvisols. In *Encyclopedia of Soil Science*; Chesworth, W., Ed.; Springer: Dordrecht, The Netherlands, 2008d; pp 440–442.

Spaargaren, O. Regosols. In *Encyclopedia of Soil Science*; Chesworth, W., Ed.; Springer: Dordrecht, The Netherlands, 2008e; pp 605, 606.

Spaargaren, O. Solonchaks. In *Encyclopedia of Soil Science*; Chesworth, W., Ed.; Springer: Dordrecht, The Netherlands, 2008f; pp 737, 738.

Stampfli, G. M.; Borel, G. D. A Plate Tectonic Model for the Paleozoic and Mesozoic Constrained by Dynamic Plate Boundaries and Restored Synthetic Oceanic Isochrons. *Earth Planet. Sci. Lett.* **2002**, *196* (1, 2), 17–33.

Stampfli, G. M.; Borel, G. D.; Cavazza, W.; Mosar, J.; Ziegler, P. A. Palaeotectonic and Palaeogeographic Evolution of the Western Tethys and Peri-Tethyan Domain (IGCP Project 369). *Episodes* **2001**, *24* (4), 222–228. https://doi.org/10.18814/epiiugs/2001/v24i4/001.

Taşoğlu, E.; Öztürk, M. Z.; Yazıcı, Ö. High-Resolution Köppen-Geiger Climate Zones of Türkiye. *Int. J. Climatol.* **2024**, *44* (14), 5248–5265. https://doi.org/10.1002/joc.8635.

TUIK. *Turkey's Statistical Yearbook-2013*; Turkish Statistical Institute: Ankara, 2014.

Türkeş, M. *Klimatoloji ve Meteoroloji*; Kriter Yayınevi: İstanbul, 2010.

van Zeist, W.; Bottema, S. Palynological Investigations in Western Iran. *Palaeohistoria* **1977**, *19*, 18–85.

van Zeist, W.; Wright, Jr. H. E. Preliminary Pollen Studies at Lake Zeribar, Zagros Mountains, Southwestern Iran. *Science* **1963**, *140* (3562), 65–67. https://www.jstor.org/stable/1710743 (accessed April 21, 2026).

Vardanian, T. The Hydro-Chemical Changes of Lake Sevan Water after the Artificial Lowering of the Water Level. In *The Role of Ecological Chemistry in Pollution Research and Sustainable Development*; Springer: Dordrecht, The Netherlands, 2009; pp 77–84.

Yılmaz, Y.; Güner, Y.; Şaroğlu, F. Geology of the Quaternary Volcanic Centers of East Anatolia. *J. Volcanol. Geotherm. Res.* **1998**, *85* (1–4), 173–210. https://doi.org/10.1016/S0377-0273(98)00055-9.

Yılmaz, Y.; Şaroğlu, F.; Güner, Y. Initiation of the Neomagmatism in East Anatolia. *Tectonophysics* **1987**, *134*, 177–199.

Zohary, M. *Geobotanical Foundations of the Middle East*; Vol. 2; Gustav Fischer: Stuttgart, 1973.

CHAPTER 2

DIVERSITY OF CYANOBACTERIAL SPECIES AND THEIR APPLICATION AREAS IN TÜRKIYE

TOLGA COŞKUN[1,2]* and ANIL ESMEYAR[3]

[1]*Technology Transfer Office, Niğde Ömer Halisdemir University, Niğde, Türkiye*

[2]*Department of Biology, Faculty of Arts and Sciences, Middle East Technical University (METU), Ankara, Türkiye*

[3]*School of Foreign Languages, Cappadocia University, Nevşehir, Türkiye*

*Corresponding authors

ABSTRACT

The increasing demand for water resources has made the protection of aquatic ecosystems and the monitoring of water quality increasingly imperative. In this context, cyanobacteria stand out as photosynthetic prokaryotic microorganisms. Cyanobacterial blooms triggered by eutrophication can produce toxins that negatively affect water quality. Cyanobacteria are morphologically divided into groups such as Chroococcales, Nostocales, Oscillatoriales, Pleurocapsales, and Stigonematales. Some of their species support plant growth by fixing nitrogen. With seasonal changes, algal blooms can occur, and these processes can disrupt ecosystem balance and reduce dissolved oxygen levels in the water. Cyanobacteria are also of economic and industrial importance, with potential use in the food, fuel, fertilizer, and pharmaceutical industries. For example, *Spirulina* is a popular food supplement, as it contains high levels of protein, vitamins, and minerals in its content. In this chapter, the effects of cyanobacteria on

Biodiversity Hotspot of the Irano-Anatolian Region. T. Pullaiah (Ed.)

DOI: 10.1201/9781779645760-2

water quality, their toxin types, biodiversity, and ecological roles, as well as their potential in different areas of application, are analyzed. The chapter argues that cyanobacteria pose a risk to ecosystem health, especially in terms of algal bloom, eutrophication, and cyanotoxin production, but can also be beneficial in various industrial and biotechnological applications.

2.1 INTRODUCTION

As the human population expands and water resource conditions shift, the demand for accessible and usable water continues to rise. The limited availability of usable water resources has heightened the significance of studies on water quality and monitoring of aquatic ecosystems. In this regard, the concern about organic pollution and eutrophication has intensified as they are important signs of cyanobacterial presence and their related toxins. Eutrophication fosters conditions for the deterioration of water quality and the formation of cyanobacteria (Koca et al., 2019).

Blue-green algae, also known as Schizophyta and/or Myxophyta, are prokaryotic organisms without true nuclei and plastids (Cirik and Gökpınar, 1993; İspirli, 2009). In this group, Cyanobacteria are among the oldest and most widespread photosynthetic microorganisms with an evolutionary history of 3.5 billion years (Koca et al., 2019). Cyanobacteria are eminent microorganisms that can occupy the area where they are located due to their metabolism, plasticity, and ability to maintain their cell structure and different ecological characteristics. They are also known to possess gram-negative bacteria capable of oxygenic photosynthesis and the photosynthetic pigments chlorophyll *a* and phycobilin's, an antenna pigment. Additionally, some species of cyanobacteria fix nitrogen and are therefore important nitrogen providers for plants (Abed et al., 2009; Çelikoğlu, 2022; Chaurasia, 2015; Filatova et al., 2020).

These organisms contribute to primary productivity in aquatic ecosystems and are widely distributed across both aquatic and terrestrial environments, including extreme habitats, due to their ecological adaptability (Hitzfeld et al., 2000; Özen 2020; Sompong et al., 2005; Taton et al., 2006). Notably, these organisms exhibit thermal tolerance, occurring in environments that span from polar ice caps to hot springs, and are widely found in various habitats, including oceans, freshwater bodies, and terrestrial surfaces (Bulut, 2023; Çelikoğlu, 2022; Genuário et al., 2019; Henson et al., 2004; Schirrmeister et al., 2015; Whitton, 2012).

2.2 GENERAL CHARACTERISTICS OF CYANOBACTERIA

Cyanobacteria, as prokaryotic organisms, are a group of microorganisms known for their photosynthetic ability (the ability to fix nitrogen and produce organic carbon and oxygen by photosynthesis). Commonly linked to aquatic habitats, cyanobacteria are widely present in both freshwater and saltwater systems (Ward et al., 2006; Yadav et al., 2017). While cyanobacteria are structurally similar to other bacteria, their possession of chlorophyll *a*, which enables them to absorb light and convert it into chemical energy, makes them functionally comparable to plants. In addition, they contain the pigments carotene, xanthophyll, phycocyanin, and phycoerythrin. Among these pigments, phycocyanin and phycoerythrin are unique to cyanobacteria (Whitton and Potts, 2000).

Due to their prokaryotic structure, they do not have true nuclei and plastids. As they do not possess a nuclear membrane, their DNA and pigment substances are dispersed in the cytoplasm (Aykın, 2023; Ballot et al., 2004; Cirik and Cirik, 2004; İspirli, 2009; Kahraman and Küplülü, 2012). In addition, pigments such as chlorophyll a and c, phycocyanin, ß+k carotenoids, and xanthophylls are present in their structure. Cyanobacterial cell walls lack cellulose and pectin, and instead store reserve materials such as glycogen, cyanophycean, and volutin (Cirik and Gökpınar, 1993; Rai et al., 2002; Rinehart et al., 1994; Whitton and Potts, 2000; Yılmaz, 2009).

Although their photosynthetic mechanisms are similar to microalgae, unlike eukaryotic microalgae, they do not have membrane-enclosed organelles such as chloroplasts. Photosynthetic pigments are located in the thylakoid membrane systems near the cell periphery. Due to these properties, they produce free oxygen during photosynthesis, which plays a critical role in adding oxygen to the atmosphere (Rinehart et al., 1994; Sancar et al., 2023; Varlı 2021; Yüksel et al., 2021). An important metabolic characteristic of cyanobacteria is their ability to fix atmospheric nitrogen. This function is commonly observed in heterocyst-forming filamentous genera such as *Anabaena, Aphanizomenon*, and *Nostoc*. However, certain non-heterocystous species, such as *Trichodesmium*, are also capable of nitrogen fixation under specific environmental conditions. They are able to convert nitrogen directly to ammonium (NH_4) under aerobic conditions using the enzyme nitrogenase (Cirik and Cirik, 2004; Mankiewicz et al., 2003).

Cyanobacteria can be unicellular, multicellular (colony forming), and filamentous forms with cell sizes ranging from 0.2–40 µm, which differentiates them from other prokaryotes morphologically, physiologically, and metabolically (Graham and Wilcox, 2000; Kahraman and Küplülü, 2012). Cyanobacteria exhibit both unicellular and filamentous forms and could be morphologically classified into five major groups (Chroococcales, Nostocales, Oscillatoriales, Pleurocapsales, and Stigonematales) (Mankiewicz et al., 2003; Metcalf and Codd, 2014; Rippka et al., 1979).

2.3 DISTRIBUTION, HABITAT, AND DIVERSITY

Cyanobacteria are an important part of freshwater, saltwater, and marine plankton and are widespread throughout the world. They can live in soil, on rocks, and symbiotically with plants or fungi. Moreover, owing to their broad ecological distribution (from terrestrial to aquatic habitats) and their capacity for oxygenic photosynthesis, cyanobacteria are believed to be responsible for the transition of Earth's atmosphere from an anoxic to an oxic state (Henson et al., 2004; İspirli, 2009; Yüksel et al., 2021). Due to their prevalence, they have the potential to significantly increase the overall biodiversity of ecosystems. They are also considered important indicators of ecosystem health. The cyanobacteria population can provide valuable information about the overall health and functioning of an ecosystem. At the same time, through its consumption of carbon dioxide, it plays a critical role in the carbon cycle and helps stabilize the level of carbon dioxide in the atmosphere. In short, they are vital for ecosystems with their photosynthetic ability, oxygen production, and role in the carbon cycle (Aydin, 2023; Eren, 2021; Kırtıloğlu, 2023; Omuzbüken and Kaçar, 2021; Özükaya and Güzey, 2024).

In terms of their ecological role, cyanobacteria contribute to the nitrogen cycle through nitrogen fixation, enabling nitrogen-dependent organisms to thrive and integrate into the food chain. This process allows for increased nutrient productivity, especially in nitrogen-deficient terrestrial and aquatic environments. Thus, by supporting the development of vegetation, it contributes to increasing biodiversity (Akdura, 2022; Aydin, 2023; Çiltepe, 2022; Çolak, 2022; Kula, 2021; Kütük and Akpınar, 2024). There are approximately 5723 species of cyanobacteria worldwide, and toxin-producing taxa are *Anabaena, Aphanizomenon, Cylindrospermum, Cylindrospermopsis, Hapalosiphon, Gleotrichia, Microcystis, Nodularia, Nostoc, Oscillatoria, Scytonema, Synechocystis,* and *Schizothrix* (Guiry, 2024; Kabziński et al., 2000). In this regard, Maraşlıoğlu and Gönülol (2025) reported that there are 5480 phytoplankton taxa (38 taxa of which are unknown) identified in

Türkiye, and cyanobacteria constitute 11% with 603 taxa (3 taxa of which are unknown). Table 2.2 shows the new registration studies for Türkiye between 2009 and 2022 by basin and water body. As is shown in Table 2.1, a total of 74 taxa belonging to the cyanobacteria group, which are new records for Türkiye, were identified in the Euphrates-Tigris (21 taxa), Sakarya (16 taxa), Western Mediterranean (8 taxa), and Akarçay (6 taxa) basins.

This section presents studies on cyanobacteria (blue-green algae) in various freshwater ecosystems in Türkiye from 1998 to 2024, including reservoirs, rivers, thermal water bodies, natural lakes, and ponds.

2.3.1 IN RIVERS

River ecosystems in Türkiye demonstrate remarkable variations in cyanobacterial taxonomic richness. The Tigris River stands out with exceptional diversity, harboring 72 taxa (Varol and Şen, 2014), the highest documented among all freshwater ecosystems included. The Munzur River in Tunceli contained 15 taxa (Kutlu and Demir, 2018), while the Seyhan River exhibited moderate diversity with 12 taxa (Aygün, 2000). In addition, the Mert River in the Kızılırmak River Basin harbored 9 taxa (Maraşlıoğlu and Bektaş, 2022), and the Tunca River contained 7 taxa (Öterler, 2003). Several rivers demonstrated lower cyanobacterial diversity: Kille Stream and Kocaavşar Stream in Balıkesir each had 6 taxa (Çelik, 2022, 2023), as did the Yeşilırmak River (Soylu and Gönülol, 2003). The lowest cyanobacterial richness was observed in the Arda River in Edirnc with only 3 taxa (Öterler 2016).

2.3.2 IN DAM LAKES

Taxonomic studies on cyanobacteria in Turkish dam lakes reveal considerable variations in diversity. Hirfanlı Dam Lake exhibited exceptional cyanobacterial richness with 39 taxa (Baykal and Açıkgöz, 2004), representing the highest diversity among all studied dam lakes. Devegeçidi Dam in Gaziantep harbored 29 taxa (Baykal et al., 2004), while four dam lakes, Demirköprü Dam Lake (Tenekecioğlu, 2011), Kayalıköy Dam Lake (Aykın, 2023), Çaygören Dam Lake (Sevindik, 2010), and Karacaören-1 Dam Lake (Ülkü (Bayoğlu), 2016), each contained 19 taxa, with the latter distributed across 9 families. In the Küçük Menderes River Basin, Köker et al. (2022) identified a total of 13 cyanobacteria species across five reservoirs and two natural lakes.

TABLE 2.1 Newly-Recorded Cyanobacteria Taxa Identified in Türkiye and Their Basin-Wise Distribution

Basin*	Number of NR	Total Number By Basin	Percent By Basin	Water Body**	Habitat***	Study Years	References
AK	6	6	8.11	L and DL	Fre	2017–2019	Maraşlıoğlu et al. (2022)
AN	1	1	1.35				
AR	1	1	1.35				
AS	1	1	1.35				
CE	1	1	1.35				
EB	3	3	4.05	L and P	Fre	2013–2016	Şahin and Akar (2019)
F-D	12	21	28.38	L and DL	Fre, Bra	2017–2019	Maraşlıoğlu et al. (2022)
	3			R and DL	Fre	2004–2005	Özer et al. (2012)
	5					2001–2003	Baykal et al. (2009)
	1			R		2008–2009	Varol and Fucikova (2015)
GE	1	1	1.35	L and DL	Fre	2017–2019	Maraşlıoğlu et al. (2022)
KM	1	1	1.35				
KO	3	3	4.05		Fre, Mar		
NA	2	2	2.7		Fre		
SA	2	16	21.62	L	Fre, Bra, Mar	2007–2009	Sevindik et al. (2010)
	2					2013	Sevindik et al. (2017)
	12			L and DL		2012–2013	Sevindik et al. (2015b)

TABLE 2.1 *(Continued)*

Basin*	Number of NR	Total Number By Basin	Percent By Basin	Water Body**	Habitat***	Study Years	References
SU	4	4	5.40	L and DL	Fre	2017–2019	Maraşlıoğlu et al. (2022)
VL	3	3	4.05		Fre, Mar		
WM	8	8	10.81		Fre, Bra		
YE	2	2	2.70		Fre		
Total		**74**	**100.00**				

**AK*: Akarçay Basin, AN: Antalya Basin, AR: Aras River Basin, AS: Asi River Basin, CE: Ceyhan Basin, EB: Eastern Blacksea Basin, F-D: Euphrates-Tigris River Basin, GE: Gediz Basin, KM: Küçük Menderes River Basin, KO: Konya Basin, NA: North Aegean Basin, SA: Sakarya Basin, SU: Susurluk River Basin, VL: Van Lake Basin, WM: Western Mediterranean Basin, YE: Yeşilırmak Basin.

***L*: Lake, DM: Dam Lake, P: Pond, R: River.

****Fre*: Freshwater, Bra: Brackish Water, Mar: Marine Water.

Several dam lakes demonstrated moderate cyanobacterial diversity, including Buldan Dam Lake with 18 species (Ustaoğlu et al., 2010), Bayındır Dam Lake in Ankara with 13 taxa (Atıcı et al., 2005), and Karacaören-1 Dam Lake in an earlier study with 13 taxa (Gülle, 2005). Kemer Dam Lake contained 10 taxa (Özyalın and Ustaoğlu 2008). Also, lower cyanobacterial diversity was observed in multiple dam lakes: Tahtalı Dam Lake and Burç Dam each harbored 7 taxa (Çelekli and Özpınar, 2021a; İspirli, 2009), Cip Dam Lake contained 6 taxa (Kazez, 2012), while both Adıgüzel and Seyhan Dam Lakes had 5 taxa each (Gürbüz, 2012; Sömek, 2011). The earliest study on Karacaören-1 Dam Lake documented only 5 taxa (Ertan et al., 2000). Sarıyar Dam Lake and Uzunçayır Dam Lake each contained 4 taxa (Dokcan, 2010; Kutlu et al., 2014), while Elmalı Dam Lake in Istanbul exhibited the lowest diversity with merely 2 taxa (Yılmaz, 2017).

2.3.3 IN LAKES AND PONDS

Natural lakes and ponds in Türkiye exhibit substantial variations in cyanobacterial diversity. Uluabat Lake in Bursa demonstrated the highest richness with 42 taxa (Karacaoğlu et al., 2004). A comprehensive study encompassing Mogan Lake, Beytepe Pond, and Delice River collectively identified 41 taxa (Batu and Akbulut, 2018). Karamuk Lake in Afyonkarahisar harbored 32 taxa (Kıvrak, 2011), and Liman Lake contained 25 taxa (Demirkalp et al., 2010). Notable diversity was also documented in Hafik Lake in Sivas (23 taxa; Kılınç, 1998), Abant Lake in Bolu (21 taxa; Çelekli et al., 2007b), and Büyükçekmece Lake in Istanbul (20 taxa; Aktan-Turan et al., 2006). Both Mogan Lake (Yerli et al., 2012) and Uluabat Lake in an earlier study (Akbulut et al., 2004) contained 19 taxa each, with the latter distributed across 10 genera.

Intermediate levels of cyanobacterial diversity were observed in several water bodies: Çağış Pond in Balıkesir and Cernek Lake in Samsun each harbored 17 taxa (Sevindik and Çelik, 2014; Taş and Gönülol, 2007), and Gölköy Lake in Bolu contained 16 taxa (Çelekli et al., 2007a). Another study in Mogan Lake documented 13 taxa (Demir et al., 2014).

Lower cyanobacterial richness was found in multiple lakes: Eğirdir Lake in Isparta, Poyrazlar Lake in Sakarya, and Beyşehir Lake each contained 9 taxa (Çiçek et al., 2017; Fakıoğlu and Demir, 2011; Sevindik et al., 2015a). Bafa Lake harbored 7 taxa (Sukatar et al., 2021), while the combined study across Aras (Aygır Lake), Çoruh (Balık and Boğa Lakes), and Eastern Black Sea (Camlı and Çakır Lakes) basins identified a total of 6 taxa (Çetin and Demir, 2023), and Öztürk and Kurt (2021) determined 6 taxa of Cyanobacteria in the Marmara Lake (Manisa).

The lowest diversity was documented in several water bodies: Tortum Lake, Homa Lagoon in İzmir, and the combined areas of Akyatan and Tuzla Lagoons in Adana, each contained only 3 taxa (Çevik et al., 2008; Fakıoğlu et al., 2018; Sabancı, 2014). Similarly, the combined study of Tahtalı, Davuldere, and Çayırköy Ponds in Kocaeli identified only 3 taxa (Yüce and Aktaş, 2020). Gökçeada Salt Lake Wetland exhibited the lowest diversity with merely 2 taxa (Aslan, 2016). In addition, Yılmaz et al. (2021) documented 34 cyanobacterial taxa distributed across 8 divisions in Lake Terkos. Additionally, Köker et al. (2017) conducted a broader survey of 18 selected waterbodies (comprising 7 natural lakes, 8 reservoirs, 1 lagoon, and 2 coastal and transitional waters), detecting cyanobacteria species in 14 of these waterbodies (78%), with algal blooms observed in 57% of the studied ecosystems. In their investigation of subterranean environments, Külköylüoğlu et al. (2014) recorded a total of 28 phytoplankton taxa, with cyanobacteria representing the dominant group (22 taxa). Gülecal and Temel (2014) identified a total of 69 phytoplankton species across six divisions from a lake and its five associated streams, with Cyanophyta contributing 6 species to this diversity.

2.3.4 *IN THERMALS AND HOT SPRINGS*

Ünal (1996) identified 74 taxa in Izmir Balçova (Hot Springs) and stated that 35 of these taxa belonged to the Cyanophyta group. Öztürk Ulçay et al. (2007) identified 19 species belonging to 8 genera distributed in İzmir Dikili (hot springs), and Gül et al. (2013) identified 30 species belonging to 7 genera in Denizli Sarayköy. Öztürk and Kurt (2024) reported a total of 46 algae taxa (43 Cyanobacteria) in seven thermal springs in Denizli, and Öztürk (2021) identified 54 taxa of cyanobacteria in Kütahya (thermal region) and reported that *Oscillatoriales* was the most dominant species in terms of taxon diversity (24 taxa) and biomass.

2.4 SEASONAL CHANGE AND BLOOM

In aquatic ecosystems, eutrophication occurs when environmental conditions are favorable and the amount of nutrients in the water exceeds a certain concentration. Algal blooms (multiplication of algae) are one of the most important biological processes leading to eutrophication, disrupting the odor, color, and ecological balance of water. This causes a decrease in the oxygen level at the bottom, which harms aquatic organisms. Eutrophication occurs in rivers, lakes, and ponds as a result of natural biological production with an increase in nutrient salts such as phosphate and nitrogen. This process leads to an increase in cyanobacteria and algae and the formation of algal masses and floating plant communities on the surface. The ability of cyanobacteria to store phosphate supports biomass growth even at low phosphate concentrations. The increase of cyanobacteria is influenced by the ratio of nitrogen (N) and phosphorus (P), which is reinforced when the ratio is 10–16 molecules N:1 molecule P (Chorus and Bartram, 1999; Gürbüz, 2008; Moss, 1988). However, the full range of factors that trigger cyanobacterial blooms has not yet been conclusively identified. Nitrogen, phosphorus, temperature, light, pH, and hydrological conditions are among the influential factors. Cyanobacteria occupy an important position among microplankton with their ability to move vertically thanks to their gas bladders (Ganf and Oliver, 1982; Gürbüz, 2008; Walsby, 1987).

Cyanobacteria are known to form algal blooms in aquatic areas in a short period of time, within a few days or weeks, with dense cell accumulation. This type of biological proliferation refers to the overgrowth of cyanobacteria and usually occurs in stagnant bodies of water in temperate climates with high nutrient inputs, where light can reach sufficient depths (Merel et al., 2013; Paerl and Otten, 2013; Yüksel et al., 2021). It is stated that algal blooms, which have been observed more frequently with the increasing impact of global warming recently, seriously threaten the health and balance of aquatic ecosystems by reducing the amount of dissolved oxygen in water (Huisman et al., 2018; Taranu et al., 2012; Visser et al., 2016).

2.5 TOXIN TYPES AND THEIR HARMS

Cyanobacteria are prokaryotic organisms that are usually found in aquatic environments and are capable of photosynthesis. These microorganisms, which play a significant role in ecosystems worldwide, exhibit considerable diversity in both natural aquatic environments and habitats influenced by anthropogenic activities. Türkiye is a habitat for various cyanobacteria species due to its geographical location and ecosystem diversity. In terms of microbial cycles and carbon and nitrogen cycles, the impacts of these species on ecosystem functionality are significant. Certain characteristics of cyanobacteria can be decisive for water quality and ecosystem health. Apart from being an important source of nutrients, some species can cause harmful algal blooms, which are critical for water resources management (Akyel, 2022; Atak, 2022; Koyuncuoğlu, 2021; Torun, 2021; Yüksel et al., 2021).

In lakes, the increase in nutrient salts such as nitrogen and phosphorus, excessive water withdrawal, climatic reasons, and wrong fishing methods cause an increase in cyanobacteria density. The mechanisms of action of cyanotoxins produced by species such as *Anabaena* spp., *Aphanizomenon* spp., *Lynbya* sp., *Cylindrospermopsis raciborskii, Microcystis aeruginosa, Nodularia spumigena, Nostoc* sp., *Planktothrix rubescens, P. agardhii,* and *Umezakia* sp. vary according to the organs they affect and their chemical structures.

1. **Neurotoxins:** Anatoxin-a (ATX-a), Anatoxin-a(s) (ATX-a(s)), Homoanatoxin-a (hATX-a), Saxitoxin (STX).
2. **Hepatotoxins:** Cylindrospermopsin (CYN), microcystins (MC), and nodularins (NOD). Additionally, among the cyanobacterial toxins, debromoaplysiatoxin, lyngbyatoxin, and aplysiatoxin are known for their dermatotoxic effects, causing skin irritation upon contact (Akçaalan Albay et al., 2015; Carmichael, 2001; Keskinkaya and Açıkgöz, 2021).

On the other hand, cyanotoxins are divided into groups as Alkaloids, Cyclic peptides, and Lipopolysaccharide endotoxins according to their chemical structures (Codd, 2000; Gürbüz, 2008; Keskinkaya and Açıkgöz, 2021). Table 2.2 lists the cyanobacteria groups according to their chemical structures and the organs they affect in mammals.

These toxins cause skin irritation, stomach cramps and vomiting, sore throat and headache, muscle and joint pain, digestive system disorders, inflammation in the mouth, and liver damage (especially colorectal and liver cancer, which is directly proportional to the concentration of macrocystis). In extreme cases, these toxins can be fatal (Albay et al., 2015; Araoz, 2005; Chorus et al., 2000; Kaebernick and Neilan, 2001; Zurawell et al., 2005). Mass fish and invertebrate deaths are observed in lakes due to excessive increases in cyanobacteria in the water. This results in major losses in water quality, and thus in its ecological status, economic and aesthetic value (Albay et al., 2015; Carmichael, 2001; Codd, 2000).

TABLE 2.2 Cyanobacterial Toxins and the Organs They Affect in Mammals*

	Toxin	Target Organ	Genera of Cyanobacteria
Cyclic Peptides	Microcystin	• Liver	*Anabaena, Anabaenopsis, Hapalosiphon, Microcystis, Nostoc, Planktothrix*
	Nodularin		*Nodularia*
Alkaloids	Anatoxin-a	• Nerve synapses	*Anabaena, Aphanizomenon, Cylindrospermopsis, Lyngbya, Planktothrix, Schizothrix, Umezakia*
	Anatoxin-a(S)	• Skin	
	Aplysiatoxin	• Liver	
	Lyngbiyatoxin-a	• Skin, intestinal system	
	Saxitoxin	• Nerve Axons	
	Cylindrospermopsin		

Source: Carmichael and Falconer (1993), Keskinkaya and Açıkgöz (2021).

2.6 APPLICATION AREAS OF CYANOBACTERIA

Cyanobacteria provide numerous benefits to living organisms and hold significant potential across various industries. They are particularly valuable in sectors such as food production, biofuel generation, agriculture (as biofertilizers), medicine, and environmental applications, including pollution control.

2.6.1 IN THE FOOD INDUSTRY

The rapid growth of the world population has prompted a global search for sustainable protein sources. In recent years, algal protein from microalgae has attracted interest as an additive and/or alternative protein source and has led to a shift in research in this direction. Some strains of cyanobacteria (*Anabaena* and *Nostoc*) are used as food in Chile, Mexico, Peru, and the Philippines. In addition, *Nostoc commune* is being used as a new supplementary source due to its high content of fiber and moderate protein. In addition, *Spirulina*'s nutritional content (60–70% protein, 20% carbohydrates, 5% lipids, 7% minerals, and 6% water) and digestibility make it an ideal food supplement. Besides, it is one of the richest sources of beta-carotene, thiamine (B1), riboflavin, and vitamin B12. It is commercially available in powder, granules, tablets, or capsules in stores (Belay et al., 1993; Pabuçcu and Yücer, 2017).

2.6.2 IN THE AGRICULTURE INDUSTRY

Cyanobacteria contribute to soil fertility in rice-producing (tropical) regions due to their nitrogen-fixing properties. Some heterocyst-forming species are able to colonize rice fields, while non-heterocyst species are able to fix atmospheric nitrogen under microaerophilic conditions. In addition to their traditional use in rice cultivation, cyanobacteria have also been found to enhance productivity in a variety of other crops, including barley, oats, tomato, radish, cotton, sugar beet, corn, red pepper, and lettuce. Symbiont *Anabaena-azollae* is estimated to fix 120–312 kg of nitrogen per hectare. Furthermore, studies conducted at the National Facility for Marine Cyanobacteria (NFMC) have demonstrated that cyanobacteria, due to their lignolytic enzymatic activity, are effective in the treatment of harmful waste materials, including pesticides, detergents, antibiotics, various organic molecules, and solid waste. Consequently, numerous countries are investing in efforts to conserve cyanobacterial biodiversity and explore their role in pollution mitigation (Pabuçcu and Yücer, 2017; Thajuddin and Subramanian, 2005).

2.6.3 IN THE PHARMACOLOGY INDUSTRY

Cyanobacteria are important sources of bioactive molecules used as protein sources in human and animal nutrition. Recent studies have revealed that cyanobacteria produce molecules with a wide range of pharmacological effects, such as antimicrobial, antiviral, anticancer, anti-inflammatory, hypocholesterolemia, and enzyme-inhibiting activities. Moreover, these molecules are not only used as pharmaceutical precursors but also serve as structural templates in the design of synthetic compounds (El-Sheekh et al., 2006; Quinn et al., 1993; Yücer, 2013). However, the emergence of resistant bacteria and pathogens against conventional chemotherapeutic agents has diminished the effectiveness of current cancer treatments. At this stage, cyanobacteria have emerged as highly valuable due to their content of pharmacologically active compounds, offering more efficient and less toxic alternatives to conventional drugs. Additionally, they serve as structural models for the development of novel therapeutics with physiological activity (Duke et al., 2002; Glombitza and Koch, 1989; Patterson et al., 1994; Yücer, 2013).

Worldwide, interest in cyanobacteria as a source of antibiotics and bioactive pharmaceutical compounds is increasing rapidly. This has enabled the isolation and characterization of novel antibiotics from cyanobacteria. In addition, the strong antimicrobial and anticancer properties of cyanobacterial compounds highlight the importance of active substance synthesis and drug design in advancing pharmacology (Yücer, 2013).

2.7 CONCLUSION

Cyanobacteria are microorganisms that play a critical role in ecosystems and have important applications in various fields. This review of cyanobacterial species and diversity in Türkiye provides a valuable foundation for understanding the biodiversity of these organisms and their ecological and functional roles within biological systems. In conclusion, this review has offered a comprehensive understanding of the distribution and ecological significance of cyanobacterial species in Türkiye. In addition to highlighting their roles in various ecosystems, it also provides several recommendations for future research and conservation strategies (Çiftçi and Güngör, 2021; Çobanoğlu, 2021; Çoşgun, 2021; Savcı, 2023; Şentürk, 2021; Zengin et al., 2023).

Given Türkiye's volatile climate and ecological landscapes, assessing the impacts of cyanobacterial species on ecosystem services is essential for maintaining ecological balance and informing sustainable management practices. The biodiversity of cyanobacteria supports the sustainability of a healthy ecosystem in natural water bodies, and therefore, conservation efforts should be strengthened by sound management of habitats and monitoring of cyanobacteria populations. Further scientific research is needed to clarify how these microorganisms interact with the physical and chemical components of their environment.

Ultimately, the organization of training sessions and public awareness campaigns is recommended as a means to foster greater community understanding of cyanobacteria, their ecological roles, and the need for their conservation. Increasing knowledge about the specific effects of cyanobacteria on agriculture, water quality, and environmental health will enable local communities to participate more effectively in efforts to protect these organisms. Collectively, these strategic recommendations represent critical steps toward the sustainable management and conservation of cyanobacterial species in Türkiye. These approaches will not only contribute to maintaining ecological balance but also pave the way for new opportunities in biotechnology and environmental protection.

KEYWORDS

- **blue-green algae**
- **cyanobacteria**
- **cyanotoxins**
- **endemism center**
- **Irano-Anatolian hotspot**
- **phytoplankton**

REFERENCES

Abed, R. M. M.; Dobretsov, S.; Sudesh, K. Applications of Cyanobacteria in Biotechnology. *J. Appl. Microbiol.* **2009**, *106* (1), 1–12. https://doi.org/10.1111/j.1365-2672.2008.03918.x.

Akbulut, E. N.; Akbulut, A.; Barlas, N.; Bayarı, S.; Karaytuğ, S.; Çelik, K.; Alper, A.; Özatlı, D.; Yılmaz, İ. *Uluabat Gölü ve Gölü Besleyen Su Kaynakları Üzerine Limnoekolojik Çalışmalar*; TÜBİTAK-YDAÇAĞ: 2004; p 113; Project No. 113Y005.

Akçaalan Albay, R. A.; Gürevin, C.; Albay, M.; Oğuz, A. *İznik Gölü'nde Toksik Siyanobakteri (Mavi-Yeşil Alg) Artışı, Siyanotoksin Üretimi ve Su Kalitesi ile Olan Etkileşiminin İncelenmesi*; TÜBİTAK-ÇAYDAG: 2015; p 90; Project No. 112Y209.

Akdura, N. *Ziraat, Orman ve Su Ürünleri Alanında Uluslararası Araştırmalar*; Eğitim Yayınevi, 2022.

Aktan-Turan, Y.; Aykulu, G.; Albay, M.; Okgerman, H.; Akcaalan, R.; Gurevin, C.; Dorak, Z. *Büyükçekmece Gölü'nde Aşırı Artış Gösteren Fitoplankterlerin Gelişimini Kontrol Eden Faktörlerin Araştırılması*; TUBITAK-ÇAYDAG: Ankara, 2006; p 112; Project No. 103Y127.

Akyel, E. *Uzaktan Algılama ile Karina Lagünü'nün Nitel ve Nicel Değişim Analizi*; Master's Thesis; İzmir Katip Çelebi Üniversitesi Fen Bilimleri Enstitüsü: 2022.

Araoz, R.; Nghiem, H.; Rippka, R.; Palibroda, N.; Marsac, N. T.; Herdman, M. Neurotoxins in Axenic Oscillatorian Cyanobacteria: Coexistence of Anatoxin-a and Homoanatoxin-a Determined by Ligand-Binding Assay and GC/MS. *Microbiology* **2005**, *151*, 1262–1273.

Aslan, H. *Gökçeada Tuz Gölü Sulak Alanının Bazı Fiziko-Kimyasal Özelliklerinin Saptanması ve Biyoçeşitliliğinin İzlenmesi*; Program Code: 1002; 2016; Project No. 115Y457.

Atak, S. *Doğancı Baraj Gölü (Bursa) Fitoplanktonu ve Trofik Seviyesinin Belirlenmesi*; Master's Thesis; Bursa Uludağ Üniversitesi Fen Bilimleri Enstitüsü: 2022; p 90.

Atıcı, T.; Obalı, O.; Çalışkan, H. Control of Water Pollution and Phytoplanktonic Algal Flora in Bayındır Dam Reservoir (Ankara). *Ege J. Fish. Aquat. Sci.* **2005**, *22* (1, 2), 79–82.

Aydın, S. Ö. *Türkiye'nin Enerji Arz Güvenliğinde Yeşil Hidrojen Potansiyeli*; Master's Thesis; MEF Üniversitesi Sosyal Bilimler Enstitüsü: 2023; p 89.

Aygün, Ş. *Seyhan Nehri Adana İç Göl Bölümünün Fiziksel, Kimyasal ve Plankton Kalitesi Özellikleri ve Mevsimsel Değişimleri*; Doktora Tezi; Çukurova Üniversitesi Fen Bilimleri Enstitüsü: Adana, 2000.

Aykın, Ç. *Kayalıköy Barajı (Kırklareli) Siyanobakteri Türleri ve Toksisite Potansiyellerinin Araştırılması*; Master's Thesis; Trakya Üniversitesi Fen Bilimleri Enstitüsü: Edirne, 2023.

Ballot, A.; Krienitz, L.; Kotut, K.; Wiegand, C.; Metcalf, J. S.; Codd, G. A.; Pflugmacher, S. Cyanobacteria and Cyanobacterial Toxins in Three Alkaline Rift Valley Lakes of Kenya—Lakes Bogoria, Nakuru and Elmenteita. *J. Plankton Res.* **2004**, *26* (8), 925–935.

Batu, A.; Akbulut, N. E. The Investigation on the Blue-Green Algae of Mogan Lake, Beytepe Pond and Delice River (Kızılırmak). *Commun. Fac. Sci. Univ. Ank. Ser. C* **2018**, *27* (2), 1–16. https://doi.org/10.1501/commuc_0000000193.

Baykal, T.; Açıkgöz, İ. Hirfanlı Baraj Gölü Algleri. *Ahi Evran Univ. Kırşehir Eğitim Fak. Derg.* **2004**, *5* (2), 115–136.

Baykal, T.; Akbulut, A.; Açıkgöz, İ.; Udoh, A. U.; Yıldız, K.; Şen, B. New Records for the Freshwater Algae of Turkey. *Turkish J. Bot.* **2009**, *33* (2), 141–152. https://doi.org/10.3906/bot-0705-10.

Belay, A.; Ota, Y.; Miyakawa, K.; Shimamatsu, H. Current Knowledge on Potential Health Benefits of Spirulina. *J. Appl. Phycol.* **1993**, *5*, 235–241.

Bulut, S. *Türkiye Sularına Ait Siyanobakteri Türlerinden İzole Edilen Ekzopolisakkaritlerin (EPS) Karakterizasyonu ve Kozmetik Sektöründe Kullanılma Potansiyellerinin Araştırılması*; Master's Thesis; Burdur Mehmet Akif Ersoy Üniversitesi Fen Bilimleri Enstitüsü: Burdur, 2023.

Carmichael, W. W. Health Effects of Toxin-Producing Cyanobacteria: The CyanoHABs. *Hum. Ecol. Risk Assess.* **2001**, *7*, 1393–1407.

Carmichael, W. W.; Falconer, I. R. Diseases Related to Freshwater Blue-Green Algal Toxins and Control Measures. In *Algal Toxins in Seafood and Drinking Water*; Falconer, I. R., Ed.; Academic Press: London, 1993; pp 187–209.

Çelekli, A.; Albay, M.; Dügel, M. Phytoplankton (Except Bacillariophyceae) Flora of Lake Gölköy (Bolu). *Turkish J. Bot.* **2007**, *31* (1), 49–65.

Çelekli, A.; Obalı, O.; Külköylüoğlu, O. The Phytoplankton Community (Except Bacillariophyceae) of Lake Abant (Bolu, Turkey). *Turkish J. Bot.* **2007**, *31* (2), 109–124.

Çelekli, A.; Özpınar, G. Ecological Assessment of Burç Reservoir's Surface Water (Turkey) Using Phytoplankton Metrics and Multivariate Approach. *Turkish J. Bot.* **2021**, *45* (5), 522–539.

Çelik, K. Kille Çayı (Balıkesir) Fitoplanktonu ve Mevsimsel Değişiminin Bazı Fiziksel ve Kimyasal Değişkenlerle Olan İlişkileri. *Osmaniye Korkut Ata Univ. Fen Bilimleri Enst. Derg.* **2023**, *6* (1), 252–260.

Çelik, K. Kocavşar Deresi (Balıkesir) Fitoplankton Ekolojisi. *J. Adv. Res. Nat. Appl. Sci.* **2022**, *8* (4), 777–784.

Çelikoğlu, E. *Farklı Ekolojik Ortamlardan İzole Edilen Siyanobakterilerin Sekonder Metabolitlerinin Tanımlanması*; Doktora Tezi; Eskişehir Teknik Üniversitesi Lisansüstü Eğitim Enstitüsü: Eskişehir, 2022.

Çetin, T.; Demir, N. Aras, Çoruh ve Doğu Karadeniz Havzalarındaki Yüksek Rakımlı Göllerde Fitoplankton Tür Kompozisyonu ve Biyokütlesi. *Anadolu Çev. ve Hay. Derg.* **2023**, *8* (3), 352–360. https://doi.org/10.35229/jaes.1274356.

Çevik, F.; Polat, S.; Dural, M. Akyatan ve Tuzla Lagünlerinin (Adana, Türkiye) Fitoplanktonu ve Mevsimsel Değişimi. *J. Fish. Sci.* **2008**, *2* (1), 19–29.

Chaurasia, A. Cyanobacterial Biodiversity and Associated Ecosystem Services: Introduction to the Special Issue. *Biodivers. Conserv.* **2015**, *24* (4), 707–710.

Chorus, I.; Bartram, J. *Toxic Cyanobacteria in Water—A Guide to Their Public Health Consequences, Monitoring and Management*; Routledge: London and New York, 1999.

Chorus, I.; Falconer, I. R.; Salas, H. J.; Bartram, J. Health Risks Caused by Freshwater Cyanobacteria in Recreational Waters. *J. Toxicol. Environ. Health* **2000**, *3*, 323–347.

Çiçek, N. L.; Ertan, Ö. O.; Erdoğan, Ö.; Didinen, H.; Boyacı, Y. Ö.; Kara, D.; Zeybek, M.; Diken, G. Distribution of Phytoplankton and Its Relationship with Physicochemical Parameters in Lake Eğirdir (Isparta/Turkey). *Biol. Divers. Conserv.* **2017**, *10* (3), 150–162.

Çiftçi, Y.; Güngör, Y. Nemrut-Süphan Öneri Jeopark Alanında (Bitlis-Türkiye) Doğal ve Kültürel Miras Bütünleşmesi ile Jeokoruma Önerileri. *Maden Tetkik ve Arama Derg.* **2021**, *165*, 191–215.

Çiltepe, F. *Adilcevaz (Bitlis) Karasal Ortamındaki Fosil Mikrobiyalitlerin Koruma Anlayışına Göre Değerlendirilmesi*; Master's Thesis; Balıkesir Üniversitesi Sosyal Bilimler Enstitüsü: 2022; p 122.

Cirik, Ş.; Cirik, S. *Su Bitkileri, Deniz Bitkilerinin Biyolojisi, Ekolojisi, Yetiştirme Teknikleri*; Ege Üniversitesi Su Ürünleri Fakültesi Yayınları: İzmir, 2004; pp 34–38.

Cirik, S.; Gökpınar, Ş. *Plankton Bilgisi ve Kültürü*; Ege Üniversitesi Su Ürünleri Fakültesi Yayınları: İzmir, 1993; p 269.

Çobanoğlu, G. Geçmişten Bugüne İstanbul Liken Çalışmaları Üzerine Bir Derleme. *Bağbahçe Bilim Derg.* **2021**, *8* (1), 259–266.

Codd, G. A. Cyanobacterial Toxins, the Perception of Water Quality and the Priorisation of Eutrophication Control. *Ecol. Eng.* **2000**, *169*, 56–60.

Çolak, S. *Deniz Seviyesi Yükselmelerinin Kıyılardaki Etkileri ve Türkiye Kıyılarında; CVI, Ulaşım, Tarihi-Kültürel-Doğal Zenginlikler ve Uyum Planlarıyla Değerlendirilmesi*; Master's Thesis; Mimar Sinan Güzel Sanatlar Üniversitesi Fen Bilimleri Enstitüsü: 2022; p 282.

Coşgun, A. Araştırma Hastanelerinde İklimlendirme ve Soğutma Alanlarında Bakım ve Servis İşlemleri. In *Mühendislik Alanında Akademik Çalışmalar*; Özdemir, Y., Ed.; Duvar Yayınları, 2021; pp 7–23.

Demir, A. N.; Fakıoğlu, Ö.; Dural, B. Phytoplankton Functional Groups Provide a Quality Assessment Method by the Q Assemblage Index in Lake Mogan (Turkey). *Turkish J. Bot.* **2014**, *38* (1), 169–179.

Demirkalp, F. Y.; Saygı, Y.; Çağlar, S. S.; Gündüz, E.; Kılınç, S. Limnological Assessment on the Brackish Shallow Liman Lake from Kızılırmak Delta (Turkey). *J. Anim. Vet. Adv.* **2010**, *9* (16), 2132–2139.

Dokcan, Ş. *Sarıyar Baraj Gölü Bentik Algleri*; Master's Thesis; Selçuk Üniversitesi Fen Bilimleri Enstitüsü: Konya, 2010.

Duke, J. A.; Bogenschutz-Godwin, M. J.; Doke, P. A.; Ducellier, J. *Handbook of Medicinal Herbs*, 2nd ed.; CRC Press, 2002.

El-Sheekh, M. M.; Osman, M. E. H.; Dyab, M. A.; Amer, M. S. Production and Characterization of Antimicrobial Active Substance from the Cyanobacterium *Nostoc muscorum*. *Environ. Toxicol. Pharmacol.* **2006**, *21*, 42–50.

Eren, Z. The Relationship of Harmful Algal Bloom and Mucilage Outbreak in the Sea of Marmara. *J. Environ. Nat. Stud.* **2021**, *3* (2), 203–213. https://doi.org/10.53472/jenas.985310.

Ertan, Ö. O.; Yüce, A.; Bilgin, Ş. Karacaören I Baraj Gölü Fitoplanktonu. In *Doğu Bölgesi IV. Su Ürünleri Sempozyumu Bildirileri*; Erzurum, 2000; pp 67–84.
Fakıoğlu, Ö.; Demir, N. Göllerin Ekolojik Durumunun Değerlendirilmesinde Fitoplankton Topluluklarının Kullanılması. *Ankara Univ. Çevrebilimleri Derg.* **2011**, *3* (1), 99–106.
Fakıoğlu, Ö.; Nuhoğlu, A.; Acar, Ç.; Eğercioğlu, M. N. *Derin Heyelan Set Gölü (Tortum Gölü)'nde Biyolojik ve Kimyasal Değişimlere Neden Olan Etkilerin Belirlenmesi*; TÜBİTAK-ÇAYDAG: 2018; p 68.
Filatova, D.; Picardo, M.; Núñez, O.; Farré, M. Analysis, Levels and Seasonal Variation of Cyanotoxins in Freshwater Ecosystems. *Trends Environ. Anal. Chem.* **2020**, *26*, e00091.
Ganf, G. G.; Oliver, R. O. Vertical Separation of Light and Available Nutrients as a Factor Causing Replacement of Green Algae in the Plankton of a Stratified Lake. *J. Ecol.* **1982**, *70*, 829–844.
Genuário, D. B.; Vaz, M. G.; Santos, S. N.; Kavamura, V. N.; Melo, I. S. Cyanobacteria from Brazilian Extreme Environments: Toward Functional Exploitation. In *Microbial Diversity in the Genomic Era*; Das, S.; Dash, H. R., Eds.; Academic Press, 2019; pp 265–284.
Glombitza, K. W.; Koch, M. Secondary Metabolites of Pharmaceutical Potential. In *Algal and Cyanobacterial Biotechnology*; Cresswell, R. C.; Rees, T. A. V.; Shah, H., Eds.; Longman: Harlow, UK, 1989; pp 161–238.
Graham, L. E.; Wilcox, L. W. Algae. In *Upper Saddle River*; Ryu, T., Ed.; Prentice-Hall, 2000; pp 97–127.
Guiry, M. D. How Many Species of Algae Are There? A Reprise. Four Kingdoms, 14 Phyla, 63 Classes and Still Growing. *J. Phycol.* **2024**, *60* (2), 214–228.
Gül, Ö.; Düşen, O.; Erduğan, H. The Preliminary Search Results of Cyanobacteria Flora of Thermal Areas in Sarayköy (Denizli), Turkey. In *Digital Proceedings of the ICOEST 2013*; Nevşehir, Turkey, 2013.
Gulecal, Y.; Temel, M. Water Quality and Phytoplankton Diversity in Büyükçekmece Watershed, Turkey. *J. Water Resour. Prot.* **2014**, *6*, 55–61. http://dx.doi.org/10.4236/jwarp.2014.61009.
Gülle, İ. *Karacaören I Baraj Gölü (Burdur) Planktonunun Taksonomik ve Ekolojik Olarak İncelenmesi*; Doktora Tezi; Süleyman Demirel Üniversitesi Fen Bilimleri Enstitüsü: Isparta, 2005.
Gürbüz, F. *Göller Bölgesindeki Bazı Su Kaynaklarında Bulunan Toksik Siyanobakteri Türlerinin Tanımlanması ve Toksin Seviyelerinin Belirlenmesi*; Doktora Tezi; SDÜ Fen Bilimleri Enstitüsü: Isparta, 2008.
Gürbüz, G. *Seyhan Baraj Gölü'nde (Adana) Fizikokimyasal Özellikler, Alg Populasyonları ve Alkalen Fosfataz Aktivitesinin Yaz Periyodunda İncelenmesi*; Master's Thesis; Çukurova Üniversitesi Fen Bilimleri Enstitüsü: Adana, 2012.
Henson, B. J.; Hesselbrock, S. M.; Watson, L. E.; Barnum, S. R. Molecular Phylogeny of the Heterocystous Cyanobacteria (Subsection IV and V) Based on nifD. *Int. J. Syst. Evol. Microbiol.* **2004**, *54*, 493–497.
Hitzfeld, B. C.; Lampert, C. S.; Späth, N.; Mountfort, D.; Kaspar, H.; Dietrich, D. R. Toxin Production in Cyanobacterial Mats from Ponds on the McMurdo Ice Shelf, Antarctica. *Toxicon* **2000**, *38* (12), 1731–1748.
Huisman, J.; Codd, G. A.; Paerl, H. W.; Ibelings, B. W.; Verspagen, J. M. H.; Visser, P. M. Cyanobacterial Blooms. *Microbiology* **2018**, *16*, 471–483.

İspirli, S. *İzmir Tahtalı Baraj Gölü'nün Toksik Siyanobakteri Türleri ve Bazı Mikrosistin Varyantları Yönünden Araştırılması*; Doktora Tezi; Ege Üniversitesi Fen Bilimleri Enstitüsü: İzmir, 2009; p 77.

Kabziński, A.; Juszczak, R.; Miękoś, E.; Tarczyńska, M.; Sivonen, K.; Rapala, J. The First Report about the Presence of Cyanobacterial Toxins in Polish Lakes. *Pol. J. Environ. Stud.* **2000**, *9* (3), 171–178.

Kaebernick, M.; Neilan, B. A. Ecological and Molecular Investigations of Cyanotoxin Production. *FEMS Microbiol. Ecol.* **2001**, *35*, 1–9.

Kahraman, S. D.; Küplülü, Ö. Siyanobakteriler ve Toksinleri. *Vet. Hekimler Derneği Derg.* **2012**, *83* (2), 36–47.

Karacaoğlu, D.; Dere, Ş.; Dalkıran, N. A Taxonomic Study on the Phytoplankton of Lake Uluabat (Bursa). *Turkish J. Bot.* **2004**, *28* (5), 473–485.

Kazez, Z. *Cip Baraj Gölü (Elazığ) Kıyı Bölgesi Algleri*; Master's Thesis; Fırat Üniversitesi Fen Bilimleri Enstitüsü: Elazığ, 2012.

Keskinkaya, H. B.; Akköz, C. Chapter 8: Siyanobakteriler ve Toksinleri. In *INSAC Academic Studies on Natural and Engineering Sciences*; Gece Kitaplığı: Ankara, 2021; pp 143–153.

Kılınç, S. A Study in the Seasonal Variation of Phytoplankton in Hafik Lake (Sivas, Turkey). *Turkish J. Bot.* **1998**, *22* (1), 35–42.

Kırtıloğlu, E. *Bafa Gölü İçin Uzaktan Algılama Teknikleri ile Siyanobakterilerin Tespiti, İzlenmesi, Zamansal Analizi ve Göle Ulaşan Kirletici Yükleriyle İlişkisinin Değerlendirilmesi*; Doktora Tezi; Konya Teknik Üniversitesi Lisansüstü Eğitim Enstitüsü: 2023; p 79.

Kıvrak, E. Karamuk Gölü (Afyonkarahisar) Fitoplankton Komünitesinin Mevsimsel Değişimi ve Bazı Fiziko-Kimyasal Özellikleri. *Ege J. Fish. Aquat. Sci.* **2011**, *28* (1), 9–20.

Koca, H. M.; Demir, A. N.; Alp, M. T. İçsularda Siyanobakteri Artışları ve Oluşturduğu Riskler. *Ziraat Mühendisliği* **2019**, *367*, 23–33.

Köker, L.; Akcaalan, R.; Oğuz, A.; Gaygusuz, O.; Gürevin, C.; Akat Köse, C.; Güçver, S.; Karaaslan, Y.; Ertürk, A.; Albay, M.; Kınacı, C. Distribution of Toxic Cyanobacteria and Cyanotoxins in Turkish Waterbodies. *J. Environ. Prot. Ecol.* **2017**, *18* (2), 425–432.

Köker, L.; Çam, A. O.; Akçaalan, R.; Albay, M. Cyanobacterial Diversity and the Presence of Microcystins in the Küçük Menderes River Basin, Türkiye. *Aquat. Sci. Eng.* **2022**, *37* (3), 129–139.

Koyuncuoğlu, P. *Kentsel Atıksu Arıtma Tesisi Kaynaklı Mikroplastik Kirlilik Profilinin Belirlenmesi: Denizli İli Örneği*; Doktora Tezi; Pamukkale Üniversitesi Fen Bilimleri Enstitüsü: 2021; p 161.

Kula, H. Ö. *Uludağ Milli Parkındaki Sarıalan ve Kirazlıyayla Piknik Alanlarındaki Antropojenik Aktivitelerin Epifitik Liken Çeşitliliği Üzerine Etkilerinin İncelenmesi*; Master's Thesis; Bursa Uludağ Üniversitesi Fen Bilimleri Enstitüsü: 2021; p 176.

Külköylüoğlu, O.; Yavuzatmaca, M.; Karacaoğlu, D.; Telli, M. The Aquatic Diversity of Ostracoda, Phytoplankton and Zooplankton from Freshwater Cave Habitats in Turkey. *Cave Karst Sci.* **2014**, *41* (3), 99–104.

Kutlu, B.; Demir, B. Munzur Çayı (Tunceli-Türkiye)'nın Algal Florasının ve Su Kalitesi Değişimi. *Turkish J. Agric. Food Sci. Technol.* **2018**, *6* (8), 985–994.

Kutlu, B.; Serdar, O.; Danabaş, D.; Aydın, R. *Uzunçayır Baraj Gölü'nün Fitoplankton Türleri ve Fiziko-Kimyasal Parametrelerinin Belirlenmesi*; ÇAYDAG: 2014; p 113; Project No. 113Y005.

Kütük, M.; Akpınar, A., Eds. *Fen ve Matematik Bilimlerinde Güncel Araştırmalar-2024*; Livre de Lyon, 2024.

Mankiewicz, J.; Tarczyńska, M.; Walter, Z.; Zalewski, M. Natural Toxins from Cyanobacteria. *Acta Biol. Cracov. Ser. Bot.* **2003**, *45* (2), 9–20.

Maraşlıoğlu, F.; Bektaş, S. Characterization of Spatiotemporal Variations in Mert Stream Water Quality by Phytoplankton Community and Biological Indices. *KSU J. Agric. Nat.* **2022**, *25*, 42–53. https://doi.org/10.18016/ksutarimdoga.vi.1055761.

Maraşlıoğlu, F.; Gönülol, A. *Turkish Algae*; Electronic Publication: Çorum, Turkey, 2025. http://turkiyealgleri.hitit.edu.tr (accessed April 21, 2026).

Maraşlıoğlu, F.; Öterler, B.; Sevindik, T. O.; Soylu, E. N.; Demir, N.; Çelekli, A.; Sömek, H.; Coşkun, T.; Solak, C. N.; Çetin, T.; Karaaslan, Y.; Temizel, B.; Koca, M.; Güzel, U.; Tunca, H.; Manavoğlu, D. C.; Yılmaz, E. New Records for the Turkish Freshwater Algal Flora in Twenty-Five River Basins of Turkey, Part II: Chlorophyta, Cyanobacteria, Euglenozoa. *Turkish J. Fish. Aquat. Sci.* **2022**, *22* (8), TRJFAS19319. https://doi.org/10.4194/TRJFAS19319.

Merel, S.; Walker, D.; Chicana, R.; Snyder, S.; Baurès, E.; Thomas, O. State of Knowledge and Concerns on Cyanobacterial Blooms and Cyanotoxins. *Environ. Int.* **2013**, *59*, 303–327.

Metcalf, J. S.; Codd, G. A. Cyanobacterial Toxins. In *The Water Environment*; FR/R0009; U.K., 2004; p 5.

Moss, B. *Ecology of Fresh Waters*; Blackwell: Oxford, 1988; p 417.

Omuzbüken, B.; Kaçar, A. Assessment of Bacteria and Archaea Levels in Çakalburnu Lagoon (İzmir) Sediments by Real-Time PCR. *Ege J. Fish. Aquat. Sci.* **2021**, *38* (2), 147–154. https://doi.org/10.12714/egejfas.38.2.02.

Öterler, B. A Short-Term Study on the Phytoplanktonic and Epilithic Algae of Arda River (Edirne, Turkey). *Ege J. Fish. Aquat. Sci.* **2016**, *33* (3), 299–306. https://doi.org/10.12714/egejfas.2016.33.3.

Öterler, B. *Tunca Nehri Fitoplanktonu ve Su Kalitesi ile Olan İlişkilerinin İncelenmesi*; Master's Thesis; Trakya Üniversitesi Fen Bilimleri Enstitüsü: Edirne, 2003.

Özen, M. *Uluabat Gölü Toksik Siyanobakteri Türleri ve Toksinlerinin Belirlenmesi*; Doktora Tezi; Bursa Uludağ Üniversitesi Fen Bilimleri Enstitüsü: Bursa, 2020.

Özer, T.; Açıkgöz Erkaya, İ.; Udoh, A. U.; Akbulut, A.; Yıldız, K.; Şen, B. New Records for the Freshwater Algae of Turkey (Tigris Basin). *Turkish J. Bot.* **2012**, *36* (6), 747–760. https://doi.org/10.3906/bot-1108-16.

Öztürk Ulcay, S.; Öztürk, M.; Kurt, O.; Taşkın, E.; Öztürk, M. Dikili İlçesi (İzmir) Kaplıcalarında Yayılış Gösteren Termal Cyanophyceae (Mavi-Yeşil Alg) Türleri. *Türk Sucul Yaşam Derg.* **2007**, *5* (8), 371–378.

Öztürk, S. Cyanobacterial Diversity and Physicochemical Characteristics of Thermal Springs in the Kütahya Province of Turkey. *Bangladesh J. Plant Taxon.* **2021**, *28* (2), 413–428.

Ozturk, S.; Kurt, O. New Records for the Freshwater Algal Flora in Turkey. *Fresenius Environ. Bull.* **2021**, *30*, 6465–6473.

Öztürk, S.; Kurt, O. The Taxonomy and Distribution of Algae in the Thermal Springs of Türkiye. *Acta Bot. Croat.* **2024**, *83* (2), 92–99.

Özükaya, Z.; Güzey, E. Production of Easily Biodegradable Bioplastic Using Micro and Macro Algae: An Experimental Research. *Eurasian Acad. Sci. Soc. Sci. J.* **2024**, *52* (52), 80–97.

Özyalın, S.; Ustaoğlu, M. R. Kemer Baraj Gölü (Aydın) Net Fitoplankton Kompozisyonunun İncelenmesi. *E.Ü. Su Ürünleri Dergisi* **2008**, *25* (4), 275–282.

Pabuçcu, K.; Yücer, T. D. Oscillatoria Türlerinin Tıbbi Özelliklerinin Araştırılması. *Gaziosmanpaşa Bilimsel Araştırma Derg.* **2017**, *6* (1), 46–59.

Paerl, H. W.; Otten, T. G. Harmful Cyanobacterial Blooms: Causes, Consequences, and Controls. *Microb. Ecol.* **2013**, *65*, 995–1010.

Patterson, G. M. L.; Larsen, L. K.; Moore, R. E. Bioactive Natural Products from Blue-Green Algae. *J. Phycol.* **1994**, *29* (39), 14.

Quinn, J.; Li, H. H.; Singer, J.; Morimoto, B.; Mets, L.; Kindle, K.; Merchant, S. The Plastocyanin-Deficient Phenotype of *Chlamydomonas reinhardtii* Ac-208 Results from a Frame-Shift Mutation in the Nuclear Gene Encoding Preapoplastocyanin. *J. Biol. Chem.* **1993**, *268*, 7832–7841.

Rai, N. A.; Bergman, B.; Rasmussen, U. *Cyanobacteria in Symbiosis*; Kluwer Academic Publishers: Netherlands, 2002.

Rinehart, K. L.; Namikoshi, M.; Choi, B. W. Structure and Biosynthesis of Toxins from Blue-Green Algae (Cyanobacteria). *J. Phycol.* **1994**, *6*, 159–176.

Rippka, R.; Deruelles, J.; Waterbury, J. B.; Herdman, M.; Stanier, R. Y. Generic Assignments, Strain Histories and Properties of Pure Cultures of Cyanobacteria. *J. Gen. Microbiol.* **1979**, *111*, 1–61.

Sabancı, F. C. Phytoplankton Distribution and Its Relationship to the Physico-Chemical Environment in a Coastal Lagoon. *Ekoloji* **2014**, *23* (90), 61–72.

Şahin, B.; Akar, B. New Records from Artabel Lakes Nature Park (Gümüşhane, Turkey) to the Freshwater Algal Flora of Turkey. *Turkish J. Bot.* **2019**, *43* (1), 135–142. https://doi.org/10.3906/bot-1805-8.

Sancar, B. Ç.; Akhan, M.; Çöl, D. G., Türkol, M. Spirulinanın Gıda Endüstrisi ve Beslenmedeki Yeri. In *International Congresses of Turkish Science and Technology Publishing*; 2023; pp 405–409.

Savcı, Y. *İznik Gölü'nü Besleyen Bazı Akarsularda Bentik Omurgasızlar Kullanılarak Su Kalitesinin Belirlenmesi*; Master's Thesis; Bursa Uludağ Üniversitesi Fen Bilimleri Enstitüsü: 2023; p 74.

Schirrmeister, B. E.; Gugger, M.; Donoghue, P. C. J. Cyanobacteria and the Great Oxidation Event: Evidence from Genes and Fossils. *Palaeontology* **2015**, *58* (5), 769–785. https://doi.org/10.1111/pala.12178.

Şentürk, B. *Ankara Üniversitesi Merkez Kampüslerinin Liken Florası*; Master's Thesis; Ankara Üniversitesi Fen Bilimleri Enstitüsü: 2021; p 72.

Sevindik, T. O. Phytoplankton Composition of Çaygören Reservoir, Balıkesir-Turkey. *Turkish J. Fish. Aquat. Sci.* **2010**, *10*, 295–304.

Sevindik, T. O.; Altundal, E.; Küçük, F. Poyrazlar Gölü (Sakarya) Fitoplanktonunun Tür Kompozisyonu. *Sakarya Univ. J. Sci.* **2015a**, *19* (3), 283–290.

Sevindik, T. O.; Çelik, K. The Effects of Certain Physical and Chemical Variables on the Succession of the Phytoplankton in the Shallow Çağış Pond (Balıkesir, Turkey). *Ekoloji* **2014**, *23* (93), 27–35. https://doi.org/10.5053/ekoloji.2014.934.

Sevindik, T. O.; Çelik, K.; Gönülol, A. Twenty-Four New Records for the Freshwater Algae of Turkey. *Turkish J. Bot.* **2010**, *34* (3), 249–259. https://doi.org/10.3906/bot-0906-56.

Sevindik, T. O.; Gönülol, A.; Önem, B.; Tunca, H.; Arabacı, S. Thirty New Records for Turkish Freshwater Algal Flora from Danamandıra Ponds Silivri, İstanbul and North Mollaköy Lake Sakarya. *Biol. Div. Conserv.* **2015**, *8* (2), 4–15.

Sevindik, T. O.; Gönülol, A.; Tunca, H.; Gürsoy, N. Y.; Küçükkaya, Ş.; Durgut Kınalı, Z. Nineteen New Records for Turkish Freshwater Algal Flora from Lake Taşkısığı and Lake Little Akgöl. *Biol. Divers. Conserv.* **2017**, *10* (1), 69–78.

Sömek, H. *Adıgüzel Baraj Gölü'nün (Güney-Denizli) Fitoplankton Kompozisyonu ve Mevsimsel Değişimi*; Doktora Tezi; Ege Üniversitesi Fen Bilimleri Enstitüsü: İzmir, 2011.

Sompong, U.; Hawkins, P. R.; Besley, C.; Peerapornpisal, Y. The Distribution of Cyanobacteria across Physical and Chemical Gradients in Hot Springs in Northern Thailand. *FEMS Microbiol. Ecol.* **2005**, *52* (3), 365–376.

Soylu, E. N.; Gönülol, A. Phytoplankton and Seasonal Variations of the River Yeşilırmak, Amasya, Türkiye. *Turkish J. Fish. Aquat. Sci.* **2003**, *3* (1), 17–24.

Sukatar, A.; Ertaş, A.; Akgül, R.; Tüney Kızılkaya, İ. Assessment of the Ecological and Trophic Status of Lake Bafa (Turkey) Based on Phytoplankton. *Ege J. Fish. Aquat. Sci.* **2021**, *38* (2), 135–146. https://doi.org/10.12714/egejfas.38.2.01.

Taranu, Z. E.; Zurawell, R. W.; Pick, F.; Gregory-Eaves, I. Predicting Cyanobacterial Dynamics in the Face of Global Change: The Importance of Scale and Environmental Context. *Glob. Change Biol.* **2012**, *18* (12), 3477–3490.

Taş, B.; Gönülol, A. An Ecologic and Taxonomic Study on Phytoplankton of a Shallow Lake, Turkey. *J. Environ. Biol.* **2007**, *28* (2), 439–445.

Taton, A.; Grubisic, S.; Balthasart, P.; Hodgson, D. A.; Laybourn-Parry, J.; Wilmotte, A. Biogeographical Distribution and Ecological Ranges of Benthic Cyanobacteria in East Antarctic Lakes. *FEMS Microbiol. Ecol.* **2006**, *57* (2), 272–289.

Tenekecioğlu, E. R. *Demirköprü Baraj Gölünde (Manisa) Mevsimsel Plankton Topluluk Yapısı Değişimleri*; Doktora Tezi; Ege Üniversitesi Fen Bilimleri Enstitüsü: İzmir, 2011.

Thajuddin, N.; Subramanian, G. Cyanobacterial Biodiversity and Potential Applications in Biotechnology. *Curr. Sci.* **2005**, *89* (1), 47–57.

Torun, Z. *Aspergillus niger ve Trichoderma virilente Mantarları Üzerine Farmakognozik Araştırmalar*; Doktora Tezi; Ankara Üniversitesi Sağlık Bilimleri Enstitüsü: 2021; p 234.

Ülkü (Bayoğlu), M. *Karacaören I Baraj Gölü Siyanobakteri (Mavi-Yeşil Alg) Florası ve Mikrosistin Düzeyinin Belirlenmesi*; Master's Thesis; Mehmet Akif Ersoy Üniversitesi Fen Bilimleri Enstitüsü: Burdur, 2016; p 154.

Ünal, T. *İzmir Balçova (Agamemnon) Kaplıcası Mikroskobik ve Makroskobik Alg Florası*; Master's Thesis; Ege Üniversitesi Fen Bilimleri Enstitüsü: 1996; p 77.

Ustaoğlu, M. R.; Balık, S.; Şipal, U. G.; Mış, D. Ö.; Aygen, C. Buldan Baraj Gölü (Denizli) Planktonu ve Mevsimsel Değişimi. *Su Ürünleri Derg.* **2010**, *27* (3), 113–120.

Varlı, Y. Contact of People with Cyanotoxins and Effects on Health. *J. Inst. Sci. Technol.* **2021**, *11* (2), 880–887.

Varol, M.; Fucikova, K. Four New Records for the Freshwater Algae of Turkey. *LimnoFish* **2015**, *1* (2), 83–88. https://doi.org/10.17216/LimnoFish-5000119624.

Varol, M.; Şen, B. Dicle Nehri'nin Planktonik Alg Florası. *J. Fish. Sci.* **2014**, *8* (4), 252–264.

Visser, P. M.; Verspagen, J. M. H.; Sandrini, G.; Stal, L. J.; Matthijs, H. C. P.; Davis, T. W.; Paerl, H. W.; Huisman, J. How Rising CO_2 and Global Warming May Stimulate Harmful Cyanobacterial Blooms. *Harmful Algae* **2016**, *54*, 145–159.

Walsby, A. E. Mechanisms of Buoyancy Regulation by Planktonic Cyanobacteria with Gas Vesicles. In *The Cyanobacteria*; Fay, P.; Van Baalen, C., Eds.; Elsevier: Amsterdam, 1987; pp 377–414.

Ward, D. M.; Bateson, M. M.; Ferris, M. J.; Kühl, M.; Wieland, A.; Koeppel, A.; Cohan, F. M. Cyanobacterial Ecotypes in the Microbial Mat Community of Mushroom Spring (Yellowstone National Park, Wyoming) as Species-like Units Linking Microbial Community Composition, Structure and Function. *Philos. Trans. R. Soc. B* **2006**, *361*, 1997–2008.

Whitton, B. A. *Ecology of Cyanobacteria II: Their Diversity in Space and Time*; Springer, 2012. https://doi.org/10.1007/978-94-007-3855-3.

Whitton, B. A.; Potts, M. *The Ecology of Cyanobacteria*; Kluwer Academic Publishers: Netherlands, 2000.

Yadav, S.; Rai, S.; Rai, R.; Shankar, A.; Singh, S.; Rai, L. C. Cyanobacteria: Role in Agriculture, Environmental Sustainability, Biotechnological Potential and Agroecological Impact. In *Plant-Microbe Interactions in Agro-Ecological Perspectives*; 2017; Vol. 2, pp 257–277.

Yerli, S. V.; Kıvrak, E.; Gürbüz, H.; Manav, E.; Mangıt, F.; Türkecan, O. Phytoplankton Community, Nutrients and Chlorophyll a in Lake Mogan (Turkey): Comparison between Current and Old Data. *Turkish J. Fish. Aquat. Sci.* **2012**, *12* (1), 95–104.

Yılmaz, M. Porsuk Çayı ve Seydisuyu'nda Bulunan Cyanobacteria'lerin Karakterizasyonu. Ph.D. Thesis, Anadolu University, Türkiye, 2009.

Yılmaz, N. Elmalı Baraj Gölü (İstanbul, Türkiye) Fitoplanktonik Alg Florası. *Erzincan Univ. Fen Bilimleri Enst. Derg.* **2017**, *10* (1), 78–89.

Yılmaz, N.; Özyiğit, I. I.; Demir, H. H.; Yalçın, I. E. Assessment on Phytoplankton Composition and Heavy Metal Pollution in a Drinking Water Resource: Lake Terkos (Istanbul, Turkey). *Desalin. Water Treat.* **2021**, *225*, 265–274.

Yüce, A.; Aktaş, M. Tahtalı, Davuldere ve Çayırköy Göletlerinin (Kocaeli) Algleri ve Su Kaliteleri Üzerine Bir Çalışma. *Iğdır Univ. Fen Bilimleri Enst. Derg.* **2020**, *10* (3), 1539–1550.

Yücer, T. D. *Bazı Cyanobacteria Türlerinin Antimikrobiyal ve Antikanser Aktivitelerinin Araştırılması*; Gazi Üniversitesi Fen Bilimleri Enstitüsü: Ankara, 2013; p 143.

Yüksel, D.; Çelik, E.; Turgay, Ö. Siyanobakteri Kaynaklı Toksin Tehlikesi. *Ecol. Life Sci.* **2021**, *16* (1), 1–17. https://doi.org/10.12739/NWSA.2021.16.1.5A0144.

Zengin, A.; Aka, B.; Gençay, S.; Uzun, M. Bor Madeni ve Endüstriyel Uygulamaları. *Avrupa Bilim Teknol. Derg.* **2023**, *46*, 161–178.

Zurawell, R. W.; Chen, H.; Burke, J. M.; Prepas, E. E. Hepatotoxic Cyanobacteria: A Review of the Biological Importance of Microcystins in Freshwater Environments. *J. Toxicol. Environ. Health* **2005**, *8*, 1–37.

CHAPTER 3

MACROFUNGI OF ANATOLIA: A COMPREHENSIVE STUDY

MIHRICAN YARATANAKUL GÜNGÖR[1], HALIL GÜNGÖR[2*], and İBRAHIM TÜRKEKUL[3]

[1]*Health Services Vocational School, Kocaeli University, Kocaeli, Türkiye*

[2]*Department of Biology, Faculty of Science, Muğla Sıtkı Koçman University, Muğla, Türkiye*

[3]*Department of Biology, Faculty of Arts and Sciences, Tokat Gaziosmanpaşa University, Tokat, Türkiye*

*Corresponding authors

ABSTRACT

Türkiye is known for its rich and diverse natural habitats, and this diversity is reflected in its fungal biodiversity as well. Türkiye, with its varied climates and soil types, hosts a wide range of fungal species, with a remarkable number of species being endemic to the region. This ecological variety makes Türkiye a hotspot for fungal biodiversity. In this study the mycota of Anatolia (Asian part of Türkiye) was analyzed utilizing recent publications. The study revealed the occurrence of 3250 taxa of macrofungi, 2060 taxa of lichens (1811 lichenized and 249 lichenicolous species) and 309 taxa of myxomycetes in Anatolia. As the macrofungi is our focus, the emphasis is given to macrofungi in the present study. Among 3250 taxa of macrofungi, 33 macrofungal taxa have been identified as endemic to Türkiye. *Cortinarius,* Russulaceae, Agaricales, and Agaricomycetes are the most speciose genus, family, order and class in the Anatolia with 154, 232, 1678, and 2722 taxa, respectively. Among the various regions, the Black Sea and Mediterranean regions of Anatolia are renowned for their extensive and

Biodiversity Hotspot of the Irano-Anatolian Region. T. Pullaiah (Ed.)

DOI: 10.1201/9781779645760-3

varied macrofungal taxa. The unique climate and ecological characteristics of these regions contribute significantly to their fungal biodiversity.

3.1 INTRODUCTION

Fungi play a vital role in ecosystems through various functions. They assist in the decomposition of organic matter, which enriches the soil and is essential for the healthy growth of plants. Additionally, fungi establish intricate relationships with other organisms, fostering biodiversity. Certain fungal species also serve as a valuable food source, providing a rich supply of vitamins, minerals, and proteins.

Anatolia is one of the world's richest and most fascinating regions in terms of its floristic diversity. This situation arises from Anatolia's diverse climate types, geographical location, geological structure, varied topography, and different soil groups. A significant portion of Anatolia lies within the Irano-Anatolian floristic region, characterized by a continental climate with hot, dry summers and cold, snowy winters (Figure 3.1).

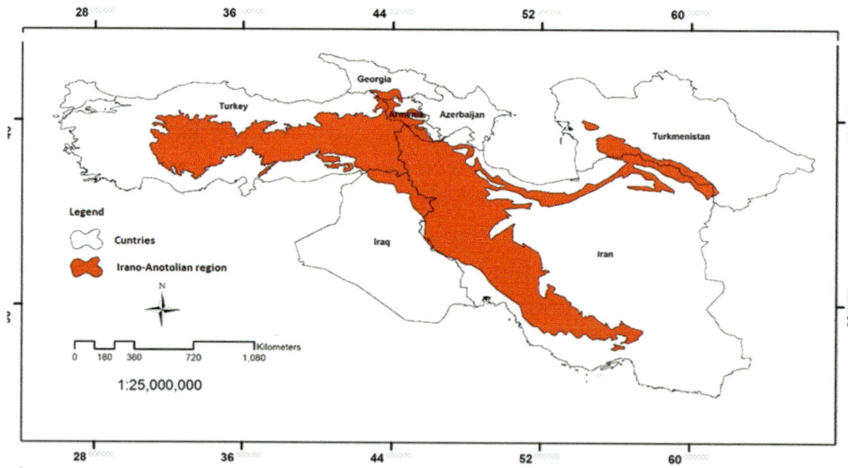

FIGURE 3.1 Geographical Location of the Irano-Anatolian biodiversity hot zone and countries covered by this area.

Source: Reprinted from Sayyad et al., 2021

The number of plant species in Türkiye, including subspecies, varieties, and newly recorded species, reaches 12,000. Of these, approximately

3,700 species (about 34%) are endemic (Güner and Aslan, 2012). This total matches the plant species count in Europe, which is also around 12,000. When compared to the flora of neighboring countries and Europe, Turkish flora stands out as remarkably rich. This biodiversity has consistently attracted foreign botanists. Consequently, many international scientists have conducted research on Turkish biodiversity, producing valuable contributions (Baytop, 2000). The diversity of flora can serve as an indicator of the richness of Turkish mycota. Numerous local and international studies have explored the richness of Turkish mycota. However, the number of taxa is expected to increase with further studies, as unexplored localities and unidentified samples are likely to contribute significantly to this growth.

3.2 STUDIES ON THE MACROFUNGI OF ANATOLIA

The first study on mycota of Anatolia was conducted by Rigler (1852), identifying 17 species in the Istanbul region. Tchihatcheff (1860) identified 33 species from around Istanbul and the Belgrad Forest. Fritsch (1873) identified 4 species from around Istanbul and the Belgrad forest, while Maire (1904) conducted studies along the Bursa-Uludağ and Ankara-Mersin roads, identifying 56 species, the majority of which were rust and parasitic fungi. In 1909, Handel-Mazzetti identified 44 species from the provinces of Istanbul, Samsun, Bursa-Uludağ, Trabzon, and Ordu. In 1932, Zwara identified 14 species belonging to the genus *Russula* from various localities. Afterwards, Albert Pilat identified a total of 118 species through field studies he carried out at various times in the Büyük Ilgaz and Küçük forests in Çankırı Province (Pilat, 1932, 1937). Following this, the German scientist Kurt Lohwag identified 88 species in the Belgrad Forest and 4 species that were wood destroyers in poplars (Lohwag, 1957, 1959). Lohwag pioneered forest phytopathologists in studying tree-parasitic fungi.

Many studies have been carried out on macrofungi in Anatolia to date, and these studies have gained significant momentum in recent years. As a result, the macrofungal diversity of many regions in our Anatolia has been determined. Today, the studies carried out to determine fungal biodiversity in Türkiye are mostly revision studies that include molecular phylogenetic researches.

3.3 CURRENT STATUS OF MACROFUNGI IN ANATOLIA

As a result of the studies carried out in Anatolia, the number of fungal taxa is increasing day by day. Numbers of Turkish macrofungi are updated with recent publications. The latest annotated checklists of Macrofungi of Türkiye were presented by Asan et al. (2022), Sesli et al. (2020), and Solak and Türkoğlu (2022). Also, the distribution of the fungal taxa is given in these checklists. Sesli et al. (2020) and Solak and Türkoğlu (2022) were taken as a base for counting the fungal taxa. It has been determined that there are around 3250 macrofungal taxa in Anatolia in the light of current literature and checklists (Acar, 2022; Akata et al., 2023; Asan et al., 2022; Güngör and Güngör, 2022; Kaygusuz et al., 2024; Kesici et al., 2022; Sadullahoğlu, 2022; Sesli et al., 2020; Solak and Türkoğlu, 2022). In Anatolia, a total of 105 hypogenous macrofungal taxa have been reported; 47 of them are ascomycetes and 58 of them are Basidiomycetes (Akata et al., 2022; Güngör and Güngör, 2022). Also, 309 myxomycete taxa have been reported from Anatolia by Baba and Sevindik (2023). The number of lichens and lichenivorous fungal taxa from Anatolia is 2060 (1811 lichenized and 249 lichenicolous species) (John and Güvenç, 2023).

3.4 BIODIVERSITY OF MACROFUNGI

The compilation revealed the occurrence of about 3,250 taxa of macrofungi in Anatolia. The macrofungal taxa are distributed under 689 genera, 183 families, 47 orders, 14 classes, and two phyla, Ascomycota and Basidiomycota (Table 3.1). Anatolia has at least 29 genera with more than 20 species; however, *Cortinarius* is the most speciose genus with 154 taxa, followed by *Russula* with 146 taxa, while *Inocybe* has 104 taxa (Table 3.2). Among various families, Russulaceae is the most speciose, with 232 taxa under 5 genera, followed by Tricholomataceae with 205 species and 28 genera (Table 3.3). Among the 44 orders of macrofungi represented in Anatolia, Agaricales is the most speciose, with 1678 taxa under 220 genera and 41 families. It can also be noted that the order Russulales is another largest order in Anatolia with 310 taxa under 32 genera and 10 families (Table 3.4). As mentioned earlier, the macrofungi of Anatolia fall under 14 classes; among them, Agaricomycetes is the most dominant with 2722 taxa belonging to 19 orders, 115 families, and 499 genera. A total of 33 macrofungal taxa are endemic to Türkiye, but it is too early to discuss

TABLE 3.1 Macrofungal Biodiversity of Anatolia and Their Hierarchical Arrangement Under Different Classes, Orders, and Families

Class	Order	Family	Genus
Agaricomycetes	Agaricales	Agaricaceae	*Agaricus* (58), *Battarrea* (1), *Chamaemyces* (1), *Chlorophyllum* (6), *Coprinus* (11), *Crucibulum* (1), *Cyathus* (3), *Cystoderma* (8), *Cystodermella* (3), *Cystolepiota* (3), *Disciseda* (1), *Echinoderma* (4), *Floccularia* (1), *Lepiota* (29), *Leucoagaricus* (15), *Leucocoprinus* (6), *Lycoperdon* (21), *Macrolepiota* (11), *Melanophyllum* (1), *Montagnea* (2), *Mycenastrum* (1), *Phaeolepiota* (1), *Staktophyllus* (1), *Tulostoma* (9), *Xerocoprinus* (1)
Agaricomycetes	Russulales	Albatrellaceae	*Albatrellopsis* (1), *Albatrellus* (5), *Byssoporia* (1), *Scutiger* (1)
Agaricomycetes	Agaricales	Amanitaceae	*Amanita* (48), *Aspidella* (1), *Limacella* (3), *Saproamanita* (3), *Zhuliangomyces* (1)
Agaricomycetes	Amylocorticiales	Amylocorticiaceae	*Amyloathelia* (1), *Amylocorticium* (1), *Anomoporia* (1), *Ceraceomyces* (1), *Phlegmacium* (1), *Plicaturopsis* (1)
Leotiomycetes	Helotiales	Arachnopezizaceae	*Arachnopeziza* (2)
Pezizomycetes	Pezizales	Ascobolaceae	*Ascobolus* (6), *Saccobolus* (1), *Thecotheus* (2)
Pezizomycetes	Pezizales	Ascodesmidaceae	*Lasiobolus* (2)
Agaricomycetes	Atheliales	Atheliaceae	*Amphinema* (1), *Athelopsis* (2), *Athelia* (6), *Byssocorticium* (2), *Fibulomyces* (1), *Leptosporomyces*(1), *Piloderma* (2)
Agaricomycetes	Auriculariales	Auriculariaceae	*Auricularia* (3), *Eichleriella* (1), *Exidia* (7)
Agaricomycetes	Russulales	Auriscalpiaceae	*Artomyces* (1), *Auriscalpium* (1), *Lentinellus* (7)
Agaricomycetes	Thelephorales	Bankeraceae	*Boletopsis* (2), *Hydnellum* (9), *Sarcodon* (7)
Sordariomycetes	Hypocreales	Bionectriaceae	*Bionectria* (1)
Agaricomycetes	Agaricales	Bolbitiaceae	*Bolbitius* (2), *Conocybe* (38), *Descomyces* (1), *Galeropsis* (1), *Pholiotina* (2), *Setchelliogaster* (1)

TABLE 3.1 *(Continued)*

Class	Order	Family	Genus
Agaricomycetes	Boletales	Boletaceae	*Aureoboletus* (2), *Boletus* (38), *Buchwaldoboletus* (1), *Butyriboletus* (6), *Caloboletus* (2), *Chalciporus* (2), *Cyanoboletus* (1), *Hemileccinum* (2), *Hortiboletus* (2), *Imleria* (1), *Imperator* (1), *Lanmaoa* (1), *Leccinellum* (2), *Leccinum* (11), *Neoboletus* (4), *Octaviania* (1), *Phylloporus* (2), *Porphyrellus* (1), *Pseudoboletus* (1), *Retiboletus* (1), *Rheubarbariboletus* (1), *Rubroboletus* (6), *Strobilomyces* (1), *Suillellus* (6), *Suillus* (17), *Tylopilus* (2), *Wakefieldia* (1), *Xerocomellus*(8)
Agaricomycetes	Russulales	Bondarzewiaceae	*Bondarzewia* (1), *Heterobasidion*(1)
Agaricomycetes	Cantharellales	Botryobasidiaceae	*Botryobasidium* (4), *Botryohypochnus* (1)
Agaricomycetes	Agaricales	Callistosporiaceae	*Callistosporium* (2), *Xerophorus*(1)
Pezizomycetes	Pezizales	Caloscyphaceae	*Caloscypha* (1)
Dothideomycetes	Catinellales	Catinellaceae	*Catinella*(1)
Leotiomycetes	Helotiales	Cenangiaceae	*Cenangium* (1), *Chlorencoelia* (1), *Encoelia* (1), *Heyderia* (1)
Agaricomycetes	Cantharellales	Ceratobasidiaceae	*Scotomyces* (1), *Thanatephorus* (2)
Agaricomycetes	Polyporales	Cerrenaceae	*Cerrena* (1), *Sistotrema* (2), *Pseudospongipellis* (1)
Leotiomycetes	Helotiales	Chlorospleniaceae	*Chlorociboria* (2)
Agaricomycetes	Agaricales	Chromocyphellaceae	*Chromocyphella* (1)
Agaricomycetes	Agaricales	Clavariaceae	*Clavaria* (4), *Clavulinopsis* (5), *Hodophilus* (1), *Ramariopsis* (1)
Agaricomycetes	Gomphales	Clavariadelphaceae	*Clavariadelphus* (5)
Sordariomycetes	Hypocreales	Cordycipitaceae	*Coniophora* (4), *Cordyceps* (1), *Isaria*(1)
Agaricomycetes	Corticiales	Corticiaceae	*Corticium* (3), *Erythricium* (1)
Agaricomycetes	Agaricales	Cortinariaceae	*Cortinarius* (154), *Dermocybe* (1), *Protoglossum* (2)

TABLE 3.1 *(Continued)*

Class	Order	Family	Genus
Agaricomycetes	Agaricales	Crassisporiaceae	*Romagnesiella* (1)
Agaricomycetes	Agaricales	Crepidotaceae	*Crepidotus* (13), *Pleuroflammula* (1), *Pleurotellus* (1), *Simocybe* (1)
Leotiomycetes	Rhytismatales	Cudoniaceae	*Cudonia* (1), *Spathularia* (2), *Spathulariopsis* (1)
Agaricomycetes	Agaricales	Cyphellaceae	*Cheimonophyllum* (1), *Chondrostereum* (1), *Cyphella* (2)
Agaricomycetes	Agaricales	Cystostereaceae	*Crustomyces* (1), *Cystostereum* (1)
Dacrymycetes	Dacrymycetales	Dacrymycetaceae	*Calocera* (3), *Dacrymyces* (7), *Guepiniopsis* (1)
Agaricomycetes	Polyporales	Dacryobolaceae	*Dacryobolus* (1), *Postia* (8)
Sordariomycetes	Xylariales	Diatrypaceae	*Diatrype*(2), *Diatrypella* (1)
Agaricomycetes	Boletales	Diplocystidiaceae	*Astraeus* (1)
Pezizomycetes	Pezizales	Discinaceae	*Discina* (5), *Gyromitra* (4)
Agaricomycetes	Russulales	Echinodontiaceae	*Amylostereum* (3)
Eurotiomycetes	Eurotiales	Elaphomycetaceae	*Elaphomyces* (9)
Agaricomycetes	Agaricales	Entolomataceae	*Alboleptonia* (1), *Clitocella* (2), *Clitopilus* (3), *Entocybe* (1), *Entoloma* (65), *Rhodocybe* (12), *Rhodophana* (1)
Pucciniomycetes	Platygloeales	Eocronartiaceae	*Eocronartium* (1)
Leotiomycetes	Erysiphales	Erysiphaceae	*Phyllactinia* (1)
Agaricomycetes	Auriculariales	Exidiaceae	*Guepinia* (1), *Pseudohydnum* (1), *Tremiscus* (1)
Exobasidiomycetes	Exobasidiales	Exobasidiaceae	*Exobasidium*(1)
Agaricomycetes	Polyporales	Fibroporiaceae	*Fibroporia* (2)
Agaricomycetes	Agaricales	Fistulinaceae	*Fistulina* (1)

TABLE 3.1 *(Continued)*

Class	Order	Family	Genus
Agaricomycetes	Polyporales	Fomitopsidaceae	*Antrodia* (11), *Buglossoporus* (1), *Daedalea* (2), *Fomitopsis* (5), *Piptoporus* (2), *Rhodofomes* (1)
Agaricomycetes	Geastrales	Geastraceae	*Geastrum* (19), *Myriostoma*(1), *Schenella* (2), *Sphaerobolus* (1)
Leotiomycetes	Helotiales	Gelatinodiscaceae	*Ascocoryne* (1), *Ascotremella* (1), *Coryne* (1), *Neobulgaria* (1)
Agaricomycetes	Polyporales	Gelatoporiaceae	*Cinereomyces* (1)
Geoglossomycetes	Geoglossales	Geoglossaceae	*Geoglossum* (2), *Leucoglossum* (1), *Trichoglossum* (3)
Agaricomycetes	Gloeophyllales	Gloeophyllaceae	*Gloeophyllum*(6), *Neolentinus* (4), *Veluticeps* (3)
Sordariomycetes	Diaporthales	Gnomoniaceae	*Apiognomonia* (1)
Agaricomycetes	Gomphales	Gomphaceae	*Gautieria* (6), *Gomphus* (1), *Phaeoclavulina* (2), *Ramaria* (21), *Turbinellus* (1)
Agaricomycetes	Boletales	Gomphidiaceae	*Chroogomphus* (6), *Gomphidius* (3)
Sordariomycetes	Xylariales	Graphostromataceae	*Biscogniauxia* (1)
Agaricomycetes	Polyporales	Grifolaceae	*Grifola* (1)
Pucciniomycetes	Pucciniales	Gymnosporangiaceae	*Gymnosporangium* (1)
Agaricomycetes	Boletales	Gyroporaceae	*Gyroporus* (1)
Pucciniomycetes	Helicobasidiales	Helicobasidiaceae	*Helicobasidium* (1)
Leotiomycetes	Helotiales	Helotiaceae	*Bisporella* (3), *Crocicreas* (2), *Cyathicula* (2), *Hymenoscyphus* (15), *Phaeohelotium* (3), *Stamnaria* (1), *Symphyosirinia* (1)
Pezizomycetes	Pezizales	Helvellaceae	*Balsamia* (1), *Barssia* (2), *Helvella* (25),
Agaricomycetes	Russulales	Hericiaceae	*Dentipellis* (1), *Hericium* (4), *Laxitextum* (1)
Leotiomycetes	Helotiales	Heterosphaeriaceae	*Heterosphaeria* (1)

TABLE 3.1 *(Continued)*

Class	Order	Family	Genus
Agaricomycetes	Auriculariales	Hyaloriaceae	*Myxarium* (1)
Leotiomycetes	Helotiales	Hyaloscyphaceae	*Cistella* (3), *Clavidisculum* (1), *Hyalopeziza* (1), *Hyaloscypha* (2)
Agaricomycetes	Cantharellales	Hydnaceae	*Cantharellus* (10), *Clavulina* (3), *Craterellus* (3), *Hydnum* (4), *Membranomyces* (1), *Merulius* (1), *Multiclavula* (1), *Pseudocraterellus* (1)
Agaricomycetes	Agaricales	Hydnangiaceae	*Hydnangium* (1), *Laccaria* (9)
Agaricomycetes	Trechisporales	Hydnodontaceae	*Litschauerella* (1), *Subulicystidium* (1), *Trechispora* (6), *Tubulicium* (1)
Agaricomycetes	Agaricales	Hygrophoraceae	*Ampulloclitocybe* (1), *Aphroditeola* (1), *Arrhenia* (11), *Cantharellula* (1), *Camarophyllus* (1), *Chrysomphalina* (2), *Cuphophyllus* (3), *Gliophorus* (3), *Humidicutis* (1), *Hygrocybe* (22), *Hygrophorus* (33), *Lichenomphalia* (2), *Pseudoarmillariella* (1)
Agaricomycetes	Boletales	Hygrophoropsidaceae	*Hygrophoropsis* (3), *Leucogyrophana* (3)
Agaricomycetes	Hymenochaetales	Hymenochaetaceae	*Coltricia* (2), *Fomitiporia* (3), *Fuscoporia* (4), *Hymenochaete* (5), *Inocutis* (2), *Inonotus* (8), *Onnia* (1), *Phellinidium* (1), *Phellinopsis* (1), *Phellinus* (14), *Phylloporia* (2), *Porodaedalea* (2), *Pseudoinonotus* (1), *Sanghuangporus* (1), *Tubulicrinis* (10), *Xanthoporia* (1)
Agaricomycetes	Agaricales	Hymenogastraceae	*Alnicola* (2), *Galerina* (19), *Gymnopilus* (7), *Hebeloma* (50), *Hymenogaster*(9), *Naucoria* (4), *Phaeocollybia* (1), *Psilocybe* (17), *Stropholoma* (1)
Agaricomycetes	Polyporales	Hyphodermataceae	*Hyphoderma* (6), *Kurtia* (2), *Metulodontia* (1)
Agaricomycetes	Hymenochaetales	Hyphodontiaceae	*Hyphodontia* (17)
Sordariomycetes	Hypocreales	Hypocreaceae	*Hypocrea* (5), *Hypomyces* (1), *Podostroma* (1)
Sordariomycetes	Xylariales	Hypoxylaceae	*Annulohypoxylon* (1), *Daldinia* (1)
Agaricomycetes	Hysterangiales	Hysterangiaceae	*Hysterangium* (7)

TABLE 3.1 *(Continued)*

Class	Order	Family	Genus
Agaricomycetes	Polyporales	Incrustoporiaceae	*Skeletocutis* (5), *Tyromyces* (2)
Agaricomycetes	Agaricales	Inocybaceae	*Inocybe* (104), *Pseudosperma* (3)
Agaricomycetes	Polyporales	Irpicaceae	*Byssomerulius* (1), *Ceriporia* (1), *Gloeoporus* (4), *Hydnopolyporus* (1), *Irpex* (1), *Leptoporus* (2), *Meruliopsis* (1)
Agaricomycetes	Polyporales	Ischnodermataceae	*Ischnoderma* (2)
Agaricomycetes	Jaapiales	Jaapiaceae	*Jaapia* (1)
Leotiomycetes	Helotiales	Lachnaceae	*Belonidium* (2), *Brunnipila* (1), *Capitotricha* (1), *Dasyscyphella* (1), *Lachnellula* (7), *Lachnum* (7), *Neodasyscypha* (1), *Perrotia* (2), *Trichopeziza* (1)
Agaricomycetes	Russulales	Lachnocladiaceae	*Vararia* (1)
Agaricomycetes	Polyporales	Laetiporaceae	*Laetiporus* (1), *Phaeolus* (1)
Sordariomycetes	Sordariales	Lasiosphaeridaceae	*Lasiosphaeria* (1), *Lasiosphaeris* (1)
Agaricomycetes	Gomphales	Lentariaceae	*Hydnocristella* (1), *Kavinia* (1), *Lentaria* (1)
Leotiomycetes	Leotiales	Leotiaceae	*Leotia* (1)
Dothideomycetes	Pleosporales	Leptosphaeriaceae	*Leptosphaeria* (1)
Agaricomycetes	Boletales	Leucogasteraceae	*Leucogaster* (4)
Agaricomycetes	Agaricales	Lycoperdaceae	*Bovista* (7), *Calvatia* (4), *Langermannia* (1)
Agaricomycetes	Agaricales	Lyophyllaceae	*Asterophora* (2), *Bryophyllum* (1), *Calocybe* (3), *Clitolyophyllum* (1), *Hypsizygus* (1), *Lyophyllum* (13), *Ossicaulis* (2), *Rugosomyces* (4), *Tephrocybe* (6), *Tricholomella* (1)
Agaricomycetes	Agaricales	Macrocystidiaceae	*Macrocystidia* (1)

TABLE 3.1 *(Continued)*

Class	Order	Family	Genus
Agaricomycetes	Agaricales	Marasmiaceae	*Baeospora* (2), *Calathella* (1), *Calyptella* (1), *Campanella* (1), *Clitocybula* (2), *Crinipellis* (2), *Gerronema* (2), *Marasmius* (22), *Megacollybia* (2)
Sordariomycetes	Diaporthales	Melogrammataceae	*Herpotrichia*(1), *Melogramma* (1)
Agaricomycetes	Polyporales	Meripilaceae	*Laeticutis* (1), *Meripilus* (1), *Rigidoporus* (2)
Agaricomycetes	Polyporales	Meruliaceae	*Aurantiporus* (1), *Bulbillomyces* (1), *Ceriporiopsis* (1), *Climacodon* (1), *Gelatoporia* (1), *Gyrophanopsis* (1), *Hypochnicium* (5), *Lawrynomyces* (1), *Mycoacia* (2), *Phlebia* (6), *Physisporinus* (1), *Sarcodontia* (4), *Scopuloides* (1)
Agaricomycetes	Hysterangiales	Mesophelliaceae	*Chondrogaster* (2)
Leotiomycetes	Helotiales	Mollisiaceae	*Mollisia* (5), *Tapesia* (2)
Pezizomycetes	Pezizales	Morchellaceae	*Disciotis* (1), *Mitrophora* (1), *Morchella* (33), *Verpa* (2)
Agaricomycetes	Agaricales	Mycenaceae	*Atheniella* (2), *Collopus* (1), *Hemimycena* (5), *Hydropus* (1), *Mycena* (70), *Panellus* (5), *Roridomyces* (1), *Scytinotus* (1), *Xeromphalina* (2)
Eurotiomycetes	Mycocaliciales	Mycocaliciaceae	*Mycocalicium* (1)
Dothideomycetes	Mycosphaerellales	Mycosphaerellaceae	*Mycosphaerella* (1), *Zasmidium* (1)
Sordariomycetes	Hypocreales	Nectriaceae	*Dialonectria* (1), *Nectria* (1), *Neonectria* (2), *Pleonectria* (1)
Agaricomycetes	Agaricales	Niaceae	*Cyphellopsis* (1), *Flagelloscypha* (1), *Lachnella* (2), *Merismodes* (1)
Agaricomycetes	Agaricales	Nidulariaceae	*Nidularia* (1)
Agaricomycetes	Agaricales	Omphalotaceae	*Connopus* (1), *Gymnopus* (22), *Lentinula* (1), *Marasmiellus* (5), *Mycetinis* (6), *Omphalotus* (3), *Rhodocollybia* (3)
Eurotiomycetes	Onygenales	Onygenaceae	*Onygena* (1)

TABLE 3.1 *(Continued)*

Class	Order	Family	Genus
Sordariomycetes	Hypocreales	Ophiocordycipitaceae	*Tolypocladium* (1)
Orbiliomycetes	Orbiliales	Orbiliaceae	*Hyalorbilia* (1), *Orbilia* (6)
Pezizomycetes	Pezizales	Otideaceae	*Otidea* (7)
Agaricomycetes	Hymenochaetales	Oxyporaceae	*Oxyporus* (1)
Agaricomycetes	Polyporales	Panaceae	*Panus* (4),
Agaricomycetes	Boletales	Paxillaceae	*Alpova* (2), *Gyrodon* (1), *Hydnomerulius* (1), *Melanogaster* (5), *Paxillus* (8)
Agaricomycetes	Russulales	Peniophoraceae	*Asterostroma* (2), *Dichostereum* (1), *Gloiothele* (2), *Peniophora* (15), *Scytinostroma* (3), *Scytinostromella* (1), *Subulicium* (1)
Pezizomycetes	Pezizales	Pezizaceae	*Adelphella* (1), *Delastria* (1), *Hydnobolites* (1), *Iodophanus* (1), *Marcelleina* (3), *Pachyella* (2), *Pachyphlodes*(2), *Peziza* (31), *Phylloscypha* (1), *Sarcopeziza* (1), *Sarcosphaera* (1), *Terfezia* (7), *Tirmania* (1)
Leotiomycetes	Helotiales	Pezizellaceae	*Calycina* (3), *Gemmina* (1), *Pezizella* (1), *Rodwayella* (1)
Leotiomycetes	Phacidiales	Phacidiaceae	*Bulgaria* (1)
Dothideomycetes	Pleosporales	Phaeosphaeriaceae	*Ophiobolus* (1)
Agaricomycetes	Phallales	Phallaceae	*Anthurus* (1), *Clathrus* (1), *Colus* (1), *Mutinus* (1), *Phallus* (2), *Pseudocolus* (1)
Agaricomycetes	Hysterangiales	Phallogastraceae	*Phallogaster* (1)
Agaricomycetes	Polyporales	Phanerochaetaceae	*Atheliachaete* (2), *Bjerkandera* (2), *Hapalopilus* (2), *Hyphodermella* (1), *Phanerochaete* (8), *Phanerodontia* (1), *Phlebiopsis* (1), *Terana* (1)
Atractiellomycetes	Atractiellales	Phleogenaceae	*Phleogena* (1)
Agaricomycetes	Agaricales	Phyllotopsidaceae	*Phyllotopsis* (1), *Pleurocybella* (1)

TABLE 3.1 *(Continued)*

Class	Order	Family	Genus
Agaricomycetes	Agaricales	Physalacriaceae	*Armillaria* (11), *Cryptomarasmius* (2), *Cylindrobasidium* (1), *Flammulina* (3), *Hymenopellis* (3), *Mucidula* (1), *Oudemansiella* (2), *Paraxerula* (1), *Rhizomarasmius* (2), *Strobilurus* (4), *Xerula* (2)
Agaricomycetes	Agaricales	Pleurotaceae	*Hohenbuehelia* (8), *Pleurotus* (12), *Resupinatus* (5)
Leotiomycetes	Helotiales	Ploettnerulaceae	*Pyrenopeziza* (3)
Agaricomycetes	Agaricales	Pluteaceae	*Pluteus* (43), *Volvariella* (17), *Volvopluteus* (2)
Agaricomycetes	Polyporales	Podoscyphaceae	*Abortiporus* (1), *Podoscypha* (1)
Agaricomycetes	Polyporales	Polyporaceae	*Cellulariella*(1), *Cerioporus* (6), *Climacocystis* (1), *Coriolopsis* (1), *Coriolus* (2), *Cerioporus* (1), *Daedaleopsis* (3), *Datronia* (3), *Dichomitus* (1), *Diplomitoporus* (1), *Erastia* (1), *Faerberia* (1), *Fomes* (2), *Ganoderma* (7), *Hexagonia* (1), *Hirschioporus* (1), *Lentinus* (5), *Lenzites* (2), *Neofavolus* (1), *Pachykytospora* (1), *Perenniporia* (4), *Picipes* (2), *Podofomes* (2), *Polyporus* (15), *Pycnoporus* (1), *Pyrofomes* (1), *Rhodonia* (1), *Royoporus* (1), *Trametes* (10), *Trichaptum* (3)
Agaricomycetes	Agaricales	Porotheleaceae	*Phloeomana* (3), *Porotheleum* (1), *Pseudohydropus* (1)
Agaricomycetes	Agaricales	Psathyrellaceae	*Coprinellus* (20), *Coprinopsis* (29), *Homophron* (2), *Lacrymaria* (4), *Panaeolina* (1), *Panaeolus* (16), *Parasola* (9), *Psathyra* (2), *Psathyrella* (49)
Agaricomycetes	Agaricales	Pseudoclitocybaceae	*Pseudoclitocybe* (3)
Pezizomycetes	Pezizales	Pseudombrophilaceae	*Pseudombrophila* (1)
Agaricomycetes	Agaricales	Pterulaceae	*Pterula* (2)
Pezizomycetes	Pezizales	Pulvinulaceae	*Pulvinula* (6)
Agaricomycetes	Corticiales	Punctulariaceae	*Dendrocorticium* (1)

TABLE 3.1 *(Continued)*

Class	Order	Family	Genus
Pezizomycetes	Pezizales	Pyronemataceae	*Acervus* (1), *Aleuria* (2), *Anthracobia* (2), *Arpinia* (1), *Byssonectria*(1), *Cheilymenia* (7), *Ciliaria* (1), *Flavoscypha* (1), *Genea* (5), *Geopora* (6), *Geopyxis* (4), *Humaria* (2), *Hydnocystis* (1), *Hypotarzetta* (1), *Inermisia* (1), *Kotlabaea* (1), *Lamprospora* (6), *Lasiobolidium* (1), *Melastiza* (2), *Neottiella* (2), *Octospora* (16), *Parascutellinia* (2), *Picoa* (2), *Plicaria* (2), *Psilopezia* (1), *Pustularia* (1), *Pyronema* (2), *Ramsbottomia* (1), *Rhodoscypha* (1), *Scutellinia* (8), *Sepultariella* (1), *Smardaea* (1), *Sowerbyella* (1), *Spooneromyces* (1), *Stephensia* (1), *Tricharina* (3), *Trichophaea* (4), *Trichophaeopsis* (1)
Agaricomycetes	Agaricales	Radulomycetaceae	*Aphanobasidium* (1), *Radulomyces* (2)
Agaricomycetes	Hymenochaetales	Repetobasidiaceae	*Fibricium* (2)
Pezizomycetes	Pezizales	Rhizinaceae	*Rhizina* (1)
Dothideomycetes	Aulographales	Rhizodiscinaceae	*Rhizodiscina* (1)
Agaricomycetes	Boletales	Rhizopogonaceae	*Rhizopogon* (8)
Leotiomycetes	Rhytismatales	Rhytismataceae	*Coccomyces* (3), *Colpoma* (1), *Hysterium* (1)
Agaricomycetes	Hymenochaetales	Rickenellaceae	*Cotylidia* (1), *Globulicium* (1), *Peniophorella* (3), *Resinicium* (1), *Rickenella* (1), *Sidera* (1)
Agaricomycetes	Russulales	Russulaceae	*Boidinia* (1), *Gymnomyces* (3), *Lactarius* (78), *Lactifluus* (4), *Russula* (146)
Leotiomycetes	Helotiales	Rutstroemiaceae	*Lanzia* (1), *Rutstroemia* (3)
Pezizomycetes	Pezizales	Sarcosomataceae	*Kompsoscypha* (1), *Microstoma* (1), *Pithya* (2), *Pseudopithyella* (1), *Plectania* (3), *Pseudoplectania* (3), *Sarcoscypha* (1)
Agaricomycetes	Agaricales	Schizophyllaceae	*Ditiola* (1), *Henningsomyces* (1), *Rectipilus* (1), *Schizophyllum* (2)

TABLE 3.1 *(Continued)*

Class	Order	Family	Genus
Agaricomycetes	Hymenochaetales	Schizoporaceae	*Alutaceodontia* (1), *Basidioradulum* (2), *Schizopora* (1), *Xylodon* (7)
Agaricomycetes	Boletales	Sclerodermataceae	*Pisolithus* (2), *Scleroderma* (8)
Agaricomycetes	Boletales	Sclerogastraceae	*Sclerogaster* (3)
Leotiomycetes	Helotiales	Sclerotiniaceae	*Botryotinia* (1), *Ciboria* (5), *Dumontinia* (1), *Sclerotinia* (2)
Agaricomycetes	Sebacinales	Sebacinaceae	*Cristella* (1), *Sebacina* (3)
Agaricomycetes	Boletales	Serpulaceae	*Serpula* (1)
Leotiomycetes	Helotiales	Solenopeziaceae	*Lasiobelonium* (2), *Trichopezizella* (1)
Agaricomycetes	Polyporales	Sparassidaceae	*Crustoderma* (2), *Sparassis* (2)
Agaricomycetes	Polyporales	Steccherinaceae	*Antrodiella* (1), *Cabalodontia* (2), *Junghuhnia* (2), *Loweomyces* (1), *Steccherinum* (2)
Agaricomycetes	Agaricales	Stephanosporaceae	*Athelidium* (1), *Cristinia* (1)
Agaricomycetes	Russulales	Stereaceae	*Aleurodiscus* (5), *Conferticium* (2), *Gloeocystidiellum* (3), *Stereum* (11), *Xylobolus* (1)
Agaricomycetes	Agaricales	Stereopsidaceae	*Stereopsis* (1)
Pezizomycetes	Pezizales	Strobiloscyphaceae	*Strobiloscypha* (1)
Agaricomycetes	Agaricales	Strophariaceae	*Agrocybe* (12), *Bogbodia* (1), *Deconica* (4), *Hypholoma* (10), *Kuehneromyces* (1), *Leratiomyces* (2), *Meottomyces* (1), *Pholiota* (26), *Protostropharia* (2), *Stropharia* (11)
Agaricomycetes	Boletales	Tapinellaceae	*Tapinella* (2)
Pezizomycetes	Pezizales	Tarzettaceae	*Tarzetta* (2)
Leotiomycetes	Thelebolales	Thelebolaceae	*Coprotus* (2)

TABLE 3.1 *(Continued)*

Class	Order	Family	Genus
Agaricomycetes	Thelephorales	Thelephoraceae	*Bankera* (2), *Lenzitella* (1), *Lenzitopsis* (1), *Odontia* (1), *Phellodon* (5), *Pseudotomentella* (2), *Thelephora* (6), *Tomentella* (13)
Agaricomycetes	Hysterangiales	Trappeaceae	*Trappea* (1)
Tremellomycetes	Tremellales	Tremellaceae	*Tremella* (7)
Agaricomycetes	Agaricales	*Tricholomataceae*	*Atractosporocybe* (1), *Bonomyes* (1), *Cellypha* (1), *Clitocybe* (46), *Collybia* (7), *Delicatula* (1), *Dermoloma* (1), *Fayodia* (1), *Gamundia* (2), *Giacomia* (1), *Infundibulicybe* (3), *Lepista* (12), *Leucocortinarius* (1), *Leucocybe* (2), *Leucoinocybe* (1), *Leucopaxillus* (9), *Melanoleuca* (31), *Myxomphalia* (1), *Neoclitocybe* (1), *Omphalina* (5), *Paralepista* (1), *Paralepistopsis* (1), *Pseudoomphalina* (2), *Pseudoporpoloma* (1), *Rhizocybe* (1), *Rimbachia* (2), *Tricholoma* (67), *Tricholomopsis* (2)
Leotiomycetes	Helotiales	Tricladiaceae	*Cudoniella* (1)
Agaricomycetes	Agaricales	Tubariaceae	*Cyclocybe* (2), *Flammulaster* (3), *Phaeomarasmius* (2), *Tubaria* (7)
Pezizomycetes	Pezizales	Tuberaceae	*Choiromyces* (1), *Reddellomyces* (2), *Tuber* (16)
Agaricomycetes	Cantharellales	Tulasnellaceae	*Tulasnella* (1)
Agaricomycetes	Agaricales	Typhulaceae	*Macrotyphula* (2), *Typhula* (8)
Ustilaginomycetes	Urocystidales	Urocystidaceae	*Urocystis* (1)
Leotiomycetes	Helotiales	Vibrisseaceae	*Vibrissea* (2)
Agaricomycetes	Corticiales	Vuilleminiaceae	*Vuilleminia* (3)
Agaricomycetes	Russulales	Xenasmataceae	*Xenasmatella* (2)
Sordariomycetes	Xylariales	Xylariaceae	*Hypocopra* (1), *Hypoxylon* (5), *Kretzschmaria* (1), *Rosellinia* (3), *Xylaria* (5)

the endemism ratio of fungal species. It is certain that Türkiye contains some rare fungal taxa. However, there is currently no specific Red Data List for macrofungi in Türkiye.

TABLE 3.2 Most Speciose Genera of Macrofungi in Anatolia

Genus	Species	Genus	Species
1. *Cortinarius*	154	6. *Tricholama*	67
2. *Russula*	146	7. *Entoloma*	65
3. *Inocybe*	104	8. *Agaricus*	58
4. *Lactarius*	78	9. *Hebeloma*	50
5. *Mycena*	70	10. *Psathyrella*	49

TABLE 3.3 Most Speciose and Dominant of Macrofungi in Anatolia

Families	Species	Genera	Families	Species	Genera
1. Russulaceae	232	5	6. Boletaceae	124	28
2. Tricholamataceae	205	28	7. Hymenogastraceae	110	9
3. Agaricaceae	199	25	8. Inocybaceae	107	2
4. Cortinariaceae	157	3	9. Pyronemataceae	95	38
5. Psathyrallaceae	132	9	10. Mycenaceae	88	9

TABLE 3.4 Most Speciose Orders of Macrofungi in Anatolia and Number of Taxa

Order	Species	Genera	Families
1. Agaricales	1678	220	44
2. Russulalaes	310	32	10
3. Pezizales	283	81	16
4. Polyporales	209	91	18
5. Boletales	186	46	12

3.5 DIVERSITY IN DIFFERENT REGIONS

Anatolia boasts a vast geography with diverse climate zones, a rich flora, and conditions that support the growth of a wide variety of fungi. The northern regions experience a Black Sea climate, while the south and west have a Mediterranean climate, and the interior regions are dominated by a continental climate. Anatolia's location at the intersection of three phytogeographic regions further enhances its fungal biodiversity. As a result, Anatolia is home to a greater variety of fungal species. However, the Euro-Siberian, Mediterranean, and Irano-Turanian floristic regions in Anatolia show significant differences in biodiversity due to their distinct climatic and geographical features.

In the Euro-Siberian region, which includes Anatolia's Black Sea coast and Northeastern Anatolia, the climate is cool, humid, rainy, and snowy, dominated by the Black Sea climate. The lush forests and rich vegetation provide an ecosystem abundant in biological diversity. Due to the high humidity, the region is also home to a great variety of fungi. In the Mediterranean region, which covers Southern and Western Anatolia, as well as the Aegean and Mediterranean coasts, the Mediterranean climate prevails, characterized by hot, dry summers and mild, rainy winters. The Mediterranean vegetation (maquis, pine forests, and shrubs) is unique to this area. While the warm winters contribute to rich fungal biodiversity, the dry summers of the Mediterranean climate can limit fungal growth.

The Irano-Turanian region, which includes Central Anatolia and parts of Eastern Anatolia, experiences a continental climate with hot, dry summers and cold, snowy winters. Fungal diversity is relatively limited due to the arid conditions and extreme temperatures of the continental climate, which restrict fungal growth.

3.6 CONSERVATION STRATEGIES OF FUNGI IN TÜRKİYE

Protecting the natural habitats of fungi is a fundamental step in maintaining fungal populations. It is important to conserve and restore various habitats such as forests, meadows, and mountainous regions. In Türkiye, fungal conservation efforts generally focus on the preservation of natural habitats and raising awareness. National parks, nature parks, and nature reserves in various regions of Türkiye play a significant role in protecting the habitats of plants, animals, and fungi. These areas provide the necessary suitable habitats for fungi to sustain their natural life cycles. Raising public awareness about the conservation of fungi and sustainable harvesting practices is crucial. To this end, various educational programs are organized in collaboration with the Ministry of Health and universities in Türkiye. Seminars are held in schools, universities, and at meetings and festivals organized by civil society organizations. Additionally, information is disseminated to the public through television, social media, and other communication channels as public service announcements (PSA).

There are legal regulations regarding the protection of fungi in Türkiye. In fact, it is necessary to obtain permission for collecting any kind of material from nature in Türkiye. These permits are issued by the relevant ministries. There are severe penalties for collecting materials from nature without permission. Hunting and collection restrictions have been imposed, particularly to protect certain species. The enforcement of these regulations is important for preserving fungal populations.

Research and monitoring activities are conducted within universities and research institutes to monitor fungal populations and develop conservation strategies. These studies are crucial for identifying which species are under threat and for guiding conservation efforts. These examples reflect various aspects of fungal conservation efforts in Türkiye. Sustaining such efforts is important for preserving fungal biodiversity and passing it on to future generations.

In Türkiye, *Boletes* (Figure 3.2a), *Morchella* spp. (Figure 3.2b), *Lactarius* spp. (Figure 3.2c), *Agaricus* spp. (Figure 3.2d), *Pleurotus* spp. (Figure 3.2e), *Russula* spp., *Tricholoma* spp., *Cantharellus* spp., *Craterellus* spp., *Amanita* spp., *Tuber* spp. (Figure 3.2f), and *Terfezia* spp. are collected from nature for consumption or trade (Metin et al., 2013). In addition to edible fungi, certain fungal species with medicinal properties, such as *Ganoderma lucidum*, are also consumed in Türkiye, either wild or cultivated forms.

Mushroom foraging in the wild is a delightful pursuit for mushroom aficionados. Gathering species like *Boletes*, *Morchella* spp., and *Tuber* spp. from natural habitats not only brings joy but also serves as a significant source of livelihood for forest communities (Metin et al., 2013). However, due to factors such as overharvesting and insufficient rainfall, the growth rates of certain fungal species have declined in some regions. Enhancing and diversifying initiatives to educate mushroom enthusiasts on the collection of mushrooms is crucial for safeguarding public health, maintaining biodiversity, and promoting sustainable livelihoods.

3.7 CONCLUSIONS

Undoubtedly, Anatolia hosts a rich macrofungal biodiversity due to its geographical location, the presence of different climate zones, and its diverse ecosystems. The number and diversity of taxa presented in the study suggest that Anatolia's mycota has great potential. Endemic 33 macrofungal taxa are globally rare species. This underscores Anatolia's importance as a significant biodiversity center on a global scale. Habitat loss, climate change, and overharvesting, in particular, are the main factors threatening the future of fungal species. The study reveals that stricter measures need to be taken against overharvesting and habitat loss in order to protect fungal biodiversity.

FIGURE 3.2 (a). *Boletus* sp. (b). *Morchella esculenta* (c). *Lactarius* sp (d). *Agaricus campestris*. (e). *Pleurotus ostreatus* (f). *Tuber aestivum*.

There are a significant number of academics researching the fungal biodiversity of Türkiye. They are investigating every part of Türkiye with research projects. Nowadays, genus-focused studies are popular in Türkiye, and the number of mycologists specializing in this subject is increasing day by day. In this context, new macrofungal taxa are being added to the mycobiota of Türkiye on a regular basis (Kaygusuz et al., 2020, 2024). Advanced researches supported by molecular taxonomy and ecological studies will contribute to the accurate identification and conservation of Turkish fungal taxa. Moreover, a better understanding of the roles of these species within ecosystems will help develop sustainable management strategies.

KEYWORDS

- **Anatolia**
- **biodiversity**
- **elevational gradients**
- **endemism**
- **macrofungi**
- **taxonomy**

REFERENCES

Acar, I. New Records from Türkiye: *Cortinarius barbatus*, *C. osmophorus*, and *C. saturninus*. *Mycotaxon* **2022**, *136* (4), 819–830.

Akata, I.; Kumru, E.; Ediş, G.; Özbey, B. G.; Şahin, E. Three New Records for Turkish Agaricales Inhabiting Ankara University Beşevler 10th Year Campus Area. *Kastamonu Univ. J. Forestry Faculty* **2023**, *23* (3), 250–263.

Akata, I.; Şen, İ.; Sevindik, M.; Kabaktepe, Ş. Truffle Checklist of Türkiye II with a New Record. *Turkish Journal of Agriculture-Food Science and Technology* **2022**, *10* (10), 1913–1920.

Asan, A.; Selçuk, F.; Giray, G.; Aydoğdu, H.; Ulukapı, M.; Ceylan, M. F. Türkiye Mantarları Listesi'ne İlaveler-1. *Bağbahçe Bilim Dergisi* **2022**, *9* (3), 65–89.

Baba, H.; Sevindik, M. Myxomycetes (Myxogastria) of Türkiye: A Checklist 2023. *Mycopath* **2023**, *21* (1), 53.

Baytop, T. *Anadolu Dağlarında 50 Yıl: Bir Bitki Avcısının Gözlemleri*; Nobel Tıp Kitabevleri: İstanbul, 2000.

Fritsch, K. Beitrag zur Flora von Constantinopel I. Kryptogamen. *Denkschriften der Kais. Akad. D. Wiss. Mathem. Naturw. Klasse* **1973**, Bd. L XVII, p 219.

Güner, A.; Aslan, S. *Türkiye Bitkileri Listesi: (Damarlı Bitkiler)*; Nezahat Gökyiğit Botanik Bahçesi Yayınları, 2012.

Güngör, M. Y.; Güngör, H. *Rhizopogon pumilionum* (Ade) Bataille; A New False Truffle (Basidiomycota) Record for Turkish Mycota. *Biodicon* **2022**, *15* (3), 303–307.

Handel-Mazzetti, H. F. Ergebnisse Einer Botanischen Reise in das Pontische Randgebirge in Sandschak Trapezunt. *Annalen des K.K. Naturhistorischen Hofmuseum* **1909**, Bd. XXIII, p 101.

Ilanloo, S. S.; Ashrafi, S.; Shabani, A. A. Modeling Habitat Suitability of the Red-Backed Shrike (*Lanius collurio*) in the Irano-Anatolian Biodiversity Hotspot. *Journal of Zoological Research* **2021**, *3* (2), 1–8.

John, V.; Güvenç, Ş. Additions to the Checklist and Bibliography of the Lichens and Lichenicolous Fungi of Türkiye II. *Archive for Lichenology* **2023**, *34*, 1–47.

Kaygusuz, O.; Knudsen, H.; Türkekul, İ.; Çolak, Ö. F. *Volvariella turcica*, A New Species from Türkiye, and a Multigene Phylogeny of *Volvariella*. *Mycologia* **2020**, *112* (3), 577–587.

Kaygusuz, O.; Reschke, K.; Kaya, A.; Dima, B.; Morozova, O.; Noordeloos, M. E.; Piepenbring, M. Morphology and Molecular Phylogeny of Four New Species of *Entoloma* (Entolomataceae, Agaricales) from Europe and Western Eurasia. *Mycological Progress* **2024**, *23* (1), 1–19.

Kesici, S.; Akçay, M. E.; Sadullahoğlu, C.; Uzun, Y. Six *Typhula* spp. Newly Recorded from Türkiye. *Mycotaxon* **2022**, *137* (3), 577–589.

Lohwag, K. Kavaklarda Odun Tahripçisi Mantarlar (Çeviren: Selik, M.). *J. Fac. For. Istanbul Univ.* **1959**, *9* (1), 7–10.

Lohwag, K. Türkiye'nin Mantar Florası Hakkında Araştırma (Çeviren: H. Ünlügil). *J. Fac. For. Istanbul Univ.* **1957**, *7* (1), 129–137.

Maire, M. Étude des Champignons Récoltés en Asie Mineure. *Bull. Société des Sciences de Nancy* **1904**, 165.

Metin, İ.; Güngör, H.; Çolak, Ö. F. Ülkemizdeki Bazı Mantar ve Mantar Ürünlerinin Dış Ticareti Üzerine Bir Araştırma ve Küresel Pazarlanmasına Yönelik Öneriler. *Mantar Dergisi* **2013**, *4* (2), 1–9.

Pilát, A. Additamenta ad Floram Asiae Minoris Hymenomycetum et Gasteromycetum. *Bull. Soc. Bot.* **1937**, *53* (3, 4), 253–264.

Pilát, A. Additamenta ad Floram Asiae Minoris Hymenomycetum. Pars Secunda: Agaricineae. *Bull. Soc. Mycol.* **1932**, *48* (3, 4), 283–302.

Rigler, L. *Die Türkei und Deren Bewohner*; Wien, Germany, 1852; p 111.

Sadullahoğlu, C. Oltu ve Narman (Erzurum) İlçelerinde Yetişen Makromantar Üzerinde Taksonomik Bir Araştırma; 2022; p 426.

Sayyad, S. I.; Sohrab, A.; Afshin, A. S. Modeling Habitat Suitability of the Red-Backed Shrike (*Lanius collurio*) in the Irano-Anatolian Biodiversity Hotspot. *Journal of Zoological Research* **2021**, *3* (2), 2. http://ojs.bilpublishing.com/index.php/jzr (accessed April 21, 2026).

Sesli, E.; Asan, A.; Selçuk, F., eds.; Abacı Günyar, Ö.; Akata, I.; Akgül, H.; Aktaş, S.; Alkan, S.; Allı, H.; Aydoğdu, H.; Berikten, D.; Demirel, K.; Demirel, R.; Doğan, H. H.; Erdoğdu, M.; Ergül, C. C.; Eroğlu, G.; Giray, G.; Haliki Uztan, A.; Kabaktepe, Ş.;

Kadaifçiler, D.; Kalyoncu, F.; Karaltı, İ.; Kaşık, G.; Kaya, A.; Keleş, A.; Kırbağ, S.; Kıvanç, M.; Ocak, İ.; Ökten, S.; Özkale, E.; Öztürk, C.; Sevindik, M.; Şen, B.; Şen, İ.; Türkekul, İ.; Ulukapı, M.; Uzun, Y. A.; Uzun, Y. U.; Yoltaş, A. *Türkiye Mantarları Listesi (The Checklist of Fungi of Türkiye)*; Ali Nihat Gökyiğit Vakfı Yayını: İstanbul, 2020; p 1177.

Solak, M.; Türkoğlu, A. *Macrofungi of Türkiye, Checklist*, 3rd ed.; Kanyılmaz Matbaacılık: İzmir, 2022; 420 p.

Tchihatcheff, P. *Asie Mineure*, Vol. III, Botanique, II; Paris, 1860; p 670.

Zwara, J. Contribution à l'étude des Russules de l'Asie Mineure. *Bull. Soc. Bot.* **1932**, *48*, 253–258.

CHAPTER 4

BRYOPHYTE DIVERSITY IN THE IRANO-ANATOLIAN REGION

PREM L. UNIYAL[1*], HIMANI YADAV[1*], PRITI GIRI[1], ANSHUL DHYANI[2], and MEENAM BHATIA[3]

[1]*Department of Botany, University of Delhi, Delhi, India*

[2]*Department of Botany, Hindu College, University of Delhi, Delhi, India*

[3]*Department of Botany, Daulat Ram College, University of Delhi, Delhi, India*

*Corresponding authors

ABSTRACT

The network of mountains and the enclosed Mediterranean Basin is known as the Irano-Anatolian Hotspot. The region harbors high endemism and diverse flora and fauna and is now recognized as a hotspot. Over 40% of the plant species in this area are thought to be indigenous. Major sections of central and eastern Turkey, southern Georgia, Azerbaijan, Armenia, northeastern Iraq, northern and western Iran, and the Northern Range of Turkmenistan are all included in the Irano-Anatolian biodiversity hotspot. This area experiences dry weather with a Mediterranean pattern of precipitation. The mountain region ranges from 1000 to 4000 m. Annual rainfall ranges from 100 to over 1000 mm in the winter and spring. It has continental, Oceanic, and Mediterranean climates. Dominated families of the study site are Pottiaceae, Bryaceae, Orthotrichaceae, Brachytheciaceae, Hypnaceae, and Grimmiaceae. *Bryum, Brachythecium, Eurhynchium, Fissidens, Grimmia,* and *Hypnum* are found to be the most species-rich genera. The Irano-Anatolian region has added numerous species to the richness of bryophytes despite this gap. Because the Irano-Anatolian region is endowed with a wide variety of bryophyte species, thorough recording is necessary in order to investigate the richness of bryofloral life.

Biodiversity Hotspot of the Irano-Anatolian Region. T. Pullaiah (Ed.)

DOI: 10.1201/9781779645760-4

4.1 INTRODUCTION

Bryophytes are considered to be the second largest group of land plants, with more than 23,000 species distributed all over the world (Buck et al., 2000). They are the pioneers, which colonized terrestrial habitats (Shaw and Renzaglia, 2004). They are an essential part of the earth's biodiversity and play a significant role in diverse terrestrial and aquatic ecosystems (Hallingback and Hodgetts, 2001). They are found on soil, rocks, and trees throughout the world, from coastal Antarctica to the tundra of the Northern Hemisphere and from the Australian deserts to the Amazon rain forests. They have remained relatively unchanged over the course of evolution and are thus generally referred to as a 'conservative group,' but have adapted themselves successfully to the changing environment (Bahuguna et al., 2013). Bryophytes are the best choice for experimental work because of their straightforward morphology, compact size, structural organization, minimal nutritional demand, great capacity for regeneration, and quick life cycle completion.

Bryophytes act as environmental indicators, which offer a better clue to the solution of phytogeographical problems than many vascular plants. They are important in controlling pore pressure during the rainy season and hence control landslides. Bryophytes are important for ecosystem nutrient dynamics, utilization as medicine, horticultural value, seedbed for higher plants, providing microhabitats for other plants and animals, and have a major role in promoting plant succession. They also have economic value, whether it is as peat for fuel, horticulture, oil absorption, or as sources for a wide variety of chemical compounds. Extracts from bryophytes are thought to have antibacterial, antifungal, and anticancer properties. They are a food source for animals in cold environments (Goffinet and Shaw, 2008) and are also used as bio-indicators (Ando and Mastuo, 1984; Bargagli et al., 1995).

4.2 IRANO-ANATOLIAN REGION

The vast, topographically intricate network of mountains and enclosed basins that divides the native populations and natural ecosystems of the Mediterranean Basin from the arid plateaus of Western Asia is known as the Irano-Anatolian Hotspot. Due to harboring high endemism and diverse flora and fauna, the area is now recognized as a new hotspot. For many centuries, the principal conduit for trade and cultural interaction between the two regions was the old Silk Road, which crossed through this area from east to west. Major sections of central and eastern Turkey, a tiny bit of southern Georgia, northeastern Iraq, a large chunk of Armenia, the Nakhichevan Province of Azerbaijan, the Northern Kopet Dag Range in Turkmenistan, and northern and western Iran are all included in the Irano-Anatolian biodiversity hotspot. It is the only global biodiversity hotspot that is completely contained inside Southwest Asia, and the area covered is 899,773 km^2. Iran has a significant degree of plant, animal, and landscape diversity, making up 54% of the Irano-Anatolian hotspot's surface area (Naroozi, 2018).

According to Tatkhtajan (1986), the Irano-Anatolian Hotspot is limited to the western portion of the Irano-Turanian Floristic Region, which also stretches eastward to encompass the arid regions surrounding the Caspian Sea, the Hindu Kush, and the western Himalayan spurs. However, other regions like the Kopet Dagh and the Zagros mountains express the mixed nature of their biogeographic links with parts of the Mediterranean and Irano-Turanian elements (Davis et al., 1971). In addition to supporting about 400 plant species, the hotspot northern region features vast ranges of arid mountains that go from west to east (the Kopet Dagh, the Southern Elburz Mountains, and the East Anatolian Mountains) and from north to south (the Zagros Mountains). These highlands stretch between the Mediterranean, Black Sea, and Caspian, the three main areas where the old Tethys Sea formerly existed. Here, the altitudes typically range from 1000 to 4000 m, with some places as low as 300 m, including the western Zagros Mountains and the foothills of the Kopet Dagh. Many peaks rise to elevations of more than 5000 m, such as the dormant volcanoes of Iran's Mt. Damavand (5671 m) and Turkey's Mt. Ararat (5165 m). The plateaus of Anatolia, Armenia, and western Iran (800 and 2000 m above sea level) are made up of primary plain steppes, which are climax steppes devoid of shrubland or forest cover (these areas are too dry and the soils poor). The surrounding mountain ranges are covered in secondary steppes or dry forests. To the north of the hotspot are three large, closed basins (Konya Closed Basin, Van Basin, and Lake Urumiyeh Basin), which are encircled by main plain steppes and include a number of tiny Tethys remnant lakes. With scorching summers and chilly winters, the climate is continental. Winters in most of the region are bitterly cold, with lows as low as –40°C. The region has the mean temperature of the dry summers is between 20 °C and 30°C, with a maximum temperature of 48°. Annual rainfall ranges from

100 to over 1000 mm in the winter and spring. At higher elevations, winter rainfall mostly takes the form of snow, which can last until June or July. With about 100 to 400 mm of precipitation annually, central Turkey and the southeast Zagros Mountains have unusually little rainfall. Mapping of species diversity is a major objective in conservation biology. The region has very high densities of endemic species and incidents of severe habitat loss, so should be taken as a conservation priority. Furthermore, hotspots for biodiversity are great places to research the diversification processes.

4.2.1 CONTINENTAL CLIMATE

In Kastamonu and northeastern Turkish regions of Erzurum, Kars, Ardahan, Posof, and Sarikamis, the summer precipitation is higher than winter precipitation. In the Kars and Ardahan region, the lowest temperatures of the coldest month range from –16°C to –17°C. The Kastamonu region, on the other hand, experiences far less cold, with the lowest temperatures of the coldest month falling between –5°C and –6°C.

4.2.2 MEDITERRANEAN CLIMATE

Along the northern slopes of the Black Sea Turkish coast, which is bordered to the west by Bulgaria and to the east by Azerbaijan, an oceanic climate predominates. Almost everything has an annual mean precipitation of more than 1000 mm, and this total consistently rises from west to east. In the western Black Sea region, the average annual precipitation ranges from 1050 to 1370 mm, and in the eastern region, it exceeds 2350 mm. The town of Kemalpasa reported the highest amount of rainfall, 2650 mm. The warmest months have mean maximum temperatures that range from 25°C to 28°C. In comparison, the temperatures inland from the Black Sea region in Yavuzkemal (600 m) and Buyukduz (1500 m) are 17°C and 22°C, respectively. The lowest average temperature of February, the coldest month, ranges from –4°C (Yavuzkemal) to –6°C (Buyukduz) inland, while it is between 1°C and 3°C by the shore.

4.2.3 MEDITERRANEAN VEGETATION

In the summertime, minimum rainfall in the majority of Turkey reveals the sorts of Mediterranean precipitation regimes. Central Anatolia is tangentially covered by the subhumid Mediterranean climate, which is primarily found in the Marmara and Aegean regions, as well as a portion of eastern Anatolia. The Mediterranean, Aegean, and Marmara coastline regions are especially characterized by humid type environments. The humid Mediterranean climate experiences between 700 and 1200 mm of mean annual precipitation, whereas the subhumid one experiences between 570 and 1100 mm of mean annual precipitation. The mean maximum temperature of the hottest month, August, is between 30° and 36°C in the coastal belts of the Mediterranean and Aegean region. It is between 27° and 34°C in the Marmara region and 25° to 28°C in the Black Sea region. On the other hand, the mean minimum temperature of the coldest month, January, is much higher but decreases inland.

4.2.4 BRYOPHYTES OF CENTRAL AND EASTERN TURKEY

Turkey is one of the richest nations in terms of biodiversity, situated in the transition zone between three distinct biogeographic regions: the Mediterranean, Irano-Turanian, and European-Siberian (Kaya and Raynal, 2001). Turkey has highly noticeable seasonal temperature variations and daily to monthly photoperiodism. It is located between 38.96° N and 35.24° E. The Turkey region is influenced by three distinct climates: oceanic, continental, and Mediterranean. The majority of the country's high montane regions, including the largest land area (480,000 out of a total of 780,600 km^2), are under the Mediterranean influence.

Prominent researchers of Turkish bryophytes were Bornmüller (1908), Czeczott (1939), Davis et al. (1971), Handel-Mazzetti (1905, 1909, 1914), Rechinger (1943a, b), and Tchihatcheff (1860). Prominent bryologists Fröehlich (1950, 1952, 1953, 1959), Henderson (1958a, b, 1961a, b, 1964), Henderson and Muirhead (1955), Henderson and Prentice (1969), Müller (1829), Schiffner (1896, 1897, 1908, 1909, 1913), and Wettstein (1889) identified their collections. Crundwell and Nyholm (1974, 1977, 1979), and Walther (1967, 1970, 1975, 1979) travelled throughout the nation and mostly discovered their valuable collections on their own. Walther and Leblebici's (1969) paper on the bryophyte vegetation of Karagöl (İzmir, Yamanlar Mts.) marked the beginning of the third phase of Turkish bryology.

Turkey's vegetation is distinct, and both vascular plants and bryophyte flora exhibit this richness. The Central Anatolian Region (Can et al., 2013; Ezer, 2006; Ezer et al., 2013, Kara et al., 2014; Manafzadeh, 2017; Oren et al., 2017; Yayıntaş, 2009; Yayıntaş and Erdağ, 1995; Yayıntaş and Yayıntaş, 2010) and the bryophyte flora of Erciyes Mountain (Kara et al., 2014) have been the subject of major bryofloristic studies, according to recent literature. Local bryologists began conducting bryofloristic investigations in Turkey (Gökler, 1992, 1993a, b; Gökler and Ozturk, 1986, 1987, 1989; Gökler et al., 1984; Kürschner and Erdağ, 2008, 2009; Kürschner and Kirmaci, 2015; Kürschner et al., 1997; Ursavaş and Çetin, 2013). These have persisted, and a large amount of data on bryophyte diversity has been gathered and the findings are available in a number of provincial checklists (Abay et al., 2016; Erdağ and Kürschner, 2017). Gökler et al. (1986) created the first liverwort checklist in Turkey. Çetin (1988) produced a more comprehensive list on liverworts and hornworts of this region. Gökler and Öztürk (1991, 1992, 1994a, b, 1996), Kürschner and Parolly (1998, 1999a, b, 2006a, b, 2008) and Özenoğlu-Kiremit et al. (2007) published a list of bryophytes found in Turkey.

Uyar and Çetin (2004) published a significant article on the Turkish moss checklist, while Kürschner and Erdağ (2005) published a second one specifically for mosses, liverworts, and hornworts in Turkey. Kirmaci and Ağcagil (2009) studied the urban bryophyte flora. Özenoglu-Kiremit and Keçeli (2009) released a comprehensive inventory of Turkey's liverworts and hornworts. Nonetheless, due to the addition of numerous new species and records (Abay et al., 2009, 2016; Alatas, 2017, 2022; Kürschner, 2014; Kürschner and Lubenau-Nestle, 2000; Kürschner et al., 2015), the number of bryophytes in the nation has continued to rise during the past few years. Taking into account all of these published data, the nation's bryophytes are represented by 960 species or roughly 6.5% of all bryophytes worldwide. In Turkey, there are 773 species of mosses, 183 species of liverworts, and 4 species of hornworts (Erdağ and Kürschner, 2017). The description of *Grimmia nutans* was published by Müller (1829).

The bryophyte flora of Turkey currently consists of 960 species (Bryophyta 773, Marchantiophyta 183, and Anthocerotophyta 4), two subspecies (Marchantiophyta), and 86 variations (10 hepatics and 76 mosses), according to the most recent checklist of Turkish bryophytes by Erdağ and Kürschner (2017). The species of bryophytes reported form the Turkey region are mentioned under section 4.9 under the code TRK.

4.3 NORTHERN AND WESTERN IRAN

With a surface area of 1,648,000 km^2 and a highly varied climate and vegetation, Iran is a desirable location for bryological research. Iran is mostly found on the Persian Plateau. Numerous large basins are surrounded by mountains encompassing 1,648,195 km^2. Over 2,000 km make up these borders, which also include almost 650 km of sea on the southern shore of the Caspian Sea.

The Armenian highlands, which topographically border parts of neighboring Turkey, Armenia, Azerbaijan, and Georgia, include portions of northwest Iran. Iran has a moderate climate overall, with October through April seeing the majority of the country's very little yearly precipitation. The majority of the nation receives 400 millimeters or less of precipitation annually. The higher Zagros Mountain valleys and the Caspian coastal plain are the main outliers, with precipitation averaging at least 600 mm and occurring mostly as snow at high elevations. Rainfall in the western portion of the Caspian surpasses 1,500 mm per year and is spread fairly evenly throughout the year. The South Caspian forests along the northern slopes of the Alborz mountains and the Zagros Oak forests in the western parts of Iran offer extensive and diverse biotopes for a wide range of bryophytes, despite the fact that large portions of the country in the Central, Eastern, and Southern regions with arid and semiarid climate provide limited habitats for moisture-requiring plants.

Throughout the 1800s, Bornmüller (1908, 1910, 1911, 1915), Buhse and Boissier (1860), Fehlner (1883, 1885), Juratzka and Milde (1870), Rechinger (1943 a, b), and Weinmann (1846) documented several bryophytes from Iran. A thorough inventory of Iran's bryophytes was supplied by Buhse and Boissier (1860), Nicholson (1920), Störmer (1963), Störmer et al. (1983). Frey and Kürschner (1977, 1983) made important additions to Iran's bryophyte flora by studying the bryophyte flora and vegetation of the North Iran National Park. The Iranian bryoflora has 15 new taxa added by Kürschner (1996, 2001, 2006, 2008, 2013). Kürschner and Frey (2011) presented an updated list of 121 taxa. Ghahreman et al. (2003) conducted a comparative analysis of the moss floras in the Golestan province and Iran. The majority of local and western botanists focused their study on this category, maybe because of Iran's rich vascular flora. Wendelbo's collection was examined by Arnell (1963) and Stormer (1963). Froehlich (1950, 1953, 1959) and Schiffner (1897, 1901, 1908, 1910, 1913) analyzed Bornmüller's collection. Frey (1974) published the first listing of Iranian hepatic species, which contained 33 species. Several field excursions in the 1970s were conducted mostly on gathering and studying Iranian bryophytes (Frey and Kürschner, 1977, 1979, 1981, 1983; Frey and Probst, 1973, 1974a, b; Tregubov and Tregubov, 1989;

Wendelbo and Rechinger, 1990). According to Frey (1986), there are 32 genera and 56 species of hepatics and 102 genera and 259 species of musci in Iran. Frey and Kürschner (1991a–d) listed 363 taxa in their "*Conspectus Bryophytorum Orientalum et Arabicorum*," which included 297 mosses, 65 liverworts, and 1 hornwort from Iran.

Akhani and Kürschner (2004) reported 447 taxa of bryophytes (two hornworts, 68 liverworts, and 367 mosses). Numerous new records were added to Iran's bryoflora by Ahmadi et al. (2004), Frey and Kürschner (2010), Ghahreman et al. (2007), Kürschner and Frey (2006), Naghinezhad et al. (2006), and Zare (2011). Iranian bryoflora overall is slightly more than 1.5% of all bryophyte species worldwide. These species include 330 species of mosses (Bryopsida) in 114 genera and 69 species of hepatics and hornworts (Hepaticopsida and Anthocerothopsida, respectively) in 36 genera (Akhani and Kürschner, 2004; Ghahreman et al., 2003; Kürschner and Frey 2011, 2020; Kürschner and Ramezani, 2005). The majority of earlier research on Iranian bryophytes was done by foreign botanists.

Kürschner (1997) provided a list of the majority of publications on the bryological works on Iran and other Southwestern Asian nations. Kürschner (2001) presented a summary of all hornworts and hepatics found in the Middle East, together with identification keys and some images. These discoveries came from the identification of newly created collections by Iranian botanists or pre-existing collections in Europe (Kürschner et al., 2000, 2015; Shirzadian, 1989, 1996, 2011; Shirzadian and Kumar, 1994; Shirzadian et al., 2000).

An updated new bryophyte checklist for Iran is serving as a foundation for the bryophyte flora as well as fostering the growing interest of botanists who have less access to outdated publications sent around Europe. "Flora Iranica," the first thorough vascular flora of Iran (Rechinger, 1963), is a valuable volume on nonvascular plants. Ghahremaninejad et al. (2016) presented the updated checklist of bryophytes in the Iran region and reported the occurrence of 534 species of bryophytes (2 hornworts, 70 liverworts, and 462 mosses) in Iran. The species of bryophytes reported from the Iran region are mentioned under section 4.9 under the code IRN.

4.4 NORTHEASTERN IRAQ

Iraq is located between latitudes 29° and 38° N and longitudes 39° and 49° E (with the exception of a tiny portion west of 39°). The Tigris and Euphrates rivers flow south through Iraq and into the Shatt al-Arab.

Iraq contains the eastern portion of the Syrian Desert and the northwest extremity of the Zagros Mountain range. Iraq features a large portion of hot, dry weather with subtropical influences. For the most part of Iraq, summer temperatures average above 40°C and frequently surpass 48°C. Wintertime highs are typically between 15 and 19°C, with nighttime lows of 2 to 5°C. Winter temperatures rarely fall below 21°C. The highest locations receive less than 250 mm of precipitation yearly, with the winter months seeing the highest rainfall.

Works on the Iraqi bryophytes so far have been concerned mainly with the moss species collected and encountered during excursions of foreign works in different parts of Iraq (Agnew and Vondracek, 1975). They recorded 54 genera and 149 species, and they referred in detail to old works on mosses collected from Mesopotamia, Iraq, Syria, and other Middle East countries, for example, Lebanon, Palestine, Persia, and Afghanistan. Maulood and Resheed (1997) have given a literature survey related to the moss flora of Iraq; they recorded a total number of 27 genera and 48 species in the northern part of Iraq, Nineveh, Erbil, and Dohok, of which 4 genera and 9 species were considered as a new record to Iraqi flora. Al-Ni'ma (2003) reported 79 species (72 mosses and 7 liverworts). There are only a few references for Iraqi mosses in the literature leading with bryophytes (Frochlich, 1959; Schiffner, 1913; Vondracek, 1965).

Bryophytorum Bibliotheca, that has been published by Frey and Kürschner (1991a–d), is a valuable contribution to a researcher in Iraq and southwest Asian countries, which has more than 125 references, scattered in various journals not all accessible to the researchers, in one catalogue, followed by few other publications (Al-Ni'ma, 2003; Al-Ni'ma and Al-Khayat, 2001, 2007; Aziz, 2010, 2011; Aziz and Al-Ni'ma, 2017; Maulood and Rasheed, 1997). There are fewer Bryophytes reported in Iraq (about 20 species of liverworts and 145 species of mosses), and that is due to the arid climate (Agnew and Vondracek, 1975). Four papers by Froehlich (1950, 1953, 1959) deal solely with Receinuer's collection from Iraq, while the others list species from Persia and Afghanistan. The species of bryophytes reported from the Iraq region are mentioned under section 4.9 under the code IRQ.

4.5 THE NAHÇEVAN PROVINCE OF AZERBAIJAN

A transcontinental nation like Azerbaijan is found at the border of Western Asia and Eastern Europe. It is located in the South Caucasus region, and it is surrounded by the Caspian Sea to the east, Russia on the north, Georgia on the northwestern side, Armenia and Turkey on the west side, and Iran on the south. It is situated at 38° and 42° N, and 44° and 51° W. Azerbaijan has a land border of 2648 km: 1007 km with Armenia, 756 km with Iran, 480 kilometers with Georgia, 390 kilometers with Russia, and 15 kilometers with Turkey. Its sea length is 800 km, and the longest part of the Caspian Sea is Azerbaijani, that is 456 km. More than half of the surface of Azerbaijan is covered by mountain ridges, crests, highlands, and plateaus scaling up to 400–1000 meters of hypsometric level. The subtropical climate lies on most of the foothills and plains of the country. The absolute minimum temperature of elongation is –33°C, and the absolute maximum temperature is what is witnessed both in Julfa and Ordubad regions. The greatest precipitation is observed in Lankaran (1,600 to 1,800 mm), and the least one in Absheron (200 to 350 mm).

The bryophyte flora of Azerbaijan has received little attention till present. The initial contribution by Weinmann (1846) listed 43 species. The first comprehensive overview of the bryophyte flora of the Caucasus countries, including Azerbaijan, was given by Brotherus (1892), who mentioned a total of 112 taxa. Ignatov and Afonina (1992) compiled information on the diversity of mosses in the countries of the former USSR. An additional list was published by Ignatova et al. (2005) and Ignatov et al. (2006), covering again the Caucasus ecoregion. A more detailed study from Azerbaijan was carried out by Lyubarskaya (1986), who recorded 414 taxa of mosses (364 species, 32 varieties, 18 forms). Gasimov and Novruzov (2017) explored the Orthotrichaceae in Azerbaijan. Qasunov et al. (2016) and Parolly et al. (2018) reported some new records of bryophytes from Azerbaijan.

Azerbaijan's existing moss flora has reached 419 taxa, including the studies conducted in the period from 1846 until the end of 2020 (368 species, 33 varieties, 18 forms). It has been established that in the local bryoflora, *Sphagnum*, *Amblystegium, Bryum, Philonotis, Mnium, Polytrichum,* and *Marchantia* are represented by the largest number. Azerbaijan has a significant resource of medicinal plants, including mosses. Mammadova et al. (2021) reported a few mosses as new for Azerbaijan. Over the years, studies of the bryoflora of Azerbaijan documented 419 taxa of bryophytes (168 genera and 65 families). The species of bryophytes reported from the Azerbaijan region are mentioned under section 4.9 under the code AZB.

4.6 SOUTHERN GEORGIA

Georgia is a high-mountainous nation, which is practically located on the territory of the South Caucasus. The nation is located between latitudes 41° and 44° N, and longitudes 40° and 47° E, covering a land that forms 67,900 km^2. The country is separated into the eastern and the western parts by the Likhi Range. The landscape of Western Georgia varies all the way between lowland marsh-forests, swamps, and temperate rainforest to eternal snows and glaciers, whereas the eastern side of the country even holds a small strip of semi-arid steppe. Climate in the region is highly inclined to altitude, and though in the lowland parts of western Georgia the climate is relatively warm all over, the foothills to the mountains are wet and cold with cool summers and cold winters covered with snow (snow cover). The forests of Western Georgia are mostly deciduous trees up to 600 m above sea level and are found with species like oak, hornbeam, beech, elm, ash, and chestnut. Species of the evergreen like box are also very much present in many regions. About 1000 out of 4000 higher plants of Georgia are in the country.

The first comprehensive overview on the bryophyte flora of the Caucasus, including Georgia, was given by Bell (1973), Brotherus (1892), Konstantinova (1992), Ignatov and Afonina (1992), and Schornherst (1945) provided state-of-the-art knowledge about bryophytes of the region. Later, Ignatov et al. (2007) gave an additional list, also covering records from the Caucasus. A first, although taxonomically and nomenclatural slightly outdated, Georgian bryophyte checklist was presented by Chikovani and Svanidze (2004), with a total of three hornworts, 171 liverworts, and 638 mosses. Most recently, the bryoflora and vegetation of the Caucasus are again in the focus of floristic and vegetation studies (Doroshina, 2011; Doroshina and Czernyadjeva, 2018; Eckstein et al., 2018; Kürschner et al., 2013; Tigishvili, 2006; Zündorf, 2011). With *Hygrohypnum eugyrium* (Schimp.) Broth. and *Hyocomium armoricum* (Brid.) Wijk and Margad., Zündorf (2011) added two species to the flora, which were collected in the western part (Adjara, Mtirala National Park) of the country. Sapegin (1910) added numerous records to the bryoflora of Caucasus. The species of bryophytes reported from the Georgia region are mentioned under section 4.9 under the code GOR.

4.7 ARMENIA

Armenia is a country in the Southern Caucasus Mountains and the lowlands in between the Black Sea and Caspian Sea, and north of the Bosphorus.

Armenia is located at latitudes of 38° and 42° north and longitudes of 43° and 47° east. It has two terrestrial ecoregions, namely the Caucasus mixed forests and Eastern Anatolian montane steppe. The territorial area of Armenia is 29,743 km^2; the majority of the territory is mountainous, and fast-flowing rivers with rare forests are present. The highest point of the land is Mount Aragats, which is located 4,090m above sea level. Armenia has a significantly highland continental climate. Summers are sunny, hot, and dry; they extend between June and mid-September. It is subject to temperatures that range between 22 and 36°C. The winters are very cold and snowy (temperature is between –10 and –5°C).

Lazarenko (1938) explored the bryoflora of Middle Asia. Dylevskya (1950) studied the moss flora of Caucasus Minor. Abramova and Abramov (1950, 1959) published on the mosses of the Caucasus and Armenia. Takhtajan (1954) brought out the flora of Armenia. Barsegyan (1959) provided the notes on *Ricciocarpus natans*. Dildaryan (1969, 1971, 1972, 1979, 1983) and Dildaryan and Avakyan (1986, 1987a, b) documented several bryophytes in various regions of Armenia.

A significant contribution to the bryoflora of Armenia was made by Manakyan (1966a, b, 1968, 1969, 1970,1971a–d, 1974,1975,1976a, b, 1986a, b, 1988, 1989a, b, 1990, 1995). He recorded several new species in this region from Armenia (Manakyan 1971a, 1976b, 1979,1983). Manakyan (1972) conducted a comparative analysis of the moss floras at different altitudinal ranges. The information on the distribution of all taxa is listed, indicating their occurrence in 12 phytogeographic areas and in three altitudinal belts (Manakyan 1989b).

Armenia is known bryological by 48 species and 1 variety of hepatics (29 genera, 20 families, 7 orders) and 347 species and 25 varieties of mosses (135 genera, 40 families, 14 orders). The species of bryophytes reported from the Armenian region are mentioned under section 4.9 under the code ARM.

4.8 TURKMENISTAN

Turkmenistan is a very landlocked state in Central Asia with one of its sides bordered by the Caspian Sea. It is situated between latitude 35° and 43° N, and longitude 52° and 67° E. The Karakum Desert covers more than 80% of the country. Bryofloristic studies in this region are very scarce and need attention. Ignatov et al. (2006) made a significant contribution to the bryophyte flora of Eastern Europe and North Asia, including Turkmenistan.

Mamatkulov (1967, 1975, 1990) and Mamatkulov et al. (1998) studied the bryophytes of the Tajik region, Middle Asia, and Kazakhstan. The species of bryophytes found in the Turkmenistan region are mentioned under section 4.9 under the code TKM.

4.9 LIST OF BRYOPHYTES OF THE IRANO-ANATOLIAN REGION

This section contains the species of bryophytes reported from the Irano-Anatolian region. The distribution of the species is given in brackets in bold with the country name in abbreviated form (TRK=Turkey; IRN=Iran; IRQ=Iraq; GOR=Georgia; AZB=Azerbaijan, ARM= Armenia, and TKM =Turkmenistan). The families under each group and the species name under each family are arranged alphabetically. The taxa are placed according to Goffinet and Shaw (2008).

4.9.1 MARCHANTIOPHYTA

- **Acrobolbaceae:** *Acrobolbus ciliatus* (Mitt.) Schiffn. (GOR).
- **Anastrophyllaceae:** *Barbilophozia attenuata* (Mart.) Loeske (TRK); *B. barbata* (Schmidel ex Schreb.) Loeske, (TRK); *B. rubescens* (R.M. Schust. & Damsh.) Kartt. & L. Söderstr., (TRK); *Gymnocolea acutiloba* (Schiffn.) Müll. Frib., (TRK); *Neoorthocaulis attenuata* (Mart.) L. Söderstr., De Roo & Hedd. (TRK).
- **Aneuraceae:** *Aneura pinguis* (L.) Dumort., (TRK, ARM); *Riccardia chamaedryfolia* (With.) Grolle, (TRK); *R. incurvata* Lindb. (ARM); *R. multifida* (L.) S. F. Gray, (TRK, IRN); *R. palmata* (Hedw.) Carruth. (TRK, IRN); *R. pinguis* (L.) S. F. Gray, (TRK); *R. sinuata* (Hook.) Trevis, (TRK).
- **Antheliaceae:** *Anthelia julacea* (L.) Dumort. (TRK, ARM).
- **Arnelliaceae:** *Southbya nigrella* (De Not.) Henriq., (TRK); *S. tophacea* (Spruce) Spruce, (TRK).
- **Aytoniaceae:** *Asterella gracilis* (F. Weber) Underw., (TRK, IRN); *A. persica* (Steph.) M. Howe, (IRN); *Mannia androgyna* (L.) A. Evans, (TRK, IRN); *M. fragrans* (Balb.) Frye & L. Clark, (TRK); *M. gracilis* (F. Weber) D.B. Schill & D.G. Long, (TRK); *M. triandra* (Scop.) Grolle, (TRK); *Plagiochasma appendiculatum* Lehm. & Lindenb. (TRK); *P. rupestre* (J.R. Forst. & G. Forst.) Steph., (TRK, IRN); *Reboulia hemisphaerica* (L.) Raddi, (TRK, IRN, IRQ).
- **Blasiaceae:** *Blasia pusilla* L., (TRK).

- **Blepharostomataceae:** *Blepharostoma trichophyllum* (L.) Dumort., (TRK).
- **Calypogeiaceae:** *Calypogeia arguta* Nees & Mont., (TRK, IRN); *C. azurea* Stotler & Crotz, (TRK); *C. fissa* (L.) Raddi, (TRK, IRN); *C. muelleriana* (Schiffn.) Müll. Frib. (TRK); *C. sphagnicola* (Arnell & J. Perss.) Warnst. & Loeske, (TRK).
- **Cephaloziaceae:** *Cephalozia catenulata* (Huebener) Lindb. (TRK); *C. macrantha* Kaal. & W.E. Nicholson, (TRK); *C. pleniceps* (Austin) Lindb., (TRK); *C. lunulifolia* (Dumort.) Dumort. (IRN); *C. rubella* (Nees) Schiffn., (TRK, IRN); *C. stellulifera* (Taylor ex Carrington & Pearson) Croz., (TRK); *C. turneri* (Hook.) Müll. Frib. (TRK); *Cephaloziella baumgartneri* Schiffn., (TRK); *C. divaricata* (Sm.) Schiffn., (TRK); *C. hampeana* (Nees) Schiffn. ex Loeske, (TRK); *Cladopodiella fluitans* (Nees) H. Buch, (TRK); *Nowellia curvifolia* (Dicks.) Mitt. (TRK).
- **Cleveaceae:** *Athalamia hyalina* (Sommerf.) S. Hatt., (TRK); *A. spathysii* (Lindenb.) S. Hatt., (TRK); *Clevea hyaline* (Sommerf.) Lindb. (TRK).
- **Codoniaceae:** *Fossombronia caespitiformis* (Raddi) De Not. ex Rabenh. (IRN).
- **Conocephalaceae:** *Conocephalum conicum* (L.) Dumort., (TRK, IRN, IRQ).
- **Corsiniaceae:** *Corsinia coriandrina* (Spreng.) Lindb., (TRK).
- **Fossombroniaceae:** *Fossombronia angulosa* (Dicks.) Raddi, (TRK); *F. caespitiformis* (Raddi) De Not. ex Rabenh., (TRK, IRQ); *F. foveolata* Lindb., (TRK); *F. husnotii* Corb., (TRK); *F. pusilla* (L.) Nees, (TRK); *Petalophyllum ralfsii* (Wilson) Nees & Gottsche ex Lehm., (TRK); *F. vulgaris* Hedw. (IRQ).
- **Frullaniaceae:** *Frullania dilatata* (L.) Dumort., (TRK, IRN); *F. cesatiana* Hampe ex Lehm., (IRN); *F. ericoides* (Nees) Mont., (TRK); *F. jackii* Gottsche, (TRK); *F. tamarisci* (L.) Dumort. (TRK, IRN).
- **Geocalycaceae:** *Chiloscyphus pallescens* (Hoffm.) Dum. (TRK); *C. polyanthos* (L.) Corda (TRK); *Lophocolea bidentata* (L.) Dum., *L. minor* Nees (ARM).
- **Gymnomitriaceae:** *Gymnomitrion concinnatum* (Lightf.) Corda, (TRK); *Marsupella emarginata* (Ehrh.) Dumort., (TRK); *M. funckii* (F. Weber & D. Mohr) Dumort., (TRK); *Nardia compressa* (Hook.) Gray, (TRK); *N. handelii* Schiffn., (TRK); *N. lignicola* Schiffn., (TRK).
- **Jubulaceae:** *Jubula hutchinsiae* (Hook.) Dumort. (IRN, ARM); *J. pennsylvanica* (Steph.) A. Evans (TRK).

- **Jungermanniaceae:** *Barbilophozia barbata* Schmid. (ARM) *B. hatcheri* Evans (ARM), *B. lycopodioides* (Wallr.) Loeske (ARM), *Jamesoniella autumnalis* (DC.) Steph., (TRK); *Jungermannia atrovirens* Dumort., (TRK); *J. atrovirens* Dum. var. rivularis Bernet. (ARM), *J. caucasica* Váňa, (TRK); *J. eucordifolia* Schljak (ARM); *J.gracillima* Sm., (IRN); *J. handelii* (Schiffn.) Amakawa, (TRK); *J. hyalina* Lyell, (TRK, IRN); *J. leiantha* Grolle, (TRK, IRN); *J. lignicola* (Schiffn.) Grolle, (TRK); *J. obovata* Nees, *J. sphaerocarpa* Hook., *J. subtilissima* (Schiffn.) Grolle: *J. subulata* A. Evans (IRN); *J. tristis* Nees (TRK); *Leiocolea alpestris* Schleich. ex F. Weber, (TRK); *L. badensis* (Gottsche ex Rabenh.) Jørg., (TRK); *L. bantriensis* (Hook.) Jørg., (TRK); *L. heterocolpos* (Thed. ex Hartm.) H. Buch, (TRK ARM); *L. turbinata* (Raddi) H. Buch, (TRK); *Lophozia ventricosa* (Dicks.) Dum. (ARM), *Schistochilopsis incisa* (Schrad.) Konst. (ARM); *Solenostoma triste* (Nees) Müll. Frib., (TRK); *Tritomaria quinquedentata* (Huds.) Buch (ARM).
- **Lejeuneaceae:** *Cololejeunea minutissima* (Sm.) Steph. (GOR); *C. rossettiana* (C. Massal.) Schiffn., (IRN); *Drepanolejeunea bidens* (Prantl) A. Evans, (GER); *Lejeunea cavifolia* (Ehrh.) Lindb., (TRK, IRN); *L. lamacerina* (Steph.) Schiffn., (TRK, IRN); *L. patens* Lindb., (TRK); *Microlejeunea ulicina* (Taylor) Steph., (TRK); *Rectolejeunea maxonii* A. Evans. (GOR).
- **Lepidoziaceae:** *Bazzania flaccida* (Dumort.) Grolle (TRK); *B. nudicaulis* A. Evans, (TRK); *B. tricrenata* (Wahlenb.) Lindb., (TRK), *B. trilobata* (L.) Gray, (TRK); *Lepidozia reptans* (L.) Dumort. (TRK).
- **Lophocoleaceae:** *Barbilophozia barbata* (Schmidel ex Schreb.) Loeske, (IRN); *Chiloscyphus pallescens* (Ehrh. ex Hoffm.) Dumort. (IRN); *Clasmatocolea rigens* (Hook. f. & Taylor) J.J. Engel (GOR); *Lophocolea bidentata* (L.) Dumort., (TRK, IRN); *Leptoscyphus cuneifolius* (Hook.) Mitt. (ARM); *L. heterophylla* (Schrad.) Dumort., (TRK, IRN); *L. minor* Nees, (TRK, IRN); *Leptoscyphus taylorii* (Hook.) Mitt. (TRK).
- **Lophoziaceae:** *Lophozia bicrenata* (Schmidel ex Hoffm.) Dumort., (TRK); *L. excisa* (Dicks.) Konstant. & Vilnet, (TRK); *L. longidens* (Lindb.) Konstant. & Vilnet, (TRK); *L. longiflora* (Nees) Schiffn., (TRK); *L. muelleri* (Nees ex Lindenb.) Dumort., (TRK); *L. porphyroleuca* (Nees) Schiffn. ex-Dalla Torre & Sarnth., (TRK); *L. sudetica* (Nees ex Huebener) Grolle, *L. turbinata* (Raddi) Steph., (TRK, IRN); *L. ventricosa* (Dicks.) Dumort., (TRK); *Leiocolea badensis* (Gottsche ex Rabenh.) Jørg., (IRN); *L. turbinata* (Raddi) H. Buch,

(IRN); *Obtusifolium obtusum* (Lindb.) S.W. Arnell, (TRK); *Tritomaria exsectiformis* (Breidl.) Schiffn. ex Loeske, (TRK); *T. quinquedentata* (Huds.) Bakalin, *T. exsectiformis* (Breidl.) Schiffn. ex Loeske, *T. exsecta* (Schmidel) Schiffn. ex Loeske., (TRK, IRN).

- **Lunulariaceae:** *Lunularia cruciata* (L.) Dumort. ex Lindb., (TRK, IRN, IRQ).
- **Marchantiaceae:** *Dumortiera hirsuta* (Sw.) Nees (IRN); *Marchantia domingensis* Lehm. & Lindenb. (GOR); *M. paleacea* Bertol., (TRK, IRN); *M. polymorpha* L., (TRK, IRN, IRQ); *M. polymorpha* var. *alpestris* (Nees) Gottsche, Lindenb. & Nees (IRQ); *Preissia quadrata* (Scop.) Nees., (TRK).
- **Metzgeriaceae:** *Apometzgeria pubescens* (Schrank) Kuwah, (TRK); *Metzgeria conjugata* Lindb. (TRK, IRN, ARM); *M. furcata* (L.) Corda (TRK, IRN, ARM); *M. pubescens* (Schrank) Raddi. (TRK, ARM).
- **Oxymitraceae:** *Oxymitra incrassata* (Brot.) Sérgio & Sim-Sim. (GOR).
- **Pelliaceae:** *Pellia endiviifolia* (Dicks.) Dumort., (TRK, IRN, ARM); *P. epiphylla* (L.) Corda, (TRK, IRN, ARM); *Pellia furcigera* (Hook.) Mass. (ARM), *P. neesiana* (Gottsche) Limpr. (TRK).
- ***Plagiochilaceae:*** *Pedinophyllum interruptum* (Nees) Kaal. (IRN); *Plagiochila asplenioides* (L.) Dumort., (TRK, IRN); *P. asplenioides* (L.) Dumort. (TRK); *P. porelloides* (Torr. ex Nees) Lindenb., (TRK, IRN); *P. tridenticulata* (Michx.) Dumort. (GOR); *Plectocolea crenulata* (Mitt.) A. Evans (TRK); *P. obovata* (Nees) Mitt. (TRK).
- **Porellaceae:** *Porella arboris-vitae* (With.) Grolle (IRN); *P. baueri* (Schiffn.) C.E.O. Jensen (TRK); *P. cordaeana* (Huebener) Moore, (TRK, IRN); *P. obtusata* (Taylor) Trevis. (TRK); *P. pinnata* L., (TRK); *Porella platyphylla* (L.) Pfeiff., (TRK, IRN); *P. rivularis* (Nees) Pfeiff., (TRK); *P. platyphylloidea* (Schwein.) Lindb. (TRK, IRN); *Porella wataugensis* (Sull.) Underw. ex M. Howe. (GOR).
- **Radulaceae:** *Radula complanata* (L.) Dumort. (TRK, IRN); *R. lindenbergiana* Gottsche ex C. Hartm. (TRK, IRN).
- **Ricciaceae:** *Riccia atromarginata* Levier (IRN); *R. bicarinata* Lindb. (TRK); *R. bifurca* Hoffm. (TRK, IRN); *R. ciliata* Hoffm. (IRN); *Riccia crytstalina* L. (IRQ); *R. crozalsii* Levier, (IRN); *R. fluitans* L., (IRN); *R. frostii* Austin, (IRN, IRQ); *R. frostii* var *major* (IRQ); *R. glauca* L., (IRN); *R. gougetiana* Durieu & Mont. (IRN); *R. lamellosa* Raddi, (IRN, AZB); *R. papillosa* Moris (TRK); *R. subbifurca* Warnst. ex Croz. (IRN); *R. sorocarpa* Bisch., (TRK, IRN, IRQ); *R. trichocarpa* M. Howe (IRN); *Ricciocarpos natans* (L.) Corda (TRK, IRN);

- **Riccardiaceae:** *Riccardia penguis* (L.) Gray (IRQ).
- **Riellaceae:** *Riella helicophylla* (Bory & Mont.) Mont. (TRK).
- **Scapaniaceae:** *Diplophyllum albicans* (L.) Dumort., (TRK); *D. nanum* Herzog (IRN); *D. taxifolium* (Wahlenb.) Dumort. (TRK, ARM); *Scapania aequiloba* (Schwägr.) Dumort., (TRK); *S. aspera* M. Bernet & Bernet, (TRK, IRN); *S. compacta* (Roth) Dumort., (TRK); *S. curta* (Mart.) Dumort., (TRK); *Scapania cuspiduligera* (Nees) K. Muell. (ARM), *S. irrigua* (Nees) Nees, (TRK, ARM); *S. nemorea* (L.) Grolle, (TRK, IRN, ARM); *S. obscura* (Arnell & C.E.O. Jensen) Schiffn., (TRK); *S. umbrosa* (Schrad.) Dumort., (TRK); *S. undulata* (L.) Dumort., (TRK, ARM); *S. verrucosa* Heeg (TRK).
- **Southbyaceae:** *Gongylanthus ericetorum* (Raddi) Nees (TRK); *Southbya nigrella* (De Not.) Henriq. (IRN).
- **Sphaerocarpaceae:** *Sphaerocarpos michelii* Bellardi, (TRK); *S. texanus* Austin (TRK);
- **Targioniaceae:** *Targionia hypophylla* L., (TRK, IRN, IRQ); *T. lorbeeriana* Müll. Frib., (IRN).
- **Trichocoleaceae:** *Blepharostoma trichophyllum* (L.) Dum. (ARM).

4.9.2 BRYOPHYTA

- **Amblystegiaceae:** *Amblystegium confervoides* (Brid.) Loeske, (TRK); *A. fluviatile* (Hedw.) Schimp. (TRK, IRN); *A. humile* (P. Beauv.) Crundw., (TRK, IRN); *A. kochii* Schimp. (IRN); *A. kurdicum* Schiffn., (TRK, IRQ); *A. juratzkanum* Schimp. (IRQ); *A. radicale* (P. Beauv.) Schimp., (IRN); *A. riparium* (Hedw.) Schimp., (IRN); *A. serpens* (Hedw.) Schimp., (TRK, IRN, IRQ, ARM); *A. serpens* (Hedw.) B.S.G. var. *juratzkanum* (Schimp.) Rau et Herv. (ARM); *A. tenax* var. *spinifolium* (Schimp.) H.A. Crum & L.E. Anderson, (IRN); *A. subtile* (Hedw.) Schimp., (TRK); *A. tenax* (Hedw.) C.E.O. Jensen, (TRK, IRN); *A. varium* (Hedw.) Lindb., (TRK, IRN, ARM); *Calliergonella cuspidata* (Hedw.) Loeske, (IRN, ARM); *C. cuspidata* var. *pungens* (Schimp.) Latzel, (IRN); *Calliergon cordifolium* (Hedw.) Kindb. (ARM); *Calliergon sarmentosum* (Wahlenb.) Kindb., (IRN); *Campylium calcareum* Crundw. & Nyholm, (TRK); *C. chrysophyllum* (Brid.) Lange, (TRK, GOR, ARM); *C. elodes* (Lindb.) Kindb. (ARM); *C. halleri* (Hedw.) Lindb. (ARM); *C. polygamum* (B.S.G.) C. Jens. (ARM); *C. sommerfeltii* (Myr.) J. Lange (ARM); *C. stellatum* (Hedw.) C.Jens. var. *protensum* (Brid.) Grout, (ARM); *C. polygamum* (Schimp.)

Lange & C.E.O. Jensen, (TRK); *C. stellatum* (Hedw.) Lange & C.E.O. Jensen, (TRK); *Campylophyllum calcareum* (Crundw. & Nyholm) Hedenäs, (IRN); *Cratoneuron commutatum* (Hedw.) Roth. (IRQ); *C. commutatum* var. *commutatum* (Hedw.) G. Roth (IRQ); *C. decipiens* (De Not.) Loeske, (TRK); *C. decipiens* var. *napaeiforme* (Schiffn.) Schiffn. (IRQ), *C. filicinum* (Hedw.) Spruce, (TRK, IRN, IRQ, ARM, TKM); *C. filicinum* var. *fallax* (Brid.) G. Roth (IRN); *Drepanocladus aduncus* (Hedw.) Warnst., (TRK, IRN, IRN, ARM); *D. aduncus* (Hedw.) Warnst. var. *kneiffii* (Schimp.) Moenk. (ARM); *D. aduncus* var. *polycarpus* (Voit) G.Roth. (ARM); *D. capillifolius* (Warnst.) Warnst., (TRK); *D. exannulatus* (Schimp.) Warnst., (TRK); *D. fluitans* (Hedw.) Warnst., (TRK, IRQ); *D. lycopodioides* (Brid.) Warnst., (TRK); *D. polycarpos* (Blandow ex Voit) Warnst., (TRK); *D. polygamus* (Schimp.) Hedenäs, (TRK); *D. revolvens* (Sw.) Warnst., (TRK, IRQ); *D. sendtneri* (C. Muell.) Warnst. (ARN); *D. uncinatus* (Hedw.) Loeske, (TRK); *D. vernicosus* (Mitt.) Warnst., (TRK); *Hamatocaulis vernicosus* (Mitt.) Hedenaes (ARM); *Hygroamblystegium fluviatile* (Hedw.) Loeske, (TRK); *H. irriguum* (Hook. & Wilson) Loeske, (TRK); *H. tenax* (Hedw.) Jenn., (TRK, ARM); *Hygroamblystegium tenax* (Hedw.) Jenn. var. *spinifolium* (Schimp.) Jenn. (ARM); *H. varium* (Hedw.) Mönk., (TRK); *H. luridum* (*Cardot) Reimers, (TRK); *Hygrohypnum luridum* (Hedw.) Jenn. (IRN, ARM); *H. molle* (Hedw.) Loeske (ARM), *H. smithii* (Sw. ex Lilj.) Broth. (ARM); *Hygrohypnum ochraceum* (Turner ex Wilson) Loeske. (IRQ); *Leptodictyum humile* (P.Beauv.) Ochyra (ARM), *L. kochii* (Schimp.) Warnst., (TRK); *L. kurdicum* (Schiffn.) Broth., (TRK); *L. riparium* (Hedw.) Warnst. (TRK, ARM, TKM); *Limprichtia cossoni* (Schimp.) Anderson et al. (ARM), *L. revolvens* (Sw.) Loeske (ARM), *Palustriella commutata* (Hedw.) Ochyra, (TRK, IRN, TKM); *P. commutata* var. *fluctuans* (Schimp.) Ochyra (IRN), *Palustriella commutata* (Hedw.) Ochyra var. *fallax* (Brid.) Ochyra (ARM); *P. decipiens* (De Not.) Ochyra (TRK, IRN, ARM); *P. decipiens* var. *napaeiformis* (Schiffn.) Ochyra (IRN); *P. falcata* (Brid.) Hedenäs (IRN, TKM); *Pseudocalliergon trifarium* (Web. et Mohr) Loeske (ARM). *Sanionia uncinata* (Hedw.) Loeske (AZB, ARM); *Warnstorfia exannulata* (Schimp.) Loeske (IRN, ARM); *W. fluitans* (Hedw.) Loeske (ARM); *W. schulzii* (Limpr.) Loeske (ARM).

- **Andreaeaceae:** *Andreaea rupestris* Hedw. (TRK, ARM).
- **Anomodontaceae:** *Anomodon acutifolius* Mitt. (IRN); A. a*ttenuatus* (Hedw.) Huebener (TRK, IRN, ARM); *A. longifolius* (Brid.) Hartm.

(IRN, ARM); *A. rostratus* (Hedw.) Schimp. (IRN); *A. rugelii* (Müll. Hal.) Keissl., (IRN, TRK); *A. viticulosus* (Hedw.) Hook. & Taylor, (TRK, IRN, ARM); *Haplohymenium triste* (Ces.) Kindb. (TRK, IRN, GOR); *Pterogonium gracile* (Hedw.) Sm. (ARM).

- **Aongestroemiaceae:** *Dichodontium flavescens* (Dicks.) Lindb., (TRK); *D. palustre* (Dicks.) M. Stech, (TRK); *D. pellucidum* (Hedw.) Schimp. (TRK).
- **Archidiaceae:** *Archidium alternifolium* (Dicks. ex Hedw.) Schimp. (TRK).
- **Aulacomniaceae:** *Aulacomnium androgynum* (Hedw.) Schwägr. (TRK); *A. palustre* (Hedw.) Schwägr. (TRK, GOR, ARM); *A. turgidum* (Wahlenb.) Schwaegr. (ARM).
- **Bartramiaceae:** *Anacolia webbii* (Mont.) Schimp., (TRK); *Bartramia aprica* Müll. Hal. (AZB) *B. halleriana* Hedw., (TRK, IRN, ARM); *B. ithyphylla* Brid., (TRK, ARM); *B. pomiformis* Brid., (TRK); *B. stricta* Brid., (TRK); *Philonotis arnellii* Husn., (TRK); *P. caespitosa* Jur., (TRK); *P. calcarea* (Bruch & Schimp.) Schimp., (TRK, IRN); *P. calcarea* var. *seriatifolia* Schiffn., (IRN); *P. capillaris* Lindb., *P. fontana* (Hedw.) Brid., (TRK, IRN, IRQ, AZB, ARM); *Philonotis fontana* (Hedw.) Brid. var. *caespitosa* (Jur.) Schimp. (ARM); *Philonotis fontana* var. *falcata* (Hook.) Brid. (ARM); *P. fontana* var. *minor* Paris (IRN); *Philonotis fontana* var. *pumila* (Turn.) Brid. (ARM); *Philonotis fontana* var. *seriata* (Mitt.) Kindb. (ARM); *P. hastata* (Duby) Wijk & Margad., (TRK, IRN); *P. laxitexta* J. Froehl. (IRN) *P. marchica* (Hedw.) Brid., (TRK, IRN, ARM); *P. rigida* Brid., (TRK, IRN); *P. rigida* var. *gracilis* Schimp., (IRN); *P. seriata* Mitt., (IRN); *P. seriata* var. *persica* Schiffn., (IRN); *P. tomentella* Molendo (TRK, IRN, AZB); *P. uncinata* (Schwägr.) Brid., (IRN); *Plagiopus oederi* (Sw.) H.A. Crum & L.E. Anderson (IRN, ARM).
- **Brachytheciaceae:** *Brachythecium acuminatum* (Hedw.) Austin (IRN); *Brachythecium albicans* (Hedw.) Schimp., (TRK, IRN, ARM, TKM); *B. campestre* (Müll. Hal.) Schimp., (TRK, IRN, ARM); *B. capillaceum* (F. Weber & D. Mohr) Giacom. (TRK, IRQ); *B. collinum* (Schleich. ex Müll. Hal.) Ignatov & Huttunen, (TRK, IRN, ARM, TKM); *B. curtum* (Lindb.) Limpr., (TRK); *B. erythrorrhizon* Schimp., (TRK); *B. geheebii* Milde, (TRK, IRN, ARM); *B. glareosum* (Bruch & Schimp. ex Fiedl.) Schimp., (TRK, ARM); *B. laetum* (Brid.) Schimp., (TRK); *B. latifolium* Kindb., (TRK); *B. mildeanum* (Schimp.) Schimp. ex Milde (TRK); *B. oedipodium* (Mitt.) A. Jaeger, (TRK); *B. olympicum* Jur., (TRK);

B. oxycladon (Brid.) A. Jaeger, (TRK, IRN); *B. plumosum* (Hedw.) Schimp., (TRK, IRQ, ARM); *B. populeum* (Hedw.) Schimp., (TRK, IRN, ARM); *B. reflexum* (Starke) Schimp., (TRK, ARM); *B. rivulare* Schimp., (TRK, IRN, IRQ, ARM); *B. rutabulum* (Hedw.) Schimp., (TRK, IRN, IRQ, ARM); *B. rutabulum* var. *meridionale* Schiffn., (IRN); *B. salebrosum* (Hoffm. ex F. Weber & D. Mohr) Schimp. (TRK, IRN, GOR, ARM); *B. starkei* (Brid.) Schimp., (TRK, ARM); *B. trachypodium* (Brid.) Schim., (TRK); *B. umbilicatum* Jur. & Milde, (TRK, IRN); *B. validum* (C.E.O. Jensen) Broth., (TRK); *B. velutinum* (Hedw.) Schimp., (TRK, IRN, IRQ, ARM, TKM); *B. velutinum* var. *salicinum* (Schimp.) Mönk., (IRN); *Bryhnia novae-angliae* (Sull. & Lesq.) Grout (IRN); *Camptothecium aureum* (Spruce) Schimp., (TRK); *C. fallax* (H. Philib. ex Schimp.) Schiffn., (TRK); *C. geheebii* (Milde) Kindb., (TRK); *C. lutescens* (Hedw.) Schimp., (TRK, ARM); *C. nitens* (Hedw.) Schimp., (TRK); *C. philippeanum* (Spruce) Kindb., (TRK); *C. sericeum* (Hedw.) Kindb., (TRK, ARM); *Cirriphyllum crassinervium* (Taylor ex Wilson) Loeske & M. Fleisch., (IRN); *C. piliferum* (Hedw.) Grout, (IRN, ARM); *C. reichenbachianum* (Huebener) Wijk & Margad. (IRN); *Eurhynchium angustirete* (Broth.) T.J. Kop. (TRK, ARM); *E. crassinervium* (Taylor ex Wilson) Schimp., (TRK, ARM); *E. euchloron* (Bruch ex Müll. Hal.) Jur. & Milde, (TRK); *E. flotowianum* (Sendtn.) Kartt., (TRK, ARM); *E. hians* (Hedw.) Sande Lac., (TRK, IRN, ARM); *E. hians* var. *rigidum* (Boulay) J. (IRN); *E. meridionale* (Schimp.) De Not., (TRK, IRN); *E. myosuroides* (Brid.) Schimp., (TRK); *E. praelongum* (Hedw.) Schimp., (TRK, IRN, IRQ); *E. praelongum* var. *stokesii* (Turner) Dixon, (IRN); *E. pulchellum* (Hedw.) Jenn., (TRK, IRN, ARM); *E. pulchellum* var. *praecox* (Hedw.) G.N. Jones, (IRN, ARM); *E. pumilum* (Wilson) Schimp., (TRK); *E. schleicheri* (R. Hedw.) Milde, (TRK, IRN, ARM); *E. speciosum* (Brid.) Jur., (TRK, IRN, ARM); *E. speciosum* var. *angustifolium* P. Størmer, (IRN) *E. stokesii* (Turner) Schimp., (TRK); *E. striatum* (Schreb. ex Hedw.) Schimp., (IRN, ARM); *E. striatulum* (Spruce) Schimp., (TRK); *E. swartzii* (Turner) Curn., (TRK); *E. swartzii* var. *rigidum* (Boulay) Thér. (IRN); *E. zetterstedtii* P. Størmer, (TRK); *Homalothecium aureum* (Spruce) H. Rob., (TRK, IRQ, TKM); *H. algerianum* Besch., (TRK); *H. fallax* H. Philib. ex Schimp., (TRK); *H. geheebii* (Milde) Wigh, (TRK); *H. lutescens* (Hedw.) H. Rob., (TRK, IRN, TKM); *H. megaptilum* (Sull.) H. Rob. (IRN); *H. nitens* (Hedw.) H. Rob. (IRN); *H. philippeanum* (Spruce) Schimp., (TRK,

IRN, IRQ, ARM, TKM); *H. sericeum* (Hedw.) Schimp., (TRK, IRN, ARM, TKM); *Isothecium alopecuroides* (Lam. ex Dubois) Isov. (IRN, ARM); *I. myurum* Brid., (IRN); *Oxyrrhynchium hians* (Hedw.) Loeske, (TRK); *O. pumilum* (Wilson) Loeske, (TRK); *O. schleicher* (R. Hedw.) Röll, (TRK); *O. speciosum* (Brid.) Warnst., (TRK); *Palamocladium euchloron* (Bruch ex Müll. Hal.) Wijk & Margad., (TRK, IRN, ARM); *P. leskeoides* (Hook.) E. Britton, (IRN); *Platyhypnidium rusciforme* M. Fleisch. (TRK); *Pseudoscleropodium purum* (Hedw.) Fleisch. ex Broth. (ARM), *Rhynchostegiella jacquinii* (Garov.) Limpr., (IRN); *R. litorea* (De Not.) Limpr. (GOR); *R. menadensis* (Sande Lac.) E.B. Bartram, (IRN); *Rhynchostegiella teesdalei* (B.S.G.) Limpr. (ARM); *R. tenella* (Dicks.) Limpr., (IRN, ARM); *Rhynchostegium megapolitanum* (Blandow ex F. Weber & D. Mohr) Schimp., (IRN, ARM); *R. riparioides* (Hedw.) Cardot, (IRN, ARM, TKM); *R. rusciforme* (Hedw.) Cardot, (IRN); *Scleropodium purum* (Hedw.) Limpr. (IRN); *Scorpiurium circinatum* (Brid.) M. Fleisch. & Loeske (IRN); *Tomentypnum nitens* (Hedw.) Loeske (AZB, ARM).

- **Bryaceae:** *Anomobryum auratum* (Mitt.) A. Jaeger (IRN); *A. concinnatum* (Spruce) A. Jaeger, (TRK); *A. julaceum* (Schrad. ex G. Gaertn., B. Mey. & Scherb.) Schimp., (TRK, ARM); *A. juliforme* Solms, (TRK); *A. lanatum* (P. Beauv.) J.R. Spence & H.P. Ramsay (IRN); *Brachymenium exile* (Dozy & Molk.) Bosch & Sande Lac. (IRN); *B. nepalense* Hook., (IRN); *Bryum algovicum* Sendtn. ex Müll. Hal., (TRK, AZB, ARM); *B. alpinum* Huds. ex With., (TRK, IRN, ARM); *B. amblyodon* Müll. Hal., (IRN); *B. argenteum* Hedw., (TRK, IRN, ARM, TKM); *B. atrovirens* Brid., (TRK, IRN); *B. badium* (Brid.) Bruch ex Milde (IRN, IRN); *B. barnesii* J.B. Wood ex Schimp., (TRK); *B. bicolor* (Hornsch.) Brid. (TRK, IRQ); *B. bornholmense* Wink. & R. Ruthe, (TRK); *B. caespiticium* Hedw., (TRK, IRN, IRQ, ARM); *B. caespiticium* Hedw. var. *badium* Brid. (ARM); *B. capillare Hedw.*, (TRK, IRQ, ARM, TKM); *B. capillare* var *meridionale* Schimp., (IRN); *B. capillare* var. *rosulatum* Mitt. (IRN); *B. cellulare* Hook., (TRK); *B. comense* Schimp., (IRN); *B. creberrimum* Taylor, (TRK, ARM); *B. cryophilum* Maort. (ARM) *B. curvatum* Kaurin & Arnell, (TRK); *B. dichotomum* Hedw., (TRK, IRN, ARM, TKM); *B. donianum* Grev., (TRK); *B. elegans Grev.*, (TRK, TKM); *B. elwendicum* Fehln., (IRN); *B funckii* Schwägr (ARM, TKM); *B. funckiioides* J. Froehl., (TRK, IRN); *B. gemmilucens* R. Wilczek & Demaret, (TRK, TMK); *B.*

gemmiparum De Not., (TRK, TMK); *B. imbricatum* (Schwägr.) Bruch & Schimp., (TRK, ARM); *B. inclinatum* (Hedw.) Dicks., (TRK); *B. intermedium* (Brid.) Turton, (TRK); *B. klinggraeffii* Schimp., (TRK); *B. knowltonii* Barnes, (TRK); *B. laevifilum* Syed (IRN); *B. moravicum* Podp. (TRK, TMK), *B. mildeanum* Jur., (TRK); *B. muehlenbeckii* Bruch & Schimp., (TRK, IRN); *B. neodamense* Itzigs., (TRK, IRN); *B. pallens* Sw., (TRK, IRN, ARM); *B. pallescens* Schleich. ex Schwägr., (TRK, IRN, ARM, TMK); *B. pallescens* var. *contextum* (Hornsch.) Huebener (IRN) *B. provinciale* H. Philib., (TRK); *B. pseudotriquetrum* (Hedw.) G. Gaertn., (TRK, IRN, ARM, IRQ); *B. pseudotriquetrum* var. *bornmuelleri* Schiffn. B. Mey. & Scherb., (TRK, IRN): *B. radiculosum* (Schwägr.) Mitt., (TRK, ARM); *B. riparium* I. Hagen, (TRK); *B. rubens* Mitt., (TRK, IRQ, ARM); *B. ruderale* Crundw. & Nyholm, (TRK); *B. rutilans* Bruch & Schimp., (TRK, TMK); *B. schleicheri* DC., (TRK, IRN, TMK); *B. schleicheri* var. *latifolium* (Schwägr.) Schimp*.* (IRN, ARM); *B. subelegans* Kindb., (TRK, ARM); *B. syriacum* Lorentz, (TRK, IRN); *B. torquescens* Bruch & Schimp., (TRK, IRN, ARM); *B. turbinatum* (Hedw.) Turner, (TRK, ARM); *B. uliginosum* (Brid.) Bruch & Schimp., (TRK, IRN, ARM); *B. ventricosum* R. Br., (TRK, IRN); *B. weigelii* Biehler, (TRK, IRN, ARM); *Gemmabryum klinggraeffii* (Schimp.) J.R. Spence & H.P. Ramsay (AZB); *Imbribryum gemmiparum* (De Not.) J.R. Spence (IRQ); *Imbribryum mildeanum* (Jur.) J.R. Spence (AZB), *Leptobryum pyriforme* (Hedw.) Wilson, (TRK, IRN, AZB, ARM); *Mielichhoferia mielichhoferiana* (Funck) Loeske, (IRN, ARM); *Plagiobryum zieri* (Dicks. ex Hedw.) Lindb., (TRK); *Ptychostomum bimum* (Schreb.) J.R. Spence (GOR) *P. compactum* Hornsch. (AZB); *P. knowltonii* (Barnes) J.R. Spence (TRK); *Rhodobryum ontariense* (Kindb.) Kindb. (ARM); *Rhodobryum roseum* (Hedw.) Limpr. (TRK, ARM); *Rosulabryum rubens* (Mitt.) J.R. Spence (IRQ).

- **Buxbaumiaceae:** *Buxbaumia viridis* (Moug. ex DC.) Brid. ex Moug. & Nestl. (TRK);
- **Calliergonaceae:** *Calliergon cordifolium* (Hedw.) Kindb., (TRK); *C. stramineum* (Dicks. ex Brid.) Kindb., (TRK); *Calliergonella cuspidata* (Hedw.) Loeske, (TRK); *Hamatocaulis vernicosus* (Mitt.) Hedenäs, (TRK);
- **Cryphaeaceae:** *Cryphaea heteromalla* (Hedw.) Brid. (TRK, IRN).
- **Climaciaceae:** *Climacium dendroides* (Hedw.) F. Weber & D. Mohr (TRK, ARM).

- **Dicrancaeae:** *Calcidicranella varia* (Hedw.) Bonfim Santos, Fedosov & Jan Kučera (IRQ); *Chorisodontium aciphyllum* (Hook. f. & Wilson) Broth. (GOR); *Cynodontium tenellum* (Schimp.) Limpr. (IRN); *Dichodontium pellucidum* (Hedw.) Schimp. (ARM); *Dicranella heteromalla* (Hedw.) Schimp., (TRK, ARM); *D. howei* Renauld & Cardot (TRK, IRN); *D. palustris* (Dicks.) Crundw. ex E.F. Warb., (TRK); *D. rufescens* (Dicks.) Schimp. (TRK); *D. squarrosa* (Schrad.) Schimp., (TRK); *D. subulata* (Hedw.) Schimp., (TRK); *D. varia* (Hedw.) Schimp., (TRK, IRQ, IRN, ARM, TMK); *Dicranodontium denudatum* (Brid.) E. Britton, (TRK, ARM); *D. cirrata* (Hedw.) Lindb. ex Milde, (TRK, IRQ, ARM); *D. crispula* (Hedw.) Lindb. (ARM); *D. intermedia* Amman (ARM), *Dicranoloma hariotii* (Müll. Hal.) Paris (GOR), *D. subimponens* (Cardot) Broth. (GOR), *Dicranoweisia crispula* (Hedw.) Milde, (TRK); *Dicranum bergeri* Bland. (ARM); *Dicranum bonjeanii* De Not., (TRK); *D. brevifolium* (Lindb.) Lindb. (ARM); *D. condensatum* Hedw., *D. congestum* Turner (TRK, ARM) *D fulvum* Hook., (TRK); *D. fuscescens* Turner, (TRK, ARM); *D. majus* Turner, (TRK); *D. montanum* Hedw., (TRK); *D. muehlenbeckii* Bruch & Schimp., (TRK, IRN, ARM); *D. oleodictyon* Dixon (GOR); *D. palustre* Bach. Pyl., (TRK); *D. polysetum* Sw., (TRK, ARM); *D. scoparium* Hedw., (TRK, GOR, IRN, ARM); *D. spadiceum* J.E. Zetterst., (TRK, ARM); *D. strictum* (Hook. f. & Wilson) Mitt., (TRK); *D. tauricum* Sapjegin, (TRK); *D. undulatum* Schrad. ex Brid., (TRK); *D. viride* (Sull. & Lesq.) Lindb., (TRK); *Kiaeria starkei* (F. Weber & D. Mohr) I. Hagen, (TRK); *Oreas martiana* (Hoppe & Hornsch.) Brid. (IRN); *Orthodicranum montanum* (Hedw.) Loeske, (TRK, ARM); *O. tauricum* (Sapjegin) Smirnova., (TRK); *Platyneuron laticostatum* (Cardot) Broth. (GOR).
- **Ditrichaceae:** *Ceratodon chloropus* (Brid.) Brid., (TRK); *C. conicus* (Spreng.) Trevis., (TRK); *C. purpureus* (Hedw.) Brid., (TRK, IRN, ARM, TMK); *Cheilothela chloropus* (Brid.) Lindb., (TRK, IRN); *Distichium capillaceum* (Hedw.) Bruch & Schimp. (TRK, IRN, IRQ, ARM); *Ditrichum cylindricum* (Hedw.) Grout, (TRK); *D. flexicaule* (Schwägr.) Brockm., (TRK, IRN, ARM); *D. heteromallum* (Hedw.) E. Britton, (TRK); *Ditrichum homomallum* (Hedw.) Brockm. (IRQ); *D. inclinatum* (Hedw.) B.S.G. (TRK, IRN ARM); *D. pallidum* (Hedw.) Brockm., (TRK); *Ditrichum pusillum* (Hedw.) Hampe (IRQ); *D. subulatum* Hobk. & Bosw., (TRK); *Garckea comosa* (Dozy & Molk.) Wijk & Margad. (IRN).

- **Diphysciaceae:** *Diphyscium foliosum* (Hedw.) D. Mohr, (TRK).
- **Drummondiaceae:** *Drummondia clavellata* (Hook. & Grev.) Hook. (GOR).
- **Encalyptaceae:** *Encalypta alpina* Sm., (TRK, ARM); *E. ciliata* Hedw., (TRK, IRN, ARM); *E. contorta* Hedw., (TRK); *E. intermedia* Jur., (TRK, IRN, TKM); *E. leptodon* Bruch ex Lindb., (TRK); *E. rhaptocarpa* Schwägr., (TRK, IRN, ARM); *E. rhaptocarpa* var. *trachymitra* (Rip.) Wijk et Marg. (ARM) *E. spathulata* Müll. Hal., (TRK, ARM); *E. streptocarpa* Hedw., (TRK, IRN, IRQ, ARM); *E. vulgaris* Hedw., (TRK, IRN, IRQ, ARM, TKM).
- **Ephemeraceae:** *Ephemerum cohaerens* (Hedw.) Hampe, (TRK); *Ephemerum cristatum* (Hook. & Wilson) Müll. Hal (IRQ); *E. minutissimum* Lindb., (TRK).
- **Entodontaceae:** *Entodon cladorrhizans* (Hedw.) C. Muell. (ARM); *E. compressus* (Hedw.) C.Muell. (ARM); *Entodon concinnus* (De Not.) Paris (TRK, IRN, ARM); *E. orthocarpos* (Brid.) Lindb., *E. schleicheri* (Schimp.) Demet. (TRK, IRN, ARM); *Pleurozium schreberi* (Willd. ex Brid.) Mitt.
- **Fabroniaceae:** *Fabronia ciliaris* (Brid.) Brid., (TRK, IRN, ARM); *F. pusilla* Raddi, (TRK, IRQ, TMK); *F. pusilla* Raddi var. *nervosa* Thér. (ARM); *Schwetschkeopsis fabronia* (Schwägr.) Broth. (IRN).
- **Fissidentaceae:** *Fissidens adianthoides* Hedw. (TRK, ARM); *F. algarvicus* Solms, (TRK); *Fissidens arnoldi* R. Ruthe, (IRQ, TMK); *F. bambergeri* Schimp., (TRK, IRN); *F. bryoides* Hedw., (TRK, IRN, ARM); *F. crassipes* Wilson ex Bruch & Schimp., (TRK, IRN, TMK); *F. curvatus* Hornsch., (TRK, IRN); *F. dubius* P. Beauv., (TRK, IRN, ARM); *F. exiguus* Sull., (IRN, TMK) *F. exilis* Hedw., (TRK, IRN); *F. gracilifolius* Brugg. Nann. & Nyholm, (TRK, IRN); *F. grandifrons* Brid. (ARM); *F. incurvus* Starke ex Röhl., (TRK, IRN); *F. intralimbatus* R. Ruthe, (TRK); *F. limbatus* Sull. (TRK); *F. mildeanus* Schimp., (TRK); *F. monguillonii* Thér., (TRK); *Fissidens obtusifolius* Wilson (GOR), *F. osmundioides* Hedw., (IRN); *F. polypodioides* Hedw. (GOR), *F. persicus* R.Ruthe (TMK); F. pusillus* (Wilson) Milde, (TRK, ARM); *F. rivularis* (Spruce) Schimp., (TRK, IRN); *F. rufulus* Schimp., (TRK, IRN); *F. serrulatus* Brid., (TRK); *F. taxifolius* Hedw., (TRK, IRN, ARM); *F. taxifolius* subsp. *pallidicaulis* (Mitt.) P. de la Varde (IRN); *F. viridulus* (Sw.) Wahlenb., (TRK, IRN, IRQ); *F. warnstorfii* M. Fleisch., (TRK).

- **Fontinalaceae:** *Fontinalis antipyretica* Hedw., (TRK, IRN, ARM); *F. duriaei* Schimp., (TRK); *F. gracilis* Lindb., (TRK); *F. hypnoides* Hartm. (TRK, IRN, ARM); *F. mesopotamica* Schiffn. (IRQ); *F. sullivantii* Lindb. (GOR).
- **Funariaceae:** *Entosthodon angustifolius* Jur. & Milde (IRQ, TKM); *E. attenuatus* (Dicks.) Bryhn, (TRK, IRN); *E. curvisetus* (Schwägr.) Müll. Hal., (TRK); *E. duriaei* Mont. (IRN, TKM); *E. fascicularis* (Hedw.) Müll. Hal., (TRK, IRN); *E. muhlenbergii* (Turner) Fife (IRN), *E. obtusus* (Hedw.) Lindb., (TRK); *E. pallescens* Jur., (TRK); *Funaria anomala* Jur., (TRK, IRN); *F. attenuata* (Dicks.) Lindb., (TRK); *F. calcarea* Wahlenb., (TRK); *F. convexa* Spruce, (TRK); *F. curviseta* (Schwägr.) Sérgio, (TRK); *F. dentata* Crome, (TRK); *F. fascicularis* (Hedw.) Lindb., (TRK); *F. handelii* Schiffn., (TRK, IRN, IRQ); *F. hygrometrica* Hedw., (TRK, IRN, IRQ, ARM, TKM); *F. microstoma* Bruch ex Schimp., (TRK, IRN); *F. muhlenbergii* Turner, (TRK, ARM); *F. obtusa* (Hedw.) Lindb., (TRK); *F. pulchella* H. Philib., (TRK); *Physcomitrella patens* (Hedw.) Bruch & Schimp. (IRN); *Physcomitrium acuminatum* Bruch & Schimp., (TRK); *P. eurystomum* Sendtn., (TRK); *P. pyriforme* (Hedw.) Brid. (TRK, ARM); *Physcomitrium sphaericum* (C. Ludw.) Brid. (IRN).
- **Grimmiaceae:** *Bucklandiella sudetica* (Funck) Bedn. Ochyra & Ochyra (GOR); *B. striatipila* (Cardot) Bedn.-Ochyra & Ochyra (GOR), *Coscinodon cribrosus* (Hedw.) Spruce, (TRK, IRN, IRQ); *Dryptodon patens* (Hedw.) Brid. (ARM); *Grimmia affinis* Hornsch., (TRK); *G. alpestris* Sommerf., (TRK, IRN, ARM, TMK); *G. anodon* Bruch & Schimp., (TRK, IRN, IRQ, ARM, TMK); *G. apiculata* Hook., (TRK); *G. caespiticia* (Brid.) Jur., (TRK, IRN, ARM); *G. caespiticia* (Brid.) Jur. var. *bornmuellerorum* Schiffn. (IRN); *G. campestris* Burch. ex Hook., (TRK); *G. capillata* De Not., (TRK, TMK); *G. commutata* Huebener (TRK); *G. crassifolia* Lindb. ex Broth., (TRK, IRN); *G. cribrosa* (Hedw.) Spruce. (TRK); *G. crinita* Brid., (TRK, ARM, TMK); *G. decipiens* (Schultz) Lindb., (TRK, ARM); *G. dissimulata* E. Maier (TRK); *G. donniana* Sm., (TRK, IRQ, ARM); *G. elatior* Bruch ex Bals. Criv. & De Not., (TRK, IRN, ARM); *G. elongata* Kaulf., (TRK, ARM); *G. exquisita* J. Muño (TMK); *G. funalis* (Schwägr.) Bruch ex De Not., (TRK, IRN, ARM, TMK); *G. gibbosa* S. Agnew (TRK); *G. hartmanii* Schimp., (TRK, IRN, ARM); *G. incurva* Schwaegr. (ARM); *G. laevigata* Mitt. (TRK, GOR, ARM), *G. lisae* De Not., (TRK,

IRN); *G. longirostris* Hook. (TRK, IRN, AZB); *G. maritima* Sm. ex R. Scott (IRQ); *G. mesopotamica* Schiffn. (TRK); *G. montana* Bruch & Schimp., (TRK, ARM); *G. nutans* Bruch, (TRK); *G. olneyi* Sull. (IRN); *G. orbicularis* Bruch ex Wilson, (TRK, IRN, ARM); *G. ovalis* (Hedw.) Lindb., (TRK, IRN, ARM, TMK); *G. patens* Hornsch. (IRQ); *G. plagiopodia* Hedw., (TRK, AZB, ARM); *G. poecilostoma* Cardot & Sebille, (TRK, TMK); *G. pulvinata* Sw., (TRK, IRN, IRQ, ARM, TMK); *G. reflexidens* Müll. Hal., (TRK); *G. singarensis* Schiffn., (TRK); *G. stirtonii* Schimp., (TRK); *G. subcaespiticia* Schiffn., (TRK); *G. tenerrima* Renauld & Cardot, (TRK); *G. tergestina* Tomm. ex Bruch & Schimp., (TRK, TMK); *G. tergestina* var. *poecilostoma* (Card. et Seb.) Loeske, (ARM); *G. trichophylla* Grev., (TRK, IRN, ARM); *Racomitrium aciculare* (Hedw.) Brid., (TRK, ARM); *R. affine* (Schleich. ex F. Weber & D. Mohr) Lindb. (IRQ); *R. canescens* (Hedw.) Brid. (ARM); *Racomitrium elongatum* Ehrh. ex Frisvoll (IRN); *R. microcarpon* (Hedw.) Brid. (IRQ, AZB); *R. sudeticum* (Funck) B.S.G. (ARM); *Schistidium agassizii* Sull. & Lesq. (IRN); *S. apocarpum* subsp. *gracile* (Röhl.) Meyl. (IRN, IRQ); *Schistidium apocarpum* var. *confertum* (Funck) Moell. (ARM); *S. atrofuscum* (Schimp.) Limpr. (IRN, IRQ); *S. brunnescens* Limpr. (IRN, TMK); *S. confertum* (Funck) Bruch & Schimp. (IRN); *S. crassipilum* H. H. Blom, (IRN); *S. flaccidum* (De Not.) Ochyra (IRN, ARM); *S. pulchrum* H.H. Blom. (TRK); *S. rivulare* (Brid.) Podp. (IRN); *S. singarense* (Schiffn.) Laz. (IRN); *S. strictum* (Turn.) Maort. (ARM); *S. trichodon* (Brid.) Poelt.

- **Habrodontaceae:** *Habrodon perpusillus* (De Not.) Lindb., (TRK);
- **Hedwigiaceae:** *Hedwigia ciliata* (Hedw.) Beauv. (TRK, IRN, IRQ, ARM); *H. stellata* Hedenäs, (TRK).
- **Helodiaceae:** *Helodium blandovii* (F. Weber & D. Mohr) Warnst., (TRK).
- **Hookeriaceae:** *Hookeria lucens* (Hedw.) Sm., (TRK).
- **Hylocomiaceae:** *Hylocomium brevirostre* (Brid.) Schimp., (TRK); *H. pyrenaicum* (Spruce) M. Fleisch. ex Broth., (TRK); *H. splendens* (Hedw.) Schimp., (TRK, ARM); *Loeskeobryum brevirostre* Brid. (ARM); *Pleurozium schreberi* (Brid.) Mitt. (ARM); *Rhytidiadelphus loreus* (Hedw.) Warnst. (IRQ); *R. squarrosus* (Hedw.) Warnst. (ARM); *R. subpinnatus* (Lindb.) T. Kop., *R. triquetrus* (Hedw.) Warnst., Fleisch. (ARM).

- **Hypnaceae:** *Cirriphyllum crassinervium* (Taylor ex Wilson) Loeske & M. Fleisch., (TRK); *C. piliferum* (Hedw.) Grout, (TRK); *C. reichenbachianum* (Huebener) Wijk & Margad., (TRK); *C. velutinoides* Loeske & M. Fleisch., (TRK); *Ctenidium molluscum* (Hedw.) Mitt., (TRK, IRN, ARM); *C. procerrimum* (Mol.) Lindb. (ARM); *Ectropothecium cyperoides* (Hook.) A. Jaeger (IRN); *Herzogiella seligeri* (Brid.) Z. Iwats. (TRK, ARM); *Homomallium incurvatum* (Brid.) Loeske (IRN, ARM), *Hypnum andoi* A.J.E. Sm., (TRK); *H. bambergeri* Schimp., (TRK); *H. callichroum* Brid., (TRK, ARM); *H. cupressiforme* Hedw., (TRK, IRN, ARM); *H. cupressiforme* Hedw. var. *lacunosum* Brid. (IRN, TKM), *H. cupressiforme* var. *tectorum* Brid., (IRN); *H. ericetorum* (Schimp.) Loeske, (TRK); *H. fastigiatum* Wibel, (TRK); *H. fertile* Sendtn., (TRK); *H. hamulosum* Schimp., (TRK); *H. imponens* Hedw., (TRK, IRN); *H. jutlandicum* Holmen & E. Warncke (TRK, IRN); *H. lacunosum* (Brid.) Hoffm. ex Brid., (TRK); *H. lindbergii* Sande Lac., (TRK, ARM); *H. macrogynum* Besch. (IRN); *H. mammillatum* (Brid.) Lindb., (TRK); *H. pallescens* (Hedw.) P. Beauv. (IRN, ARM), *H. ravaudii* Boulay, (TRK); *H. recurvatum* (Lindb. & Arnell) Kindb., (TRK); *H. resupinatum* Taylor, (TRK); *H. revolutum* (Mitt.) Lindb. (TRK, IRN, ARM); *H. tectorum* Funck ex Brid., (TRK); *H. uncinulatum* Jur., *H. vaucheri* Lesq. (IRN, ARM), *Platygyrium repens* (Brid.) Schimp. (IRN); *H. vaucheri* Lesq., (TRK); *Isopterygiopsis muelleriana* (Schimp.) Z. Iwats., (TRK); *I. pulchella* (Hedw.) Z. Iwats., (TRK, ARM); *Isopterygium elegans* (Brid.) Lindb. (IRQ); *Isopterygium depressum* (Brid.) Mitt. (TRK); *I. muellerianum* (Schimp.) Z. Iwats., (TRK); *I. pulchellum* (Hedw.) A. Jaeger, (TRK); *I. seligeri* (Brid.) Dixon ex C.E.O. Jensen, (TRK); *Orthothecium chryseon* (Schwaegr.) B.S.G. (ARM); *O. intricatum* (C. Hartm.) B.S.G. (ARM); *Platydictya confervoides* (Brid.) Crum (ARM); *P. subtilis* (Hedw.) Crum (ARM); *Platygyrium repens* (Brid.) B.S.G. (ARM); *Ptilium crista-castrensis* (Hedw.) De Not. (ARM); *Pylaisia polyantha* (Hedw.) Schimp. (TRK, IRN); *Pylaisiella polyantha* (Hedw.) Grout, (ARM); *Stereodon revolutus* Mitt. (TKM); *S. vaucheri* (Lesq.) Lindb. ex Broth. (TKM), *Taxiphyllum wissgrillii* (Garov.) Wijk et Marg. (ARM).
- **Lembophyllaceae:** *Isothecium alopecuroides* (Lam. ex Dubois) Isov. (TRK); *I. filescens* (Brid.) Mönk., (TRK); *I. myosuroides* (Lam. ex Dubois) Isov., (TRK); *I. myurum* (Lam. ex Dubois) Isov., (TRK); *I. stoloniferum* Brid., (TRK); *I. viviparum* Lindb., (TRK);

- **Leskaeaceae:** *Lescuraea decipiens* (Limpr.) Loeske, (TRK); *L. denudata* (Kindb.) Mönk., (TRK); *L. incurvata* (Hedw.) E. Lawton, (TRK); *L. mutabilis* (Brid.) Lindb. ex I. Hagen, (TRK, ARM); *L. patens* (Lindb.) Arnell & C.E.O. Jensen, (TRK, IRN); *Leskea perstricta* Dixon (IRN); *L. polycarpa* Hedw. (IRN, ARM); *L. radicosa* (Mitt.) Mönk., (TRK); *L. saxicola* (Schimp.) Molendo, (TRK); *L. striata* (Schwägr.) Schimp., (TRK); *Leskea catenulata* (Brid. ex Schrad.) Mitt., (TRK); *L. incurvata* Hedw., (TRK); *L. polycarpa* Hedw., (TRK); *L. tectorum* (Funck ex Brid.) Lindb., (TRK); *Leskeella incrassate* (Lindb. ex Broth.) Broth., (TRK, ARM); *L. nervosa* (Brid.) Loeske, (TRK, ARM); *L. tectorum* (Funck ex Brid.) I. Hagen., (TRK); *Lindbergia brachyptera* (Mitt.) Kindb., (ARM); *Pseudoleskea incurvata* (Hedw.) Loeske (IRN, ARM), *Pseudoleskeella catenulata* (Schrad.) Kindb. (IRN, ARM); *P. laxiramea* (Schiffn.) Broth. (IRN); *P. nervosa* (*Brid.) Nyholm (IRN); *P. radicosa* (Mitt.) Kindb. (ARM); *P. tectorum* (Funck ex Brid.) Kindb. ex Broth. (IRN, ARM, TMK); *Ptychodium plicatum* (Schleich. ex F. Weber & D. Mohr) Schimp. (IRN, ARM).
- **Leucobryaceae:** *Campylopus atrovirens* De Not., (TRK); *C. fragilis* (Brid.) Bruch & Schimp., (TRK); *C. pilifer* Brid., (TRK); *C. schimperi* Milde (IRQ); *Leucobryum albidum* (Brid. ex P. Beauv.) Lindb., (TRK, GOR); *L. glaucum* (Hedw.) Ångstr., (TRK, IRQ); *L. juniperoideum* (Brid.) Müll. Hal., (TRK); *Oncophorus virens* (Hedw.) Brid., (TRK, ARM); *O. wahlenbergii* Brid. (ARM); *Paraleucobryum enerve* (Thed.) Loeske, (TRK); *P. longifolium* (Ehrh. ex Hedw.) Péterfi, (TRK); *P. sauteri* (Bruch & Schimp.) Loeske (TRK).
- **Leucodontaceae:** *Antitrichia breidleriana* Schiffn. (IRQ); *A. californica* Sull. ex Lesq., (TRK, IRN); *A. curtipendula* (Hedw.) Brid., (TRK, ARM); *Leucodon balcanicus* (Hedw.) Schwägr., (TRK); *L. caucasicus* Jur. & Milde, (TRK); *L. flagellaris* Broth., (TRK); *L. immersus* Lindb., (TRK, IRN, ARM); *L. sciuroides* (Hedw.) Schwägr., (TRK, IRN, ARM);
- **Meesiaceae:** *Amblyodon dealbatus* (Hedw.) P. Beauv. (TRK, IRN, ARM); *Meesia triquetra* (Richter) Aongstr. (ARM); *M. uliginosa* Hedw. (ARM); *Paludella squarrosa* (Hedw.) Brid. (ARM).
- **Mniaceae:** *Cinclidium arcticum* B.S.G. (ARM) *Epipterygium tozeri* (Grev.) Lindb. (TRK); *Mniobryum albicans* (Wahlenb.) Limpr. (TRK); *Mnium affine* Blandow (TRK, GOR); *M. ambiguum* H. Müll., (TRK, ARM); *M. confertidens* (Lindb. & Arnell) Kindb. (TRK); *M. cuspidatum* Neck. ex Lindb., (TRK); *M. ellipticum* Brid. (TRK); *M. hornum* Hedw.

(TRK); *Mnium immarginatum* Broth. (IRN); *M. longirostrum Brid.* (TRK); *M. lycopodioides* Schwägr. (TRK); *M. magnifolium* Horik. (TRK); *M. marginatum* (With.) P. Beauv. (IRN, ARM); *M. medium* (Besch.) Müll. Hal. (TRK); *M. menziesii* (Hook.) Müll. Hal. (IRQ); *M. orthorrhynchum* Müll. Hal., (TRK); *M. punctatum* Hedw., (TRK); *M. riparium* Mitt., (TRK); *M. rostratum* Schrad., (TRK); *M. rugicum* Laurer, (TRK); *M. seligeri* Jur. ex Warnst., (TRK); *M. spinosum* (Voit) Schwägr. (TRK, IRN, ARM); *M. spinulosum* Bruch & Schimp., (TRK); *M. stellare* Hedw., (TRK, IRN, ARM); *M. thomsonii* Schimp., (TRK, IRN, AZB, ARM); *M. undulatum* Hedw., (TRK); *Plagiomnium affine* (Blandow) T.J. Kop., (TRK, IRN, ARM); *P. confertidens* (Lindb. & Arnell) T.J. Kop., (TRK); *P. cuspidatum* (Hedw.) T.J. Kop., (TRK, IRN, ARM); *P. elatum* (Bruch & Schimp.) T.J. Kop., (TRK, IRN, ARM); *P. ellipticum* (Brid.) T.J. Kop., (TRK, IRN, ARM); *P. medium* (Bruch & Schimp.) T.J. Kop., (TRK, ARM); *P. rostratum* (Schrad.) T.J. Kop., (TRK, IRN, IRQ, ARM); *P. undulatum* (Hedw.) T.J. Kop. (IRN, ARM), *Pohlia annotina* (Hedw.) Lindb., (IRN, ARM); *P. cruda* (Hedw.) Lindb., (IRN, ARM); *P. elongata* Hedw. var. greenii (Brid.) Shaw (ARM); *P. filum* (Schimp.) Maort. (ARM); *P. flexuosa* Harv., (IRN); *P. ludwigii* (Spreng. ex Schwägr.) Broth., (IRN); *P. melanodon* (Brid.) Shaw, (ARM, TKM); *P. nutans* (Hedw.) Lindb. (IRN, ARM); *Pohlia obtusifolia* (Vill. ex Brid.) L.F. Koch., (TRK); *P. pentasticha* (Schiffn.) Schiffn. ex Podp., (IRN); *P. wahlenbergii* (F. Weber & D. Mohr) A.L. Andrews, (IRN, AZB, ARM); *Rhizomnium punctatum* (Hedw.) T.J. Kop. (IRN, ARM); *Trachycystis ussuriensis* (Maack & Regel) T.J. Kop. (IRN, ARM).

- **Myuriaceae:** *Hyocomium armoricum* (Brid.) Wijk & Margad., (TRK).
- **Neckeraceae:** *Homalia besseri* Łobarz., (TRK, ARM); *H. lusitanica* Schimp., (TRK); *H. trichomanoides* (Hedw.) Brid., (TRK, IRN, ARM); *H. webbiana* (Mont.) Mitt., (TRK); *H. woronowii* Thér., (TRK); *Metaneckera menziesii* (Drumm.) Steere, (TRK); *N. complanata* (Hedw.) Huebener, (TRK, IRN, ARM); *N. complanata* var. *tenella* Schimp. (IRN); *Neckera crispa* Hedw., (TRK, ARM); *N. besseri* (Łobarz.) Trevis., (TRK); *N. mediterranea* H. Philib., (TRK); *N. pennata* Hedw. (IRN, ARM); *N. pumila* Hedw. (TRK, ARM); *Neckeradelphus menziesii* (Drumm.) Steere., (TRK).
- **Orthodontiaceae:** *Orthodontium lineare* Schwägr. (IRQ).

- **Orthotrichaceae:** *Amphidium lapponicum* (Hedw.) Schimp. (ARM); *A. mougeotii* (Bruch & Schimp.) Schimp., (TRK, IRQ, ARM); *Lewinskya affinis* (Brid.) F. Lara, Garilleti & Goffinet, (TRK); *L. striata* (Hedw.) F. Lara, Garilleti & Goffinet, (TRK); *L. transcaucasica* Eckstein, Garilleti & F. Lara (GOR); *Orthotrichum acuminatum* R. Br. bis, (TRK); *O. affine* Brid., (TRK, IRN, IRQ, ARM, TKM); *O. alpestre* Hornsch. ex Bruch & Schimp., (TRK, IRN, ARM); *O. anomalum* R. Br. (TRK, IRN, ARM, TKM); *O. anomalum* Hedw. subsp. *saxatile* (Milde) Kho. (IRN), *O. crenulatum* Mitt. (TKM); *O. cupulatum* Hoffm. ex Brid., (TRK, IRN, IRQ); *O. cupulatum* var. *bistratosum* Schiffn. (IRQ), *O. cupulatum var. riparium* Huebener (TMK, ARM); *O. diaphanum* Schrad. ex Brid., (TRK, IRQ, ARM); *O. fallax* Bruch ex Brid., (TRK); *O. fastigiatum* Bruch ex Brid., (TRK, ARM); *O. franzonianum* De Not., (TRK); *O. gymnostomum* Bruch ex Brid., (TRK); *O. laevigatum* J.E. Zetterst., (TRK); *O. leiocarpum* Bruch & Schimp., (TRK); *O. lyellii* Hook. & Taylor, (TRK, IRQ); *O. macrocephalum* F. Lara Garilleti & Mazimpaka (IRN); *O. microcarpum* De Not. (IRN); *O. obtusifolium* Brid., (TRK, IRN, ARM, TKM); *O. octoblepharis* Brid. (IRQ); *O. pallens* Bruch ex Brid., (TRK, IRN, ARM, TKM); *O. patens* Bruch ex Brid., (TRK, ARM); *O. philibertii* Venturi (IRN); *O. pumilum* Bruch & Schimp., (TRK, ARM); *O. rivulare* Turner, (TRK, IRN, AZB); *O. rupestre* Schleich. ex Schwägr. (TRK, IRN, IRQ, ARM); *O. saxatile* Brid., (TRK); *O. schimperi* Hammar, (TRK, IRQ); *O. sordidum* Sull. et Lesq. (ARM); *O. speciosum* Nees, (TRK, IRQ, ARM); *O. sprucei* Mont., (TRK); *O. stramineum* Hornsch. ex Brid. (TRK, ARM, TKM); *O. striatum* Hedw. (ARM); *O. sturmii* Hoppe & Hornsch., (TRK); *O. tenellum* Bruch ex Brid., (TRK, GOR, ARM); *O. tortidontium* F. Lara, Garilleti & Mazimpaka, (TRK); *O. urnaceum* C. Muell. (ARM); *O. urnigerum* Myrin (ARM), *Pulvigera lyellii* (Hook. & Taylor) Plášek, Sawicki & Ochyra, (TRK); *Ulota crispa* (Hedw.) Brid. *U. curvifolia* (Wahlenb.) Lilj. (IRN).
- **Plagiotheciaceae:** *Herzogiella seligeri* (Brid.) Z. Iwats. (IRN); *Orthothecium intricatum* (Hartm.) Schimp., (TRK); *Plagiothecium denticulatum* (Hedw.) Schimp. (IRN, IRQ, ARM); *P. denticulatum* var. *obtusifolium* (Turner) Moore (IRN); *P. laetum* Schimp. (IRN); *P. nemorale* (Mitt.) Jaeg. (ARM); *P. platyphyllum* Mönk. (IRN); *P. succulentum* (Wilson) Lindb. (IRN, AZB); *P. undulatum* (Hedw.) Schimp. (IRN); *Stereophyllum wightii* (Mitt.) A. Jaeger. (IRN).

- **Polytrichaceae:** *Atrichum angustatum* (Brid.) Bruch & Schimp. (TRK, IRN, GOR); *A. flavisetum* Mitt. (ARM); *A. tenellum* (Röhl.) Bruch & Schimp., (TRK); *A. undulatum* (Hedw.) P. Beauv. (TRK, IRN, ARM); *Oligotrichum hercynicum* (Hedw.) DC., (TRK); *Pogonatum aloides* (Hedw.) P. Beauv. (IRN); *Pogonatum neesii* (Müll. Hal.) Dozy (GOR); *P. urnigerum* (Hedw.) P. Beauv. (IRN, ARM); *Polytrichastrum alpinum* (Hedw.) G.L. Sm. (ARM); *P. formosum* Hedw., (TRK); *P. sexangulare* (Flörke ex Brid.) G.L. Sm., (TRK); *Polytrichum commune* Hedw., (TRK, GOR, ARM); *P. formosum* Hedw. (ARM); *P. juniperinum* Hedw., (TRK, ARM); *P. longisetum* Sw. ex Brid. (ARM); *P. piliferum* Hedw. (ARM); *P. sexangulare* Brid. (ARM);
- ***Pottiaceae:*** *Acaulon mediterraneum* Limpr., (TRK); *A. muticum* (Hedw.) Müll.Hal (TKM); *A. piligerum* (De Not.) Limpr., (TRK); *A. triquetrum* (Spruce) Müll. Hal., (TRK, IRQ); *Aloina aloides* (W.D.J. Koch ex Schultz) Kindb., (TRK, AZB); *A. ambigua* (Bruch & Schimp.) Limpr., (TRK, IRN, IRQ); *A. bifrons* (De Not.) Delgad., (TRK); *A. brevirostris* (Hook. & Grev.) Kindb., (TRK, TKM); *Anoectangium aestivum* (Hedw.) Spruce (IRQ, TKM); *A. handelii* Schiffn. (IRN, IRQ, TKM); *A. rigida* (Hedw.) Limpr., (TRK); *Astomum crispum* (Hedw.) Hampe (ARM); *A. levieri* (Limpr.) Kindb. (ARM); *Barbula acuta* (Brid.) Brid. (IRQ); *B. convoluta* Hedw. (IRN, ARM); *Barbula convoluta* var. *commutata* (Jur.) Husn. (IRQ); *B. fallax* Hedw. (IRQ); *B. hornschuchiana* Schultz, (IRQ); *B. hornschuchiana* var. *pseudorevoluta* Reimers (IRQ); *B. icmadophila* Schimp. ex Müll. Hal. (IRQ); *B. revoluta* Brid. (IRQ); *B. rigidula* (Hedw.) Mitt. (IRQ); *B. tophacea* (Brid.) Mitt. (IRQ); *B. trifaria* (Hedw.) Mitt. (IRQ); *B. trifaria* var. *desertorum* (J. Froehl.) S. Agnew (IRQ); *B. unguiculata* Hedw., (TRK, IRN, IRQ, ARM, TKM); B. *unguiculata* var. *cuspidata* (Schultz) Brid. (IRQ); *B. vinealis* Brid., (TRK); B. *vinealis* var. *cylindrica* (Taylor) Boulay (IRQ); *Bryoerythrophyllum ferruginascens* (Stirt.) Giacom., (TRK); *B. recurvirostrum* (Hedw.) P.C. Chen, (TRK, IRN, ARM); *Cinclidotus aquaticus* (Hedw.) Bruch & Schimp., (TRK); *C. bistratosus* Kürschner & Lüb. -Nestle, (TRK); *C. fontinaloides* (Hedw.) P. Beauv., (TRK, IRN, IRQ); *C. mucronatus* (Brid.) Guim., (TRK); *C. nigricans* (Brid.) Wijk & Margad. (TRK, IRQ); *C. nyholmiae* Çetin, (TRK); *C. riparius* (Host ex Brid.) Arn., (TRK, IRQ, ARM); *Crossidium crassinerve* (De Not.) Jur. (IRN TKM); *C. squamiferum* (Viv.) Jur., (IRN, ARM); *C. squamiferum* var. *pottioideum* (De Not.) Mönk. (IRN, TKM); *Crossidium griseum* (Jur.) Jur., (TRK);

C. squamiferum (Viv.) Jur., (TRK); *Crumia latifolia* (Kindb.) Schof. (ARM); *Desmatodon convolutus* (Brid.) Grout, (TRK); *D. heimii* (Hedw.) Mitt., (TRK); *D. latifolius* (Hedw.) Brid., (TRK, ARM); *D. leucostoma* (R. Br.) Berggr., (TRK); *D. obtusifolius* (Schwägr.) Schimp., (TRK); *D. systylius* Schimp., (TRK); *Dialytrichia mucronata* (Brid.) Broth., (TRK); *Didymodon aronis* (Lorentz) J. Guerra, (TRK); *D. acutus* (Brid.) K. Saito, (TRK, ARM); *D. ceratodonteus* (Müll. Hal.) Dixon, (TRK); *D. cordatus* Jur. (TRK, TKM); *D. fallax* (Hedw.) R.H. Zander, (TRK, ARM, TKM); *D. ferrugineus* (Schimp. ex Besch.) M.O. Hill, (TRK, TKM); *D. glaucus* Ryan, (TRK); *D. icmadophilus* (Schimp.) K. Saito (IRN, ARM); *D. insulanus* (De Not.) M.O. Hill (TKM); *D. luridus* Hornsch. ex Spreng., (TRK, IRN, TKM); *D. aronis* (Lorentz) J. Guerra, (TRK); *D. acutus* (Brid.) K. Saito, (TRK); *D. nicholsonii* Culm., (TRK); *D. planotophaceus* J. Froehl. (IRN); *D. reflexus* (Brid.) Zander (ARM); *D. revolutus* Broth. (IRQ); *D. rigidicaulis* (Müll. Hal.) K. Saito, (TRK); *D. rigidulus* Hedw. (ARM.); *D. rubellus* (Huebener) Bruch & Schimp., (TRK); *D. sinuosus* (Mitt.) Delogne, (TRK); *D. spadiceus* (Mitt.) Limpr., (TRK, IRN); *D. tophaceus* (Brid.) Lisa, (TRK, IRN, ARM, TKM); *D. tophaceus* (Brid.) Lisa var. *excurrens* (Broth.) Mont. (IRN), *D. umbrosus* (Müll. Hal.) R.H. Zander, (TRK); *Didymodon validus* Limpr. (GOR); *D. vinealis* (Brid.) R.H. Zander, (TRK, IRN, ARM, TKM); *D. vinealis* var. *flaccidus* (Schimp.) Zander (ARM), *D. vinealis* var. *luridus* (Hornsch.) Zander (ARM), *Eucladium verticillatum* (Brid.) Bruch & Schimp. (IRN, TMK); *Gymnostomum aeruginosum* Sm., (TRK, IRQ, ARM); *Gymnostomum calcareum* Nees & Hornsch. (IRN, IRQ, ARM); *G. calcareum* var. *muticum* Boulay IRN); *G. mosis* (Lorentz) Jur. & Milde IRN, IRQ); *G. lanceolatum* M.J. Cano, Ros & J. Guerra, (TRK); *G. luisieri* (Sérgio) Sérgio ex Crundw., (TRK); *G. recurvirostrum* Hedw. (IRQ), *G. rupestre* Schleich. ex Schwägr., (TRK, IRN, IRQ); *G. tenue* Hedw., (TRK); *G. viridulum* Brid. (TRK); *Gyroweisia luisieri* Sérgio (TRK); *G. reflexa* (Brid.) Schimp., (TRK); *G. tenuis* (Hedw.) Schimp., (TRK, IRN, IRQ); *Hydrogonium ehrenbergii* (Lorentz) A. Jaeger, (TRK, IRN, IRQ, TKM); *Hymenostomum brachycarpum* Paris, (TRK); *H. tortile* (Schwägr.) Bruch & Schimp. (IRQ); *Hymenostylium recurvirostrum* (Hedw.) Dixon, (TRK, IRN, IRQ); *H. recurvirostre* var. *latifolium* (J. E. Zetterst.) Kho. (IRN, ARM); *Hyophila involuta* (Hook.) A. Jaeger (TRK); *Mildeella bryoides* (Dicks.) Limpr., (TRK); *Molendoa sendtneriana* (Bruch & Schimp.) Limpr. (TRK); *Leptobarbula berica*

(De Not.) Schimp., (TRK); *Microbryum curvicollum* (Hedw.) R.H. Zander, (TRK); *M. davallianum* (Sm.) R.H. Zander, (TRK, IRN); *M. floerkeanum* (F. Weber & D. Mohr) Schimp., (TRK); *M. rectum* (With.) R.H. Zander, (TRK); *M. starckeanum* (Hedw.) R.H. Zander, (TRK); *Microbryum vlassovii* (Laz.) R.H. Zander (TKM), *Molendoa handelii* (Schiffn.) Brinda & R.H. Zander (IRQ), *Oxystegus cylindricus* (Bruch ex Brid.) Hilp., (TRK, IRN); *O. hibernicus* (Mitt.) Hilp., (TRK); *O. tenuirostris* (Hook. & Taylor) A.J.E. Sm., (TRK, ARM); *Phascum bryoides* Dicks., (TRK); *P. curvicollum* Hedw., (TRK, IRQ); *P. cuspidatum* Schreb. ex Hedw., (TRK, ARM); *P. cuspidatum* Hedw. var. *piliferum* (Schreb. ex Hedw.) Hook. & Taylor (IRQ); *P. floerkeanum* F. Weber & D. Mohr, (TRK); *P. piliferum* Schreb. ex Hedw., (TRK); *P. recurvifolium* Dicks., (TRK); *P. schreberianum* Dicks., (TRK); *P. vlassovii* (Laz.) Ros & Herrnst., (TRK, ARM); *Pleurochaete squarrosa* (Brid.) Lindb., (TRK, IRN, IRQ, TKM); *Pottia bryoides* (Dicks.) Mitt. (ARM); *P. commutata* Limpr. (IRQ); *Pottia crinita* Wilson ex Bruch & Schimp. (IRN); *P. davalliana* (Sm.) C.E.O. Jensen (IRQ); *P. lanceolata* (Hedw.) Müll. (IRN, IRQ, ARM); *P. mutica* Venturi (IRQ); *P. truncata* (Hedw.) Bruch & Schimp. (IRQ); *P. wilsonii* (Hook.) Bruch & Schimp. (IRQ); *P. wilsonii* var. *crinita* (Wilson ex Bruch & Schimp.) Warnst., *Pseudocrossidium hornschuchianum* (Schultz) R.H. Zander (TKM, ARM); *Pseudocrossidium revolutum* (Brid.) R. H. Zander (IRN); *Pterygoneurum ovatum* (Hedw.) Dixon (IRN, IRQ, ARM), *Pterygoneurum subsessile* (Brid.) Jur. (TKM); *Stegonia latifolia* (Schwägr.) Venturi ex Broth. (IRN, ARM); *Streblotrichum convolutum* (Hedw.) P. Beauv. (IRQ); *S. calcicola* J. J. Aman (IRN); *Syntrichia caninervis* Mitt. (AZB, TKM); *S. caninervis* Mitt. var. *caninervis* (IRN); *S. caninervis* var. *gypsophila* (J. J. Amann ex G. Roth) Gil. (IRN); *S. caninervis* var. *pseudodesertorum* (Vondr.) M.T. Gallego. (IRN); *S. fragilis* (Taylor) Ochyra (IRN); *S. handelii* (Schiffn.) S. Agnew & Vondr. (IRN, TKM); *Syntrichia inermis* (Brid.) Bayrh. (IRN); *S. laevipila* Brid. (IRQ); *S. latifolia* (Bruch ex Hartm.) Huebener (IRQ); *S. minor* (Bizot) M.T. Gallego (IRN); *S. montana* Nees (IRN, IRQ); *S. montana* var. *calva* (Durieu & Sagot ex Schimp.) J. J. Amann (IRQ); *S. papillosissima* (Copp.) Loeske (IRN, IRQ); *S. princeps* (De Not.) Mitt. (TRK, IRN, IRQ); *S. princeps* subsp. *parnassica* (Schiffn.) Podp. (IRN); *S. pseudodesertorum* (J. Froehl.) S. Agnew & Vondr. (IRQ); *S. pseudohandelii* (J. Froehl.) S. Agnew & Vondr. (IRQ, TKM), *S. pulvinata*

(Jur.) Jur. (IRQ); *S. ruralis* (Hedw.) F. Weber & D. Mohr, (TRK, IRN, IRQ, TKM); *S. ruralis* var. *ruraliformis* (Besch.) Delogne (IRN, TKM), *S. ruraliformis* (Besch.) Mans. (IRN), *S. virescens* (De Not.) Ochyra, (TRK, IRN, TKM); *S. virescens* var. *iranica* (W.A. Kramer) R. H. Zander Far. (IRN); *Timmiella anomala* (Bruch & Schimp.) Limpr. (IRN), *T. barbuloides* (Brid.) Mönk. (IRN, IRQ); *T. crassinervis* (Hampe) L.F. Koch (IRN); *T. astoma* Schiffn. (IRN, AZB); *T. atrovirens* (Sm.) Lindb., *T. brevissima* Schiffn. (IRN); *T. canescens* Mont. (IRN); *T. demawendica* Schiffn. (IRN, AZB); *T. eucalyptrata* Lindb. *T. inclinata* (R. Hedw.) Limpr. (IRN); *T. flavovirens* (Bruch) Broth., (TRK); *Tortella fragilis* (Hook. & Wilson) Limpr. (TKM, ARM), *T. inflexa* (Bruch) Broth. (IRN); *T. nitida* (Lindb.) Broth. (IRN); *T. tortuosa* (Schrad. ex Hedw.) Limpr. (TRK, IRN, IRQ, ARM, TKM); *Tortula acaulon* (With.) R.H. Zander (TMK), *T. atrovirens* (Sm.) Lindb. (TKM, ARM); *T. calcicolens* Kramer (ARM), *T. caninervis* (Mitt.) Broth. (ARM), *T. grandiretis* Broth. (TKM), *Tortula intermedia* (Turner) Lindb. (IRQ, ARM); *T. inermis* (Brid.) Mont. (IRN, AZB, ARM); *T. laevipila* (Brid.) Schwägr. (IRQ, ARM), *T. lanceola* R.H. Zander, (TKM); *T. lingulata* Lindb., (TRK); *T. marginata* (Bruch & Schimp.) Spruce (IRQ); *T. mucronifolia* Schwägr. (IRN, TKM, ARM); *T. muralis* Hedw. (IRN, IRQ, TKM); *T. muralis* var. *aestiva* Brid. ex Hedw. (IRN, IRQ, ARM, TKM); *T. muralis* var. *muralis* Hedw. (IRQ); *T. norvegica* (Web.f.) Wahlenb. ex Lindb. (ARM) *T. obtusifolia* (Schwägr.) Mathieu (IRN, ARM); *T. princeps* De Not. (ARM); *T. protobryoides* R.H. Zander (TKM); *T. revolutifolia* Laz. (TKM); *T. revolvens* (Schimp.) G. Roth var. *obtusata* Reim. (IRN); *T. ruraliformis* (Besch.) Ingh. (ARM); *T. ruralis* (Hedw.) Gaertn et al. (ARM); *T. sinensis* (C. Muell.) Broth. (ARM); *T. subulata* Hedw. (TRK, IRN, ARM); *T. subulata* var. *graeffii* (Warnst.) J. J. Amann (IRN), *T. systylia* (Schimp.) Lindb. (TRK, IRN); *T. subulata* Hedw. *T. thianschanica* Broth. (TKM); *T. trachyphylla* Broth. (TKM); *T. transcaspica* Broth. (TKM); *T. virescens* (De Not.) De Not. (ARM); *Trichostomopsis haussknechtii* (Jur. & Milde) S. Agnew & Far. (IRN); *Trichostomum brachydontium* Bruch (IRN, AZB); *T. crispulum* Bruch (IRN, IRQ, ARM); *T. crispulum* var. *elatum* Schimp. (IRQ); *T. mildeanum* Jur. (IRN); *T. persicum* Jur. & Milde (IRN); *T. tenuirostre* (Hook. & Taylor) Lindb. (IRN); *Trichostomum triumphans* De Not. (TKM), *Weissia brachycarpa* (Nees & Hornsch.) Jur. (IRN, ARM); *W. condensa* (Voit) Lindb. (IRN, ARM, TKM); *W. controversa* Hedw., (TRK, IRN, ARM,

TKM); *W. controversa* var *amblyodon* (Brid.) Sendtn. (IRQ); *W. fallax* Sehlm. (IRQ); *W. krassavinii* (Laz.) Laz. ex Ochyra (TKM); *W. papillosissima* Laz (TKM), *W. squarrosa* (Nees & Hornsch.) Müll. (IRN); *W. triumphans* (De Not.) M.O. Hill. (IRN); *W. tyrrhena* M. Fleisch. (TKM), *W. wimmeriana* (Sendtn.) Bruch et al. (ARM, TKM).

- **Pterigynandraceae:** *Myurella apiculata* (Sommerf.) Schimp., (TRK); *M. julacea* (Schwägr.) Schimp., (TRK, ARM); *M. tenerrima* (Brid.) Lindb., (TRK, TMK); *Pterigynandrum filiforme* Hedw. (TRK, ARM); *Pterygoneurum subsessile* (Brid.) Jur. (AZB).
- **Ptychometriaceae:** *Campylostelium pitardii* (Corb.) E. Maier, (TRK, TMK).
- **Rhabdoweisiaceae:** *Cynodontium bruntonii* (Sm.) Bruch & Schimp., (TRK); *C. inclinatum* (Hedw.) Brid., (TRK); *C. jenneri* (Schimp.) Stirt., (TRK); *C. polycarpon* (Hedw.) Schimp., (TRK); *C. strumiferum* (Hedw.) Lindb., (TRK); *C. tenellum* (Schimp.) Limpr., (TRK); *C. virens* (Hedw.) Schimp., (TRK); *Eucladium verticillatum* (With.) Bruch & Schimp., (TRK, IRQ); *E. verticillatum* var. *angustifolium* Lindb. (IRQ).
- **Rhytidiaceae:** *Rhytidium rugosum* (Hedw.) Kindb. (IRN, ARM); *Rhytidiadelphus triquetrus* (Hedw.) Warnst. (IRN).
- **Schistostegaceae:** *Schistostega pennata* (Hedw.) F. Weber & D. Mohr. (IRQ).
- **Sematophyllaceae:** *Brotherella harveyana* (Mitt.) Dixon (IRN); *Clastobryum capillaceum* (Griff.) Broth. (IRN).
- **Seligeriaceae:** *Seligeria pusilla* (Hedw.) Bruch et al. (TKM).
- **Sphagnaceae:** *Sphagnum centrale* C.E.O. Jensen (TRK); *S. contortum* Schultz, (TRK); *S. fallax* (H. Klinggr.) H. Klinggr., (TRK); *S. fimbriatum* Wils (ARM); *S. flexuosum* Dozy & Molk., (TRK); *S. fuscum* (Schimp.) Klinggr. (ARM); *S. girgensohnii* Russow (GOR, ARM); *S. imbricatum* Hornsch. ex Russow (GOR); *S. magellanicum* Brid., (TRK); *S. molle* Sull., (TRK); *S. riparium* Aongst. (ARM); *S. rubellum* Wilson, (TRK); *S. squarrosum* Crome (ARM); *S. subsecundum* Nees. (GOR).
- **Tetraphidaceae:** *Georgia browniana* (Dicks.) Müll. Hal., (TRK); *G. pellucida* (Hedw.) Rabenh., (TRK, IRN, ARM).
- **Timmiaceae:** *Timmia austriaca* Hedw. (ARM), *T. megapolitana* Hedw. var. *bavarica* (Hessl.) Brid. (ARM).
- **Thamnobryaceae:** *Thamnium alopecurum* (Hedw.) Schimp. (IRN); *T. alopecurum* var. *caucasicum* Kindb. Gil. (IRN); *T. alopecurum* var. *corticola* (IRN); *Thamnobryum alopecurum* (Hedw.) Nieuwl. (IRN, ARM).

- **Theliaceae:** *Thelia asprella* (Schimp.) Sull.
- **Thuidiaceae:** *Abietinella abietina* (Hedw.) M. Fleisch., (TRK, ARM); *A. hystricosa* (Mitt.) Broth., (TRK); *Bryohaplocladium virginianum* (Brid.) Wat. et Iwats. (ARM); *Haplocladium microphyllum* (Sw. ex Hedw.) Broth., (TRK); *Heterocladium dimorphum* (Brid.) Ignatov & Fedosov, (TRK); *Thuidium abietinum* (Hedw.) Schimp. (IRN); *Thuidium delicatulum* (Hedw.) Schimp., (TRK, ARM); *T. philibertii* Limpr. (IRN, ARM); *T. tamariscinum* (Hedw.) B.S.G. (ARM).
- **Timmiaceae:** *Timmia megapolitana* Hedw. subsp. *bavarica* (Hessl.) Brassard (IRN); *T. norvegica* J. E. Zetterst. (IRN).
- **Seligeriaceae:** *Blindia acuta* (Hedw.) Bruch & Schimp. (TRK).

4.9.3 ANTHOCEROPHYTA

- **Anthocerotaceae:** *Anthoceros caucasicus* Steph. (IRN); *Anthoceros punctatus* L., (TRK); *A. dichotomus* Raddi, (TRK); *Phaeoceros laevis* (L.) Prosk. (IRN).
- **Notothyladaceae:** *Phaeoceros bulbiculosus* (Brotéro) Proskauer, (TRK); *P. laevis* (L.) Proskauer, (TRK).

4.10 DISCUSSION

The Irano-Anatolian region is one of the world's 36 hotspots for biodiversity. The area gets a lot of rainfall, which keeps things damp and moist and encourages the growth of many different plant species, particularly bryophytes. As a result of the study, 200 species of liverworts, 833 mosses, and three species of hornworts were recorded from the Irano-Anatolian hotspot region. The Turkey region is found to be the most species-rich region and appears to be the center of bryophyte diversity. The habitat diversity, especially highlands and montane vegetation, provides more congenial and varied microhabitats for bryophyte diversity. *Porella platyphylla*, *Frullania dilatata*, *Radula complanata*, *Leucodon sciuroides*, *Lewinskya striata*, *Orthotrichum pallens*, and *Pterigynandrum filiforme* were found to be widely distributed species occurring as epiphytes. The presence of many species of *Sphagnum* and the members of Amblystegiaceae indicate the availability of a good amount of moisture and wetland areas in the region; however, there were many dry regions where the members of Pottiaceae were dominant. Additionally, four endemic and vulnerable species of vipers can be found in the hotspot. Two of these bryophyte species—*Jungermannia caucasica* and *Jungermannia lignicola*—are native to Turkey.

Dominated families of the study site are Pottiaceae, Bryaceae, Orthotrichaceae, Brachytheciaceae, Hypnaceae, and Grimmiaceae. According to the humidity preference of taxa, mesophytes (59%) are dominant in the study site. The data show that mesophytes are represented by 51%, followed by xerophytes (27%), hygrophytes (15%), and epiphytes (11%). The distribution of taxa indicate that the region has mostly semi-arid habitats.

Among mosses, *Bryum*, *Brachythecium*, *Eurhynchium*, *Fissidens*, *Grimmia*, and *Hypnum* are found to be the most species-rich genera, followed by *Mnium*, *Orthotrichum*, and *Didymodon*. Presence of the unique genera like *Encalypta*, *Fontinalis*, *Hedwigia*, and *Sphagnum*, and several other species, indicate that the region possesses unique habitats and microclimatic conditions and definitely deserves to be a hot spot for plant diversity. The dominant liverwort families are Jungermanniaceae, Lophoziaceae, Metzgeriaceae, Lejeuniaceae, Porellaceae, and Scapaniaceae.

Phytosociological study revealed that the Orthothotrichaceae, Frullaniaceae, Radulaceae, Hypnaceae, and Leucodontaceae form the epiphytic bryophytes, which were recorded frequently on *Fagus orientalis*, *Quercus vulcanica*, *Pinus sylvestris*, *P. brutia*. The dry habitat taxa were found in open areas and also on the lower base of the trees.

The Irano-Anatolian region is endowed with a wide variety of bryophyte species; thorough recording is necessary in order to investigate the richness of bryoflora life. The wide usage of DNA data in systematics in the last decade has brought a lot of changes in the understanding of moss phylogeny, resulting in alterations in moss taxonomy and nomenclature, especially in those of pseudocarps.

ACKNOWLWDGEMENTS

The authors are thankful to the Institution of Eminence (IoE), University of Delhi, Delhi, for the financial support (IoE/2023-24/12/FRP). Himani Yadav acknowledges the University Grants Commission (UGC) for providing a fellowship.

KEYWORDS

- **biodiversity hotspot**
- **bryophytes**
- **climatic factors**
- **epiphytic taxa**
- **habitat diversity**
- **leafy liverworts**

REFERENCES

Chapter 4

References

Abay, G.; Batan, N.; Özdemir, T. Bryophyte Checklist of Rize, North-East Turkey. *Arctoa* **2016**, *25* (2), 386–392.

Abay, G.; Uyar, G.; Keçeli, T. Contributions to the Bryoflora of Kaçkar Mts (NE Anatolia, Turkey). *Phytol. Balcanica* **2009**, *15* (3), 317–329.

Abramova, A. L.; Abramov, I. I. New Mosses for the Flora of the Caucasus. *Bot. Zhurn.* **1950**, *35* (5), 514–516.

Abramova, A. L.; Abramov, I. I. On the Mosses of Armenia. *Trudy Bot. Inst. Akad. Nauk SSSR, Ser. 2* **1959**, *12*, 360–366.

Agnew, S.; Vondráček, M. A Moss Flora of Iraq. *Feddes Repertorium* **1975**, *86* (6–8), 341–489.

Ahmadi, Sh.; Shirzadian, S.; Tavasoli, A. Four New Species for Iran Bryoflora. *Rostaniha* **2004**, *5* (1), 109, 110.

Akhani, H.; Kürschner, H. An Annotated and Updated Checklist of the Iranian Bryoflora. *Cryptogamie, Bryologie* **2004**, *25* (4), 315–347.

Alataş, M.; Batan, N.; Ezer, T.; Uyar, G. The Epiphytic Bryophyte Flora and Vegetation of Boraboy and Destek Forests (Amasya, Turkey). *Pak. J. Bot.* **2017**, *49* (5), 1779–1786.

Alataş, M.; Ezer, T.; Erata, H.; Batan, N. Epiphytic Bryophyte Vegetation of *Fagus orientalis* Trees in Hıdırnebi Plateau (Trabzon, Turkey). *Botanica Pacifica* **2022**, *11* (2), 65–70.

Al-Ni'ma, B. A. B. Study of Some Bryophytes from Areas in North of Iraq. Ph.D. Thesis, College of Science, Mosul University, Iraq, 2003.

Al-Ni'ma, B. A. B.; Al-Khay'yat, A. H. Checklist of Iraqi Bryoflora. *Bull. Iraq Nat. Hist. Mus.* **2001**, *9* (3), 7–18.

Al-N'ima, B. A. B.; Al-Khyat, A. H. Record of New Bryophytes for Iraqi Bryoflora. *J. Educ. Sci.* **2007**.

Ando, H.; Matsuo, A. Applied Bryology. In *Advances in Bryology*; Cramer, J., Ed.; International Association of Bryologists: Berlin, 1984; Vol. 2, pp 133–224.

Arnell, S. Liverworts (Hepaticae). In *Iranian Plants Collected by Per Wendelbo in 1959. VI. Mosses (Musci)*; Størmer, P., Ed.; *Aarbok Universitetet i Bergen, Matematisk-Naturvitenskapelig Serie* 1963, No. 11, p 35.

Aziz, F. H. Seventeen spp. New Records for the Moss Flora of Iraq. *Phyton-Int. J. Exp. Bot.* **2011**, *80*, 35–46.

Aziz, F. H. Six New Records of Hepatophyta Liverworts to Iraqi Flora. *Sci. J. Duhok Univ.* **2010**, *13* (1), 86–98.

Aziz, F. H.; Al-Niama, B. A. B. Species Checklist of Bryophyte, Pteridophyta, Lichens of Iraq and Kurdistan. *ZANCO J. Pure Appl. Sci.* **2017**, *29* (2), 53–73.

Bahuguna, Y. M.; Gairola, S.; Semwal, D. P.; Uniyal, P. L.; Bhatt, A. B. Bryophytes and Ecosystem. In *Biodiversity of Lower Plants*; Gupta, R., Kumar, M., Eds.; 2013; pp 279–296.

Bargagli, R.; Brown, D. H.; Nelli, L. Metal Biomonitoring with Mosses: Procedures for Correcting for Soil Contamination. *Environ. Pollut.* **1995**, *89* (2), 169–175.

Barsegyan, A. M. On the Finding of *Ricciocarpus natans* (L.) Corda in Water Bodies of Erevan Depression (Armenian SSR). *Bot. Zhurn.* **1959**, *44* (4), 568–570.

Bell, B. G. A Synoptic Flora of South Georgian Mosses: II. *Chorisodontium*, *Dicranoloma*, *Dicranum*, *Platyneurum* and *Conostomum*. *Br. Antarct. Surv. Bull.* **1973**, *37*, 33–52.

Bornmüller, J. Bearbeitung der von J. A. Knapp im nordwestlichen Persien gesammelten Pflanzen. *Verhandlungen der Zoologisch-Botanischen Gesellschaft* **1910**, 61–194.

Bornmüller, J. Beiträge zur Flora des Elbursgebirge Nord-Persiens. *Bull. Herb.* **1908**, *8*, 832, 833.

Bornmüller, J. Collectiones Straussianae Novae—Weitere Beiträge zur Kenntnis der Flora West-Persiens. Hepaticae, Musci. *Bot. Zentralbl., Abt. II* **1911**, *28*, 522, 523.

Bornmüller, J. Plantae Brunsianae: Aufzählung der von F. Bruns im nördlichen Persien gesammelten Pflanzen. *Bot. Zentralbl., Abt. II* **1915**, *33*, 270–324.

Brotherus, V. F. Enumeratio Muscorum Caucasi. *Acta Soc. Sci. Fenn.* **1892**, *19*, 1–170.

Buck, W. R.; Goffinet, B.; Shaw, A. J. Testing Morphological Concepts of Orders of Pleurocarpous Mosses (Bryophyta) Using Phylogenetic Reconstructions Based on trnL-trnF and rps4 Sequences. *Mol. Phylogenet. Evol.* **2000**, *16* (2), 180–198.

Bushe, F.; Boissier, E. Aufzählung der auf einer Reise durch Transkaukasien und Persien gesammelten Pflanzen. *Nouveaux Mémoires de la Société Impériale des Naturalistes de Moscou* **1860**, *12*, 234–238.

Can, S. M.; Kara, R.; Ezer, T. Bryophyte Flora of Melendiz Mountain in Turkey. *Turk. J. Bot.* **2013**, *37*, 575–583.

Çetin, B. Checklist of the Liverworts and Hornworts of Turkey. *Lindbergia* **1988**, *14*, 12–14.

Chikovani, N.; Svanidze, T. Checklist of Bryophyte Species of Georgia. *Braun-Blanquetia* **2004**, *34*, 97–116.

Crundwell, A. C.; Nyholm, E. *Dicranella howei* Ren. & Card. and Its Relationships to *D. varia* (Hedw.) Schimp. *Lindbergia* **1977**, *4*, 35–38.

Crundwell, A. C.; Nyholm, E. *Funaria muhlenbergii* and Related European Species. *Lindbergia* **1974**, *2*, 222–229.

Crundwell, A. C.; Nyholm, E. Some Additions to the Bryophyte Flora of Turkey. I. Hepaticae. *J. Bryol.* **1979**, *10*, 479–489.

Czeczott, H. A Contribution to the Knowledge of the Flora and Vegetation of Turkey. *Feddes Repertorium, Beih.* **1939**, *107*, 1–281.

Davis, P. H.; Happer, P. C.; Hedge, I. C. *Plant Life of Southwest Asia*; The Botanical Society of Edinburgh: Edinburgh, 1971.

de Tchihatcheff, P. *Asie Mineure*, Vol. III, Tome 2: Botanique; Paris, 1860; p 676.

Dildaryan, B. I. *Bryoflora of Forests of North-East Armenia.* Ph.D. Thesis, Erevans. Gos. Univ., Erevan, 1971; p 298.

Dildaryan, B. I. *Bryoflora of Forests of Upper Mountain Belt of North-East Armenia.* Gos. Univ. **1969**, Vol. 2 (10), pp 91–96.

Dildaryan, B. I. Mosses of Soil Banks in Sevan Lake Basin. In *Materialy VI Konferenzii po Sporovym Rasteniyam*; Nakhuzrishvili, I. G., Ed.; Izd. Akad. Nauk. Gruz. SSR: Tbilisi, 1983; p 136.

Dildaryan, B. I. The Development of Epigeic Mosses in *Quercus–Carpinus* Forests of Central Armenia. In *Materialy V Konferenzii po Nizshim Rasteniyam Zakavkazia*; Uliyanishchev, V. I., Ed.; Elm: Baku, 1979; pp 123, 124.

Dildaryan, B. I. The Distribution of Mosses in Altitudinal Belts in Forests of North-East Armenia. In *Materialy*; Teterevnikova-Babayan, D. N., Ed.; Erevans. Gos. Univ.: Erevan, 1972; pp 69, 70.

Dildaryan, B. I.; Avakyan, K. G. Some Distributional Patterns of Mosses on Southern Slopes of Tekhenisskij Range. In *Materialy VII Zakavkazskoj Konferenzii po Sporovym Rasteniyam*; Osipyan, L. L., Ed.; Izd. Erev. Gos. Univ.: Erevan, 1986; p 41.

Dildaryan, B. I.; Avetisyan, M. S. Contributions to the Bryoflora of Khosrovskij Reserve. *Voprosy Biologii (Erevan)* **1987b**, *4*, 69–74.

Dildaryan, B. I.; Avetisyan, M. S. Contributions to the Moss Flora of Southern Slopes of Pambakskij Range. *Voprosy Biologii (Erevan)* **1987a**, *4*, 44–48.

Doroshina, G. Y. New Moss Records from Georgia. 1. *Arctoa* **2011**, *19*, 281.

Doroshina, G. Y.; Czernyadjeva, I. V. New National and Regional Bryophyte Records, 54. *J. Bryol.* **2018**, *40* (1), 74.

Dylevskya, I. V. Contributions to Moss Flora of Caucasus Minor. *Zametki Sist. Geogr. Rast. Bot. Inst. Akad. Nauk. Gruz. SSR* **1950**, *21*, 4–6.

Eckstein, J.; Garilleti, R.; Lara, F. *Lewinskya transcaucasica* (Orthotrichaceae, Bryopsida) sp. nov. A Contribution to the Bryophyte Flora of Georgia. *J. Bryol.* **2018**, *40* (1), 31–38.

Erdağ, A.; Kürschner, H. A Reference List of Turkish Bryophytes: The State of Knowledge from 1829 until 2017. *Anatolian Bryol.* **2017**, *3* (2), 81–102.

Ezer, T. Topographic Influence on Overflow Dynamics: Idealized Numerical Simulations and the Faroe Bank Channel Overflow. *J. Geophys. Res.: Oceans* **2006**, *111*.

Ezer, T.; Kara, R.; Alataş, M. *Scapania gracilis* Lindb. (Hepaticae, Scapaniaceae), New to Bryophyte Flora of Turkey. *Folia Cryptogamica Estonica* **2013**, *50*, 117–119.

Fehlner, C. *Bryum elwendicum* n. sp. *Verh. Zool. Bot. Ges. Wien* **1883**, *33*, 435, 436.

Fehlner, C. Hepaticae. Musci. In *Die botanischen Ergebnisse der Polak'schen Expedition nach Persien im Jahre 1882*; Stapf, O., Ed.; *Denkschr. Akad. Wiss. Math.-Nat. Kl., Abt. 2* **1885**, *50*, 4–6.

Frey, W. Bryophyte Flora and Vegetation of South-West Asia. *Proc. R. Soc. Edinburgh, Sect. B: Biol. Sci.* **1986**, *89*, 217–227.

Frey, W. Die Lebermoose des Iran und Afghanistan. *Bryologist* **1974**, *77*, 48–56.

Frey, W.; Kürschner, H. Contributions towards a Bryophyte Flora of Iran: New Records from Iran. *Iran. J. Bot.* **1983**, *2*, 13–19.

Frey, W.; Kürschner, H. Das Fossombronio-Gigaspermetum mouretii in der Judäischen Wüste. 2. Ökosoziologie und Lebensstrategien. *Cryptogam. Bot.* **1991c**, *2*, 73–84.

Frey, W.; Kürschner, H. Die Epiphytische Moosvegetation im Hyrkanischen Waldgebiet (Nord-Iran). *Beih. Tübinger Atlas Vorderen Orients, Reihe A (Naturwissenschaften)*; Wiesbaden, 1979.

Frey, W.; Kürschner, H. Lebensstrategien Epiphytischer Bryophyten im Tropischen Tieflands- und Bergregenwald am Mt. Kinabalu (Sabah, Nord-Borneo). *Nova Hedwigia* **1991d**, *53*, 307–330.

Frey, W.; Kürschner, H. Lebensstrategien von Terrestrischen Bryophyten in der Judäischen Wüste. *Bot. Acta* **1991a**, *104*, 172–182.

Frey, W.; Kürschner, H. Morphologische und Anatomische Anpassungen der Arten in Terrestrischen Bryophytengesellschaften entlang eines Ökologischen Gradienten in der Judäischen Wüste. *Bot. Jahrb. Syst.* **1991b**, *112*, 529–552.

Frey, W.; Kürschner, H. New and Noteworthy Records to the Bryophyte Flora of Iran. *Nova Hedwigia* **2010**, *90* (3, 4), 503–512.

Frey, W.; Kürschner, H. Studies on the Bryophyte Flora and Vegetation of the Mohammad Reza Shah National Park, N. Iran. *Iran. J. Bot.* **1977**, *1*, 137–153.

Frey, W.; Kürschner, H. The Bryological Literature of South-West Asia. *J. Hattori Bot. Lab.* **1981**, *50*, 217–229.

Frey, W.; Probst, W. Hängeformen von *Pseudoleskeella laxiramea* (Schiffn.) Broth. und *Leucodon immersus* Lindb. (Musci) im Südkaspischen Waldgebiet (Iran): Ein Beitrag zur Morphologie und zu den Lebensbedingungen. *Bot. Jahrb. Syst.* **1974b**, *94*, 267–282.

Frey, W.; Probst, W. The Pulvinate Moss Vegetation in the Karadj Valley (Elburs Mountains, Iran). *Bot. Jahrb. Syst.* **1973**, *93* (3), 404–423.

Frey, W.; Probst, W. Vegetationsanalytische Untersuchungen im Dünengebiet bei Babolsar, Kaspisches Meer, Iran. *Bot. Jahrb. Syst.* **1974a**, *94*, 96–113.

Froehlich, J. Berichtigung zu Bryophyten aus Iran. *Ann. Naturhist. Mus. Wien* **1953**, *59*, 118.

Froehlich, J. Bryophyten aus Iran. *Ann. Naturhist. Mus. Wien* **1950**, *57*, 37–41.

Froehlich, J. Bryophyten aus Vorderasien. *Ann. Naturhist. Mus. Wien* **1959**, *63*, 31, 32.

Gasimov, T.; Novruzov, V. Species of Family Orthotrichaceae Arn. (Bryopsida) in Bryophyte Flora of Azerbaijan. In *3rd International Symposium on Euro-Asian Biodiversity*; Minsk, 2017.

Ghahreman, A.; Faridi, M.; Shirzadian, S.; Attar, F. New and Interesting Moss Records for Iran. *Turk. J. Bot.* **2007**, *31* (1), 41–48.

Ghahreman, A.; Mehdigholi, K.; Nejad-Sattari, T.; Attar, F. A Preliminary Study of Moss Flora of Golestan Province and Comparison with Iranian Mosses. *Iran. J. Bot.* **2003**, *10* (1), 63–81.

Ghahremaninejad, F.; Shirzadian, S. An Updated List of the Bryological Literature on Iran. *Ann. Naturhist. Mus. Wien* **2016**, *118*, 181–188.

Goffinet, B.; Shaw, A. J., Eds. *Bryophyte Biology*; Cambridge University Press: Cambridge, 2008.

Gökler, İ. A Taxonomical Investigation of the Liverworts of the Aegean Region. *Dokuz Eylül Üniversitesi Yayınları* **1993a**, *6*, 33–44.

Gökler, İ. An Investigation on the Liverworts of West Anatolia. *Doğa, Türk Botanik Dergisi* **1992**, *16*, 1–8.
Gökler, İ. Studies on the Taxonomy and Ecology of Some Western Anatolian Liverworts. *Eğitim Bilimleri Dergisi* **1993b**, *2*, 79–85.
Gökler, İ.; Inoue, H.; Öztürk, M. A New Record for Turkey, *Pellia neesiana* (Gott.) Limpr. *J. Fac. Sci. Ege Univ., Ser. B* **1984**, *7*, 85–89.
Gökler, İ.; Öztürk, M. An Investigation on the Liverworts (Hepaticae) of Black Sea Region. *Doğa, Türk Botanik Dergisi* **1989**, *13*, 242–248.
Gökler, İ.; Öztürk, M. Liverworts (Marchantiopsida) from the Artvin Province (A4, A5). In *XI. Ulusal Biyoloji Kongresi*; 1992; pp 185–194.
Gökler, İ.; Öztürk, M. Liverworts (Marchantiopsida) from the Istanbul Province (A1). In *XII. Ulusal Biyoloji Kongresi*; Temmuz, 1994a; pp 174–178.
Gökler, İ.; Öztürk, M. Liverworts of Turkey and Their Position in South-West Asia. *Candollea* **1991**, *46*, 359–366.
Gökler, İ.; Öztürk, M. Liverworts of Turkish Thrace. *Bocconea* **1996**, *5*, 319–323.
Gökler, İ.; Öztürk, M. New Concepts Concerning the Methods Used in the Identification of Liverworts (Hepaticae). *Doğa, Türk Botanik Dergisi* **1987**, *11*, 306–312.
Gökler, İ.; Öztürk, M. Studies on the Taxonomy and Ecology of Liverworts of the Kütahya Province. *Ege Üniversitesi Fen Fakültesi Dergisi* **1994b**, *16*, 1525–1529.
Gökler, İ.; Öztürk, M. Taxonomical Investigations on Some Liverworts (Hepaticae) Distributed in Turkey. 1. Jungermanniales Anacrogynae and J. Acrogynae. *Doğa, Türk Biyoloji Dergisi* **1986**, *10*, 163–170.
Gökler, İ.; Öztürk, M.; Kesercioglu, T. Checklist of Liverworts (Hepaticae) Recorded from Turkey. *J. Fac. Sci. Ege Univ.* **1986**, *8*, 1–10.
Hallingbäck, T.; Hodgetts, N. Mosses, Liverworts and Hornworts: Status Survey and Conservation Action Plan for Bryophytes. *Belgian Journal of Botany* **2001**, *134*, 95, 96.
Handel-Mazzetti, H. V. Hepaticae. In *Ergebnisse einer Naturwissenschaftlichen Reise zum Erdschias-Dagh*; Penther, A.; Zederbauer, E., Eds.; *Ann. Naturhist. Mus. Wien* **1905**, *20*, 359–464.
Handel-Mazzetti, H. V. Musci. In *Ergebnisse einer Botanischen Reise in das Pontische Randgebirge im Sandschak Trapezunt, unternommen im Jahre 1909*; pp 124–132.
Handel-Mazzetti, H. V. Studies on the Asiatic Species of the Genus *Porella* (Hepaticae). VI. *J. Hattori Bot. Lab.* **1914**, *40*, 121–138.
Henderson, D. M. Contributions to the Bryophyte Flora of Turkey: II. *Notes R. Bot. Gard. Edinburgh* **1958a**, *22*, 189–193.
Henderson, D. M. Contributions to the Bryophyte Flora of Turkey: III. *Notes R. Bot. Gard. Edinburgh* **1958b**, *22*, 611–620.
Henderson, D. M. Contributions to the Bryophyte Flora of Turkey: IV. *Notes R. Bot. Gard. Edinburgh* **1961a**, *23*, 263–278.
Henderson, D. M. Contributions to the Bryophyte Flora of Turkey: V. Summary of Present Knowledge. *Notes R. Bot. Gard. Edinburgh* **1961b**, *23*, 279–301.
Henderson, D. M. Contributions to the Bryophyte Flora of Turkey: VI. *Notes R. Bot. Gard. Edinburgh* **1964**, *25*, 279–291.
Henderson, D. M.; Muirhead, C. W. Contributions to the Bryophyte Flora of Turkey. *Notes R. Bot. Gard. Edinburgh* **1955**, *22*, 29–43.

Henderson, D. M.; Prentice, H. T. Contributions to the Bryophyte Flora of Turkey: VIII. *Notes R. Bot. Gard. Edinburgh* **1969**, *29*, 235–262.

Ignatov, M. S.; Afonina, O. M. Checklist of Mosses of Former USSR. *Arctoa* **1992**, *1*, 1–85.

Ignatov, M. S.; Afonina, O. M.; Ignatova, E. A.; et al. Checklist of Mosses of East Europe and North Asia. *Arctoa* **2006**, *15*, 1–130.

Ignatov, M. S.; Gardiner, A. A.; Bobrova, V. K.; Milyutina, I. A.; Huttunen, S.; Troitsky, A. V. On Relationships of Mosses of the Order Hypnales, with Special Reference to Taxa Traditionally Classified in Leskeaceae. In *Pleurocarpous Mosses: Systematics and Evolution*; Newton, A. E.; Tangney, R. S., Eds.; Systematics Association Special Volume 71; 2007; pp 177–213.

Ignatova, E. A.; Ignatov, M. S.; Seregin, A. P.; Akatova, T. V.; Konstantinova, N. A. Bryophyte Flora of the Projected Utrish Nature Reserve (North-West Caucasus, Russia). *Arctoa* **2005**, *14*, 39–48.

Juratzka, J.; Milde, J. Beitrag zur Moosflora des Orients: Kleinasien, das westliche Persien und den Caucasus betreffend. *Verh. Zool.-Bot. Ges. Wien* **1870**, *20*, 589–602.

Kara, R.; Ezer, T.; Gözcü, M. C.; Bozdoğan, Ş. G. Bryophyte Flora of Erciyes Mountain in Turkey, with Six Bryophyte Records from the Country. *Turk. J. Bot.* **2014**, *38* (4), 763–781.

Kaya, Z.; Raynal, D. J. Biodiversity and Conservation of Turkish Forests. *Biol. Conserv.* **2001**, *97* (2), 131–141.

Kırmacı, M.; Ağcagil, E. The Bryophyte Flora in the Urban Area of Aydın (Turkey). *Int. J. Bot.* **2009**, *5* (3), 216–225.

Konstantinova, N. Protection of Bryophytes in the Murmansk Region (Russia). *Biol. Conserv.* **1992**, *59* (2, 3), 191–195.

Kürschner, H. A Key to the Pleurocarpous Mosses (Bryophytina) of the Near and Middle East. Towards a Bryophyte Flora of the Near and Middle East, 5. *Nova Hedwigia* **2006**, *83*, 353–386.

Kürschner, H. An Annotated, Corrected, and Updated List of the Bryological Literature of Southwest Asia. *Cryptogam. Bryol. Lichénol.* **1997**, *18*, 1–46.

Kürschner, H. Biogeography of South-West Asian Bryophytes—With Special Emphasis on the Tropical Element. *Turk. J. Bot.* **2008**, *32*, 433–446.

Kürschner, H. Bryophytes of Northern Origin in Southern Turkey (Amanos Mts., Eastern Aladağları)—Relicts or Indicators for Long-Range Dispersal. *Nova Hedwigia* **2014**, *99*, 13–26.

Kürschner, H. *Jubula hutchinsiae* subsp. *caucasica* Konstant. & Vilnet (Marchantiophyta: Jubulaceae)—New for the Bryophyte Flora of Iran and Turkey. *Polish Bot. J.* **2013**, *58*, 211–216.

Kürschner, H. Towards a Bryophyte Flora of the Near and Middle East: New Records from Iran, Jordan, Kuwait, Lebanon, Oman, Saudi Arabia, Syria, and Turkey. *Nova Hedwigia* **1996**, *63*, 261–271.

Kürschner, H. Towards a Bryophyte Flora of the Near and Middle East. 3. An Artificial Key to the Anthocerotophytina and Hepaticophytina in the Near and Middle East. *Nova Hedwigia* **2001**, *72*, 161–200.

Kürschner, H.; Batsatsashvili, K.; Parolly, G. Noteworthy Additions to the Bryophyte Flora of Georgia. *Herzogia* **2013**, *26*, 213–216.

Kürschner, H.; Erdağ, A. Bryophytes of Turkey: An Annotated Reference List of the Species with Synonyms from the Recent Literature and an Annotated list of Turkish Bryological Literature. *Turkish J. Bot.* **2005**, *29* (2), 95–154.

Kürschner, H.; Erdağ, A. The Grimmietum Commutato-Campestris in Turkey. Ecology and Life Syndromes of a Saxicolous Bryophyte Community with the Description of Two New Subassociations. *Nova Hedwigia* **2009**, *88*, 441–463.

Kürschner, H.; Erdağ, A. The Saxicolous Homalothecio-Neckeradelphetum Menziesii New to Western Turkey: Phytosociology, Ecology and Life Strategies. *Nova Hedwigia* **2008**, *87*, 97–112.

Kürschner, H.; Frey, W. Liverworts, Mosses and Hornworts of Southwest Asia: Marchantiophyta, Bryophyta, Anthocerotophyta. *Nova Hedwigia* **2011**, *139*, 1–240.

Kürschner, H.; Frey, W. *Liverworts, Mosses and Hornworts of Southwest Asia (Marchantiophyta, Bryophyta, Anthocerotophyta)*; *Nova Hedwigia, Beihefte*, Beih. 149; J. Cramer: 2020.

Kürschner, H.; Frey, W. *Tortula grandiretis* Broth. (Pottiaceae), an Irano-Turanian Moss New to Iraq and Further Novelties from the Near and Middle East. Towards a bryophyte flora of the Near and Middle East, 4. *Nova Hedwigia* **2006**, *83*, 143–149.

Kürschner, H.; Kirmaci, M. *Pohlia sphagnicola* (Bruch & Schimp.) Lindb. & Arnell. In: New National and Regional Bryophyte Records, 43. *J. Bryol.* **2015**, *37*, 138, 139.

Kürschner, H.; Kirmaci, M.; Erdağ, A. On the Hygrophytic Element in the Bryoflora of Turkey—a First Attempt to High Mountain Montio-Cardaminetea Communities. *Nova Hedwigia* **2015**, *100*, 407–421.

Kürschner, H.; Lubenau-Nestle, R. *Cinclidotus bistratosus* (Cinclidotaceae, Musci), a New Species to the Hygrophytic Moss Flora of Turkey. *Nova Hedwigia* **2000**, *70*, 471–478.

Kürschner, H.; Papp, B.; Akhani, H. New Records to the Bryophyte Flora of Iran. Studies on the Flora and Vegetation of the Golestan National Park, NE Iran IV. *Nova Hedwigia* **2000**, *71* (3, 4), 509–518.

Kürschner, H.; Parolly, G. *Eremonotus myriocarpus* (Carrington) Lindb. & Kaal. ex Pearson. In: New National and Regional Bryophyte Records, 12. *J. Bryol.* **2006a**, *28*, 69.

Kürschner, H.; Parolly, G. On the Occurrence of *Grimmia reflexidens* (Musci, Grimmiaceae) in Turkey. *Fragmenta Floristica et Geobotanica* **1999b**, *44*, 287–290.

Kürschner, H.; Parolly, G. *Syntrichia montana* Nees var. calva (Durieu & Sagot ex Bruch & Schimp.) J. J. Amann. In: New National and Regional Bryophyte Records, 19. *J. Bryol.* **2008**, *30*, 235.

Kürschner, H.; Parolly, G. The Epipterygio-Riccietum frostii ass. nov.: Ecology and Life Strategies of an Ephemeral Bryophyte Community in Western Turkey. *Lindbergia* **1999a**, *24*, 84–92.

Kürschner, H.; Parolly, G. *Tortula brevissima* Schiffn., a New Record for the Moss Flora of Turkey. *Lindbergia* **1998**, *23*, 110–112.

Kürschner, H.; Parolly, G. *Warnstorfia sarmentosa* (Wahlenb.) Hedenäs. In: New National and Regional Bryophyte records, 13. *J. Bryol.* **2006b**, *28*, 154.

Kürschner, H.; Ramezani, K. E. *Helodium blandowii*, a Noteworthy New Record to the Bryophyte Flora of Iran. *Rostaniha* **2005**, *6* (1), 33–36.

Kürschner, H.; Shumilovskikh, L.; Djamali, M.; De Beaulieu, J. L. A Late Holocene Subfossil Record of *Sphagnum squarrosum* Crome (Sphagnopsida, Bryophyta) from NW Iran. *Nova Hedwigia* **2015**, *100* (3, 4), 373–381.

Kürschner, H.; Tonguç, Ö.; Yayintaş, A. First record of *Hedwigia stellata* (Musci, Hedwigiaceae) from Turkey. *Fragmenta Floristica et Geobotanica* **1997**, *42*, 586–588.

Lazarenko, A. S. Materials on brioflora of Middle Asia. *Zhurn. Inst. Bot. Akad. Nauk Ukr. SSR* **1938**, *26–67*, 191–216.

Lyubarskaya, L. B. *Concept of Mosses of Azerbaijan*; Inst. Bot. Akad. Nauk Azerb. USSR: Baku, 1986.

Mamatkulov, U. K. *Mosses of Darvaz Range*; Donish: Dushanbe, 1975; p 100.

Mamatkulov, U. K. Mosses of the Ramit Reserve. *Izv. Otd. Biol. Nauk Akad. Nauk Tadjikskoj SSR* **1967**, *3* (28), 83–86.

Mamatkulov, U. K. *The Flora of the Bryophytes of the Tadjik SSR*; Vol. 1; Donish: Dushanbe, 1990; p 320.

Mamatkulov, U. K.; Baitulin, I. O.; Nesterova, S. G. *Bryophytes of Middle Asia and Kazakhstan*; Almaty, 1998; p 232.

Mammadova, A.; Erata, H.; Atamov, V.; Batan, N. New Moss Records from Azerbaijan. *Anatolian Bryol.* **2012**, *7* (2), 70–76.

Manafzadeh, S.; Staedler, Y. M.; Conti, E. Visions of the past and dreams of the future in the Orient: The Irano-Turanian Region from Classical Botany to Evolutionary Studies. *Biol. Rev. Camb. Philos. Soc.* **2017**, *92* (3), 1365–1388.

Manakyan, V. A. A Review of Bryoflora of Armenia. In *Problemy Bryologii v SSSR*; Abramov, I. I., Ed.; Nauka: Leningrad, 1989b; pp 157–168.

Manakyan, V. A. Contributions to the Bryoflora of Forests of Zangezur. In *Tezisy Dokladov Nauchn. Konf. Sotrudnikov Aspirantov*; Mulkidjanyan, Ya. I., Ed.; Izd. Akad. Nauk Arm SSR: Erevan, 1966a; pp 18–20.

Manakyan, V. A. Contributions to the bryoflora of Platanum forest in Armenia. *Biol. Zhurn. Armenii* **1966b**, *19* (4), 92–96.

Manakyan, V. A. Contributions to the Bryoflora of Zangezur. II. *Biol. Zhurn. Armenii* **1969**, *22* (11), 78–86.

Manakyan, V. A. Contributions to the Bryoflora of Zangezur. III. *Biol. Zhurn. Armenii* **1971b**, *24* (2), 84–90.

Manakyan, V. A. Contributions to the Bryoflora of Zangezur. IV. *Biol. Zhurn. Armenii* **1971c**, *24* (7), 33–39.

Manakyan, V. A. Contributions to the Bryoflora of Zangezur. V. *Biol. Zhurn. Armenii* **1971d**, *24* (11), 115–117.

Manakyan, V. A. Contributions to the Moss Flora of the Caucasus Minor. *Novosti Sist. Nizsh. Rast.* **1986a**, *23*, 241–244.

Manakyan, V. A. Genus *Sphagnum* L. in Armenia. In *Materialy VII*; Osipyan, L. L., Ed.; Erevansk. Gos. Univ.: Erevan, 1986b; p 54.

Manakyan, V. A. Mosses of Armenian Forests. *J. Bot. Taxon. Geobot.* **1990**, *101* (5, 6), 305–309.

Manakyan, V. A. Mosses of Forests of Megri. *Biol. Zhurn. Armenii* **1975**, *28* (9), 70–76.

Manakyan, V. A. Mosses of High Mountains of Megri. *Biol. Zhurn. Armenii* **1976a**, *29* (4), 30–36.

Manakyan, V. A. *Mosses of South Zangezur*; Ph.D. Thesis; Bot. Inst. Akad. Nauk Arm SSR: Erevan, 1970; p 28.

Manakyan, V. A. *Mosses of South-East Armenia*; Izd. Akad. Nauk Arm. SSR: Erevan, 1989a; p 313.

Manakyan, V. A. Mosses of the Lower Mountain Belt of Megri. *Uchenye Zapiski Erevanskogo Gos. Univ.* **1974**, *n 1*, 137–142.

Manakyan, V. A. Mosses of the Main Forest Areas in Armenia. In: *Flora, Rastitelnost i Rastitelnye Resursy Arm SSR*; Arm. Otd. Vses. Bot. Obshchestva: Erevan, 1988; Vol. 11, pp 64–95.

Manakyan, V. A. New Data of Bryoflora of Armenia. *Uchenye Zapiski Erevanskogo Gos. Univ.* **1976b**, *n 3*, 127–131.

Manakyan, V. A. New Data on Bryoflora of Zangezur and Armenia. *Biol. Zhurn. Armenii* **1983**, *36* (10), 859–862.

Manakyan, V. A. New Genera and Species for Moss Flora of Armenia. *Uchenye Zapiski Erevanskogo Gos. Univ.* **1971a**, *1*, 96–102.

Manakyan, V. A. New Records of Bryophytes in Armenia. *Biol. Zhurn. Armenii* **1979**, *32* (6), 588–591.

Manakyan, V. A. On Altitudinal Distribution of Mosses of South Zangezur. In: Teterevnikova-Babayan, D. N., Ed.; *Tezisy Dokladov IV*; Erevansk. Gos. Univ.: Erevan, 1972; pp 77–81.

Manakyan, V. A. Results of Bryological Studies in Armenia. *Arctoa* **1995**, *5*, 15–34.

Manayyan, V. A. Preliminary Data on Bryoflora of Khustup Mountains. In *Materialy III Zakavkazskoj Konferenzii po Sporovym Rasteniyam*; Nukhizrishvili, I. G., Ed.; Mezniereba: Tbilisi, 1968; pp 272–276.

Maulood, B. K.; Rasheed, Y. A. A Study on the Moss Flora of Iraq. In *Proceedings of the 3rd Scientific Conference, University of Salahaddin*; Arbil, Iraq, June 3–4, 1997; *Zanco*, Special Issue 2, pp 67–86.

Müller, F. A. Erstes Verzeichniss ardinischer Laubmoose, Wie Auch Derjenigen Welche Von Meinem Freunde Herrn Fleischer Bei Smyrna Aufgefunden Worden Sind, nebst Beschreibungen und Abbildungen einiger neuer Arten. *Flora* **1829**, *12*, 385–396.

Naghinezhad, A. R.; Saeidi, M. S.; Norouzi, M.; Faridi, M. Contribution to the Vascular and Bryophyte Flora as well as Habitat Diversity of the Boujagh National Park, N. Iran. *Rostaniha* **2006**, *7* (2), 83–105.

Nicholson, W. E. Mosses from the Caspian and Black Sea regions. *J. Bryol.* **1920**, *23*, 90, 91.

Noroozi, J.; Talebi, A.; Doostmohammadi, M.; Rumpf, S. B.; Linder, H. P.; Schneeweiss, G. M. Hotspots within a Global Biodiversity Hotspot—Areas of Endemism are Associated with High Mountain Ranges. *Scientific Reports* **2018**, *8* (1), 10345.

Oren, M.; Uyar, G.; Ezer, T.; Gözcü, M. C. New and Noteworthy Bryophyte Records for Turkey and Southwest Asia. *Telopea* **2017**, *20*, 97–104.

Özenoğlu-Kiremit, H.; Keçeli, T. An Annotated Check-List of the Hepaticae and Anthocerotae of Turkey. *Cryptogam. Bryol.* **2009**, *30*, 343–356.

Özenoğlu-Kiremit, H.; Sukatar, A.; Gökler, İ. Studies on the Hornworts and Liverworts Flora of Antalya. *Turk. J. Bot.* **2007**, *31*, 529–537.

Parolly, G.; Qasımov, T.; Kürschner, H. *Riccia lamellosa* Raddi (Marchantiophyta): A New Record for Azerbaijan. *Cryptogam. Bryol.* **2018**, *39* (2), 163–168.

Qasımov, T.; Kürschner, H.; Parolly, G. Three New Moss Records from the Bozqır Yaylası Area of Azerbaijan. *Herzogia* **2016**, *29* (2), 805–809.

Rechinger, K. H. *Flora Aegaea: Flora der Inseln und Halbinseln des Agaischen Meeres*; Denkschriften der Akademie der Wissenschaften, Wien, Mathematisch-Naturwissenschaftliche Klasse: Wien, 1943a; Vol. 105 (1).

Rechinger, K. H. *Flora Iranica: Flora des Iranischen Hochlandes und der Umrahmenden Gebirge*; Graz, Austria, 1963.
Rechinger, K. H. Neue Beiträge zur Flora von Kreta. *Denkschriften der Akademie der Wissenschaften, Wien, Mathematisch-Naturwissenschaftliche Klasse* **1943b**, *105* (1, 2).
Sapegin, A. Contributions to the Bryoflora of Caucasus. *Vestnik Tiflisskogo Bot. Sada* **1910**, *16*, 17–21.
Schiffner, V. Beiträge zur Kenntnis der Bryophyten von Persien und Lydien. *Österreichische Botanische Zeitschrift* **1908**, *58*, 225–232.
Schiffner, V. Bryologische Fragmente LXII–LXV. *Österr. Bot. Zeitschr.* **1910**, *60*, 431–436.
Schiffner, V. Bryophyta aus Mesopotamien und Kurdistan, Syrien, Rhodos, Mytilini und Prinkipo. *Ann. Naturhist. Mus. Wien* **1913**, *27*, 472–504.
Schiffner, V. Einige Materialien zur Moosflora des Orients. *Österr. Bot. Zeitschr.* **1901**, *51*, 156–161.
Schiffner, V. Hepaticae. In *Ergebnisse einer botanischen Reise in das Pontische Randgebirge im Sandschak Trapezunt*; Handel-Mazzetti, H. v., Ed.; Naturhistorisches Museum: Wien, 1909; pp 133–141.
Schiffner, V. Musci Bornmülleriani. *Österreichische Botanische Zeitschrift* **1897**, *47*, 125–132.
Schiffner, V. Ueber die von Sintenis in Türkisch-Armenien gesammelten Kryptogamen. *Österreichische Botanische Zeitschrift* **1896**, *46*, 274–278.
Schornherst, R. O. Mosses of Stone Mountain, Georgia. *Bryologist* **1945**, *48* (1), 29–33.
Shaw, J.; Renzaglia, K. Phylogeny and Diversification of Bryophytes. *Am. J. Bot.* **2004**, *91* (10), 1557–1581.
Shirzadian, S. Additions to the Moss Flora of Iran. *Yushania* **1989**, *6* (4), 17–20.
Shirzadian, S. Five New Records of Mosses to the Bryophyte Flora of Iran. *Phytomorphology* **2011**, *61* (3, 4), 68–71.
Shirzadian, S. Systematic Studies on Moss Flora (Bryophytes) of Iran with an Introduction to their Applied Aspects. *Res. Reconstr.* **1996**, *30*, 50–54.
Shirzadian, S.; Darzikolaei, S. A.; Uniyal, P. L. Seven New Records of Mosses in Iran. *Telopea* **2014**, *17*, 393–401.
Shirzadian, S.; Kumar, S. S. A Report on The Moss Flora of Northern Iran. *Iran. J. Bot.* **1994**, *6* (2), 177–184.
Shirzadian, S.; Sotoodehnia, S.; Kianmehr, H. Four New Records for the Moss Flora of Iran. *Rostaniha* **2000**, *1* (4), 11–22.
Størmer, P. Iranian Plants Collected by Per Wendelbo in 1959. VI. Mosses (Musci). *Aar-bok Univ. Bergen Mat.-Naturvitensk. Ser.* **1963**, *11*, 1–34.
Størmer, P.; Wendelbo, P.; Arnell, S. V. *Iranian Plants Collected by Per Wendelbo in 1959: VI. Mosses (Musci)*; Norwegian Universities Press, 1983.
Takhtajan, A. *Floristic Regions of the World*; Univ. California Press, 1986.
Takhtajan, A. L., Ed. *Flora of Armenia*, Vol. 1; Izd. Akad. Nauk Arm. SSR: Yerevan, 1954.
Tigishvili, K. Ecocoenotic Distribution of Bryoflora in Lower Forest Belt of Lagodekhi Reserve. *Proc. Georgian Acad. Sci. Biol. Ser. B* **2006**, *4*, 61–66.
Tregubov, A.; Tregubov, V. Les Bryophytes des Forêts Caspiennes du Nord de l'Iran. *Bulletin de la Faculté Forestière, Université de Tehran* **1989**, *17*, 1–30.
Ursavaş, S.; Çetin, B. Contribution to the Moss Flora of Kizildağ (Isparta) National Park in Turkey. In *Current Progress in Biological Research*; Silva-Opps, M., Ed.; Intech: 2013.

Uyar, G.; Çetin, B. A New Check-List of The Mosses of Turkey. *J. Bryology* **2004**, *26* (3), 203–220.

Vondracek, M. Some New Mosses from Iraq Collected by Hadac, E. *Rull. Soc. Amis Sci. Lett. Poznan, Ser. D* **1965**, *6*, 117–122.

Walther, K. Beiträge zur Moosflora Westanatoliens I. *Mitteil. Staatsinst. Allgemeine Bot. Hamburg* **1967**, *12*, 129–186.

Walther, K. Beiträge zur Moosflora Westanatoliens II. *Mitteil. Staatsinst. Allgemeine Bot. Hamburg* **1970**, *13*, 167–180.

Walther, K. Die epiphytischen Moosgesellschaften des Nif Dağ bei Izmir, Westanatolien. *Documents Phytosociologiques* **1979**, *4*, 943–950.

Walther, K. Zur Moosvegetation der *Liquidambar*-Wälder Südwest-Anatoliens. *Phytocoenologia* **1975**, *2*, 13–18.

Walther, K.; Leblebici, E. Die Moosvegetation des Karagöl-Gebietes im Yamanlar Dağ nördlich Izmir. *Monograph Facult. Sci. Ege Univ.* **1969**, *10*, 3–48.

Weinmann, J. A. Supplement II ad Syllabus Muscorum Frondosorum Hucusque in Imperio Rossico Collectorum. *Bull. Soc. Imp. Nat. Moscou* **1846**, *18*, 429–489.

Wendelbo, P.; Rechinger, K. H. In *Flora Iranica*; Rechinger, K. H., Ed.; Akademische Druck- und Verlagsanstalt: Graz, 1990; Vol. 165, pp 13–57.

Wettstein, R. V. Beitrag zur Flora des Orientes. Bearbeitung der von Dr. A. Heider im Jahre in Pisidien und Pamphylien gesammelten Pflanzen. *Sitzungsberichte Akad. Wissensch. Wien* **1889**, *98*, 348–398.

Yayıntaş, A. N.; Erdağ, A. Some Mosses from Ihlara Valley. *J. Faculty of Science, Ege University* **1995**, *18* (1), 1–7.

Yayintaş, O. T. A Contribution to the Bryophyte Flora of Southwestern Turkey: Bryophyte flora of Yılanlı Mountain (Muğla, Turkey). *Eur. Asian J. Biosci.* **2009**, *3*, 29–39.

Yayıntaş, Ö. T.; Yayıntaş, A. N. Bryofloristic Records of Mosses in Kayseri–Yahyalı Hacer Forest. *Phytol. Balcanica* **2010**, *16*, 3.

Zare, H.; Akbarinia, M.; Hedenäs, L.; Maassoumi, A. A. Eighteen Mosses from the Hyrcanian Forest Region New to Iran. *J. Bryol.* **2011**, *33* (1), 62–65.

Zündorf, H. J. *Hygrohypnum eugyrium* und *Hyocomium armoricum*, zwei bemerkenswerte Arten neu für Georgien und das Kaukasus-Gebiet. *Herzogia* **2011**, *24*, 163–166.

CHAPTER 5

PTERIDOPHYTES OF THE IRANO-ANATOLIAN HOTSPOT

HASAN YAŞAYACAK*

Department of Biodiversity, General Directorate of Nature Conservation and National Parks, Ankara, Türkiye

*Corresponding authors

ABSTRACT

This chapter focuses on the distribution models and problems of Pteridophytes in the Irano-Anatolian hotspot. This region has a low diversity of Pteridophytes in comparison to other hotspots. The most important factor affecting the distribution and abundance of pteridophytes in the region is the relatively dry climate of the region and the resulting moisture deficit. Because of its location, the Irano-Anatolian hotspot is closely related to the Mediterranean and Caucasus hotspots, which are its neighboring regions. Pteridophytes are predominantly distributed in the peripheral regions near the Mediterranean and Caucasus hotspots. This situation can be explained by the possible effects of the Caucasus humid climate and the Mediterranean temperate climate on the hotspot. In order to create the Pteridophyte distributions and categories of the hotspot, a regional pteridophyte inventory was made and lists were created using literature, scientific publications, and databases. Climate change, habitat loss, and human impacts were evaluated to understand regional issues for pteridophytes in the hotspot. The possible destructive effects of climate change in the region and the scarcity of pteridophyte endemics for the region, their problems, and possible potentials are discussed.

Biodiversity Hotspot of the Irano-Anatolian Region. T. Pullaiah (Ed.)

DOI: 10.1201/9781779645760-5

5.1 INTRODUCTION

The British ecologist Norman Myers first published the biodiversity hotspot thesis in 1988. Myers, Conservation International, and collaborators later revised estimates of remaining primary habitat and defined the hotspots formally as biogeographic regions with at least 1500 species of "endemic" vascular plants native to the hotspot, where natural vegetation has at least 70% loss in the hotspot (Myers et al., 2000). There are currently 36 recognized biodiversity hotspots (Figure 5.1).

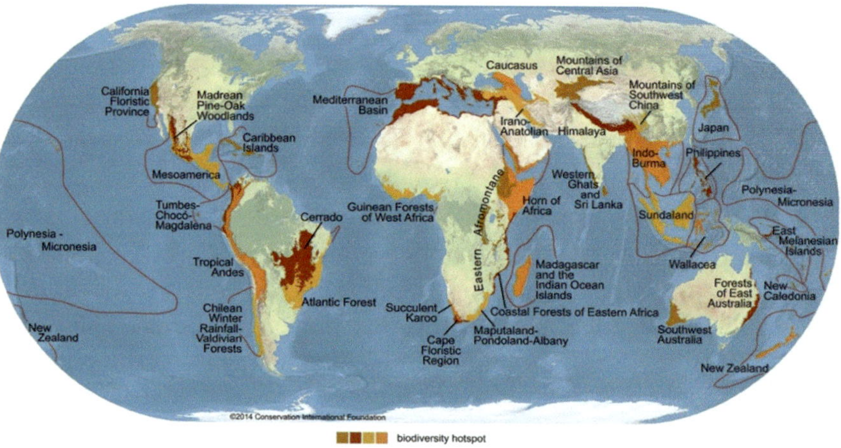

Conservation International (conservation.org) defines 35 biodiversity hotspots — extraordinary places that harbor vast numbers of plant and animal species found nowhere else. All are heavily threatened by habitat loss and degradation, making their conservation crucial to protecting nature for the benefit of all life on Earth.

FIGURE 5.1 The world's biodiversity hotspots.

Source: Conservation International. https://creativecommons.org/licenses/by-sa/4.0/

5.2 A BRIEF HISTORY OF PTERIDOPHYTES IN IRANO-ANATOLIAN HOTSPOT

Members of Pteridophytes in this region have been studied by several botanists since the 17th century; Alexandr A. Grossheim, Alexandr Fomin, A. Wilhelm Roth, J.F. Nicolaus Bornmüller, Pierre Edmond Boissier, Benjamin Balansa, Augistin Pyramus DC., J. Baptiste Bory, Olof Swartz, Thomas Walter, Manoog Haradjan, John Sibthorp, H. Handel-Mazzetti, Edward Newman, H. Wilhelm Schott, C. Christoph Schmidel, J.H. Friedrich Link, Peter Forsskal, William Hudson, C.A. Julius Milde, and Carl Tobey are notable botanists. P.H. Davis made a great contribution to the flora of the region by compiling these studies with his work "Flora of

Turkey and the East Aegean Islands" (Davis, 1965–1985). The last volume of this book was prepared by Güner et al. (2000). Khoshraves et al. (2009) and Fraser-Jenkins and Parris (2020) have significantly contributed to the pteridophyte floras of Iran and Türkiye, respectively, by presenting updated checklists for each region.

5.3 BIODIVERSITY IN THE IRANO-ANATOLIAN HOTSPOT

The Irano-Anatolian hotspot occupied an area of 899,773 km². The region encompasses a significant portion of central and eastern Turkey, a minor portion of southern Georgia, portions of Azerbaijan and Armenia, the northeast of Iraq, the Zagros Mountains region, the GNU Biosphere Reserve, and the Alborz Mountains region, extending to the Kopetdag Mountains in Turkmenistan. The total number of plant species of the area is 6000. This hotspot has 2500 endemic plant species, 10 species of mammals, 12 species of reptiles, 30 species of fish, and 2 species of endemic amphibians. There is a total of 90 species of fish, 116 species of reptiles, 18 species of amphibians, and 140 species of mammals. The protected area of this region is 56,193,000 km² and its population density is 58 inhabitants per square kilometer (Sayyad et al., 2021).

The region includes huge dry mountain steppes stretching from west to east in the northern part (Eastern Anatolian Mountains, Southern Alborz Mountains, and Kopet Mountain) and from north to south in the southern part (Zagros Mountains). These plateaus lie between the three major remnants of the ancient Tethys Sea: the Mediterranean, the Black Sea, and the Caspian. In terms of its geological structure, the hotspot has a mixture of ancient massifs, folded sedimentary rocks, and recent igneous rocks. The southern part is dominated by limestone blocks mostly from the Paleozoic Age. Between the mountain ranges, there are broken plateaus covered with alluvial soil. Various parts of the hotspot are seismically active. Here, the altitudes are generally 1,000 m to 4,000 m. Some areas, such as the foothills of Mount Kopet and the western Zagros Mountains, are as low as 300 m. Some peaks exceed 5,000 m, including the extinct volcanoes of Mount Ararat in Türkiye (5,165 m) and Mount Damavand (5,671 m) in Iran. The plateaus of Anatolia, Armenia, and Western Iran range from 800 to 2000 m and are characterized by primary lowland steppes, which are peak steppes that are never covered by any forest or shrubland (because these areas are very dry and have poor soils); the mountainous areas around the plateaus are covered with dry forests or secondary steppes.

There are three large closed basins (Konya Closed Basin, Van Basin, and Lake Urumiyeh Basin) in the north of the hotspot; these include several small Tethys lakes surrounded by primary lowland steppes. The mountains served as both a refuge and a corridor between the Eastern Mediterranean and Western Asia during the movements of the ancient Tethys Sea and its remnants. Approximately 10 million years ago, the Turkish and Arabian plates advancing northwards collided with Asia, forming the mountains and plateaus of the Iran-Anatolia region. This separated the Tethys Sea from the Indian Ocean, creating the Mediterranean Sea. The remnants of the Tethys on the Iranian Anatolian plateaus dried up and refilled several times. These movements made the hotspot an area of continuous dispersal and local speciation, and many endemic species, especially native plant species, appeared at this time (Ekim et al., 2000). The Salt Lake Basin, Lake Van, Lake Urumiyeh, and many other small lakes in Türkiye and Iran, which are centers of endemism for aquatic species of the hotspot, are particularly important in representing the biodiversity of the Tethys. During long periods of glaciation, this region was invaded by many invasive species from Northern Europe and Asia. Subsequently, periods of drought disrupted the ranges of many species and effectively isolated European-Siberian forest and meadow species, so much so that they became restricted to riparian forests along mountain valleys or forests in the mountain lakes of the hotspot.

The climate is continental; summers are hot, and winters are cold. Much of the region experiences extremely cold winters, with extreme temperatures falling as low as –40°C. Summers are dry with an average temperature of 20°–30°C, with temperatures reaching up to 48°C. Annual rainfall varies between 100 mm and 1000 mm, and most of it falls in winter and spring. Winter precipitation occurs mostly in the form of snow and can remain until June–July at high altitudes. Precipitation is quite low in central Türkiye and the southeastern Zagros Mountains; it varies between 100 to 400 mm per year. The Iran-Anatolia Hotspot is limited to the western part of the Iran-Turanian Floristic Region, which extends eastward to include the Caspian Sea, the Hindu Kush, and the arid lands surrounding the spurs of the Western Himalaya (Takhtajan, 1986). The main habitat in the hotspot is montane forest-steppe, supporting oak-dominated (*Quercus* spp.) deciduous forests in the west and south (Anatolia and Zagros mountains) and juniper (*Juniperus* spp.) forests in the east (Earth's southern slopes, Alborz Mountains, and Mount Kopet). Forest steppe mainly occurs at altitudes up to 700 m and the forest border (1900 to 2200 m). Primary forests have been largely destroyed, and mountainous secondary steppes now dominate the landscape, impoverished by dry forests. A relatively broad zone of subalpine and alpine vegetation covers the mountain peaks above the forest boundary (Eken et al., 2004; Zohary, 1973).

5.4 ECOREGIONS IN IRANO-ANATOLIAN HOTSPOT

According to Irano-Anatolian hotspot, there are 7 distinct ecoregions presented in Table 5.1. These ecoregions constitute the most sensitive areas of the region. Deforestation, human pressure, and therefore habitat loss, constitute the most important threat factors. Economic and agricultural measures to reduce human-nature conflict will reduce threats. It would be beneficial to increase protected areas containing buffer zones.

TABLE 5.1 Ecoregions of Irano-Anatolian (Atlas for the End of the World)

1. Central Anatolian steppe	9,703 km^2 remnant habitat	6,942 km^2 protected areas
2. Central Anatolian steppe and woodlands	64,767 km^2 remnant habitat	27,640 km^2 protected areas
3. Eastern Anatolian steppe and woodlands	76,961 km^2 remnant habitat	22,671 km^2 protected areas
4. Eastern Anatolian montane steppe	160,277 km^2 remnant habitat	33,435 km^2 protected areas
5. Elbruz Range Forest and steppe	47,372 km^2 remnant habitat	7,692 km^2 protected areas
6. Kopet Dag woodlands and forest and steppe	35,592 km^2 remnant habitat	8,783 km^2 protected areas
7. Zagros mountains forest and steppe	37,710 km^2 remnant habitat	71,165 km^2 protected areas

5.5 PTERIDOPHYTE TAXA IN IRANO-ANATOLIAN HOTSPOT

5.5.1 EURO-SIBERIAN REGION (WEST PALEARCTIC SECTION) (26)

Asplenium adiantum-nigrum subsp. *woronowii* (An.: Anatolia, Ar.: Armenia, Ir.: Iran), *Asplenium cuneifolium* (An.: Anatolia, Ar.: Armenia), *Asplenium septentrionale* (An., Ar., Ir.), *Athyrium distentifolium* (An., Ar.), *Athyrium filix-foemina* (An., Ar., Ir.), *Cystopteris fragilis* subsp. *alpina*

(An.: Anatolia, Ir.: Iran), *Dryopteris caucasica* (An., Ir.), *Dryopteris filix-mas* (An., Ar., Ir.), *Dryopteris affinis* subsp. *affinis* (An., Ir.), *Dryopteris dilatata* (An., Ir.), *Equisetum arvense* (An., Ar., Ir.), *Equisetum fluviatile* (An., Ar.), *Equisetum hyemale* (An., Ar.), *Equisetum palustre* (An., Ar., Ir.), *Gymnocarpium dryopteris* (An.: Anatolia), *Huperzia selago* (An., Ar.), *Onoclea struthiopteris* (An., Ar., Ir.), *Polypodium vulgare* (An., Ar., Ir.), *Polystichum aculeatum* (An., Ar., Ir.), *Polystichum braunii* (An., Ir.), *Polystichum lonchitis* (An., Ar., Ir.), *Polystichum woronowii* (An., Ir.), *Salvinia natans* (An., Ar., Ir.), *Selaginella helvetica* (An., Ar.), *Thelypteris palustris* (An., Ar., Ir.), *Woodsia alpina* (An., Ar., Ir.).

5.5.2 MEDITERRANEAN REGION (2)

Asplenium sagittatum (An.), *Dryopteris pallida* (An., Ir).

5.5.3 IRANO-ANATOLIAN REGION (2)

Asplenium lepidum subsp. *haussknechtii* (An., Ir.), *Hemionitis persica* (An., Ar., Ir.).

5.5.4 COSMOPOLITAN (13)

Adiantum capillus-veneris (An., Ar., Ir.), *Asplenium adiantum-nigrum* subsp. *adiantum-nigrum* (An., Ar., Ir.), *Asplenium ruta-muraria* (An., Ar., Ir.), *Asplenium trichomanes* subsp. *trichomanes* (An., Ar., Ir.), *Asplenium trichomanes* subsp. *quadrivalens* (An., Ir.), *Asplenium viride* (An., Ar., Ir.), *Azolla filiculoides* (An., Ir.), *Botrychium lunaria* (An., Ar., Ir.), *Cystopteris fragilis* subsp. *fragilis* (An., Ar., Ir.), *Cystopteris fragilis* subsp. *dickieana* (An., Ar., Ir.). *Equisetum ramosisismum* (An., Ar., Ir.), *Ophioglossum vulgatum* (An., Ar., Ir.), *Pteridium aquilinum* (An., Ar., Ir.).

5.5.5 PLURIREGIONAL/UNKONWN REGION (8)

Anogramma leptophylla (An., Ir.), *Asplenium ceterach* (An., Ar., Ir.), *Asplenium scolopendrium* (An., Ar., Ir.), *Equisetum telmatei* (An., Ar., Ir.), *Hemionitis marantae* (An., Ar., Ir.), *Hemionitis pteridioides* (An., Ar., Ir.), *Pteris cretica* (An., Ir.), *Pteris vittata* (An., Ir.).

The taxa number of pteridophytes in the phytogeographic regions of Irano-Anatolian are given in Figure 5.2.

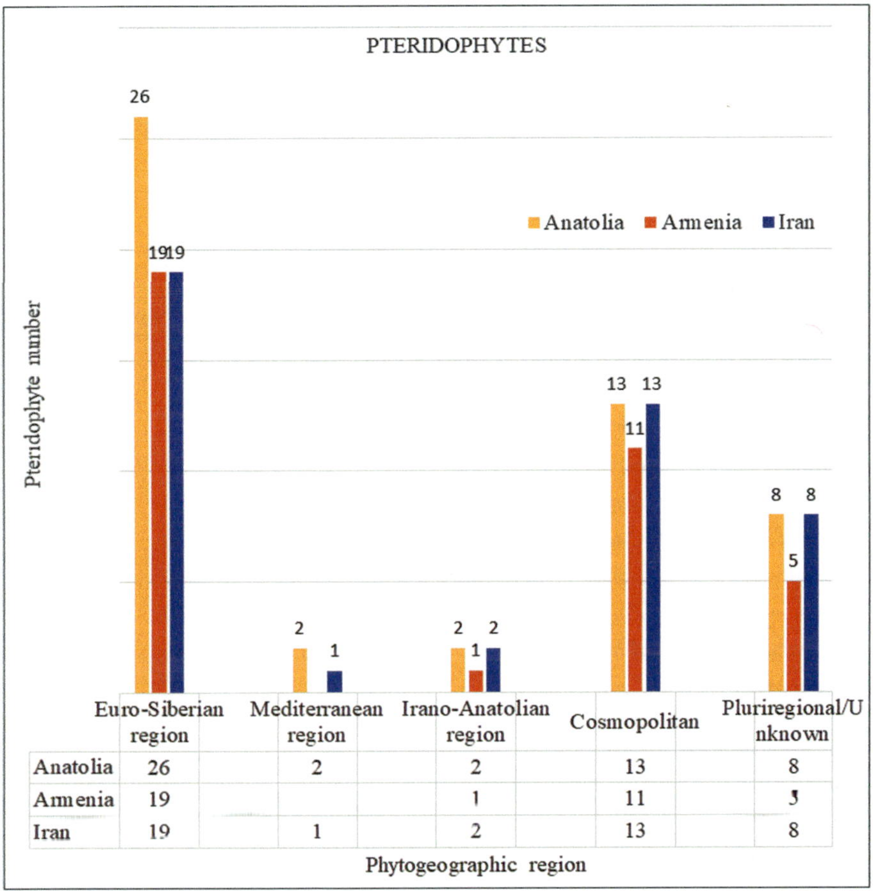

	Euro-Siberian region	Mediterranean region	Irano-Anatolian region	Cosmopolitan	Pluriregional/Unknown
Anatolia	26	2	2	13	8
Armenia	19		1	11	5
Iran	19	1	2	13	8

FIGURE 5.2 Pteridophyte taxon diversity in the phytogeographic regions of Irano-Anatolian hotspot.

5.6 FLORISTIC COMPOSITION OF PTERIDOPHYTES IN THE IRANO-ANATOLIAN HOTSPOT

The pteridophyte flora of the hotspot comprises 51 taxa, distributed across 13 families and 22 genera (Table 5.2). Notably, Aspleniaceae, Dryopteridaceae, Pteridaceae, Woodsiaceae, and Equisetaceae dominate in terms of family size, collectively representing 80% of the region's ferns. The species exhibiting high distribution and relative abundance primarily belong to the dominant genera such as *Asplenium*, *Dryopteris*, *Polystichum*, *Equisetum*. Out of the total of 51 taxa, 15 exhibit exceptionally high relative abundance and distribution in the hotspot. These are the 15 species: *Adiantum capillus-veneris*, *Asplenium adiantum-nigrum*, *Asplenium ceterach*, *Asplenium ruta-muraria*, *Asplenium scolopendrinum*, *Asplenium trichomanes*, *Athyrium filix-foemina*, *Cystopteris fragilis*, *Dryopteris filix-mas*, *Equisetum arvense*, *Equisetum ramosissimum*, *Equisetum telmateia*, *Polypodium vulgare*, *Polystichum aculeatum*, *Pteridium aquilinum*.

TABLE 5.2 Pteridophyte Composition in Hotspot

Families	Genera	Number of taxa
Aspleniaceae	*Asplenium*	12
Dennstaedtiaceae	*Pteridium*	1
Dryopteridaceae	*Dryopteris, Polystichum*	9
Equisetaceae	*Equisetum*	6
Lycopodiaceae	*Huperzia*	1
Onocleaceae	*Onoclea*	1
Ophioglossaceae	*Botrychium, Ophioglossum*	2
Polypodiaceae	*Polypodium*	1
Pteridaceae	*Adiantum, Anogramma, Hemionitis, Pteris*	7
Salviniaceae	*Azolla, Salvinia*	2
Sellaginellaceae	*Selaginella*	1
Thelypteridaceae	*Thelypteris*	1
Woodsiaceae	*Athyrium, Cystopteris, Gymnocarpium, Woodsia*	7
Total number of taxa		51

Irano-Anatolian hotspot, one of the 36 recognized hot spots on Earth, accounts for only 0.4% of the reported global pteridophyte species richness, which stands at 13,117 according to the latest data. This low percentage suggests that the Irano-Anatolian hot spot exhibits relatively limited pteridophyte diversity compared to other designated hotspots. Figure 5.3 presents the species diversity of Pteridophyte families across each region in Irano-Anatolian hotspot.

5.7 HABITAT DIVERSITY OF PTERIDOPHYTES IN THE IRANO-ANATOLIAN HOTSPOT

In pteridophytes, habitat diversity plays an important role in the number and distribution of the species. Habitat differences in phytogeographic regions provide the opportunity for different genetic resources to coexist. This is beneficial in measuring biological diversity. Whether it is climate change or human impact, the biggest factor that will affect future biodiversity will be habitat degradation.

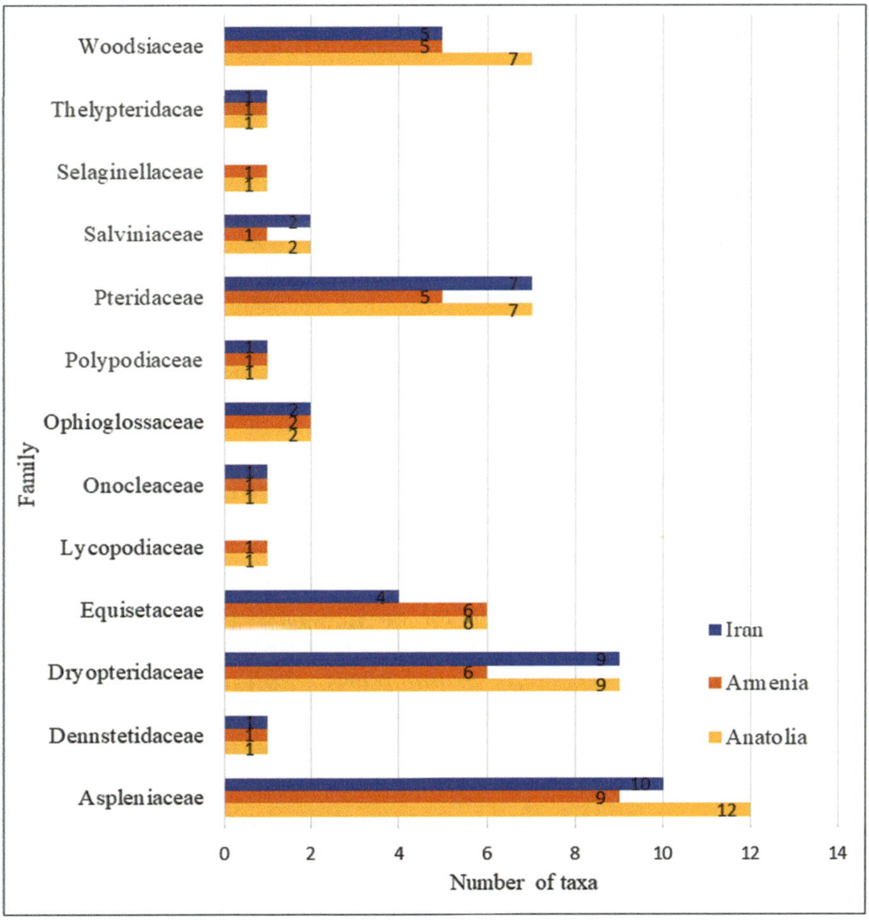

FIGURE 5.3 Species diversity of Pteridophyte families across each region in Irano-Anatolian hotspot.

5.7.1 HYDROPHILUS (WATER-LOVING) AQUATIC PTERIDOPHYTES: 2 (4%)

Azolla filiculoides, *Salvinia natans*

5.7.2 HYGROPHILUS (MOISTURE-LOVING) PTERIDOPHYTES: 12 (24%)

Adiantum capillus-veneris, Botrychium lunaria, Equisetum arvense, Equisetum fluviatile, Equisetum hyemale, Equisetum palustre, Equisetum ramosissimum, Equisetum telmatei, Huperzia selago, Ophioglossum vulgatum, Selaginella helvetica, Thelypteris palustris

5.7.3 PETROPHILUS (ROCK-LOVING) PTERIDOPHYTES: 19 (37%)

Anogramma leptophylla, Asplenium adiantum-nigrum subsp. *woronowii, Asplenium ceterach, Asplenium cuneifolium, Asplenium lepidum* subsp. *haussknechtii, Asplenium ruta-muraria, Asplenium septentrionale, Asplenium trichomanes* subsp. *trichomanes, Asplenium trichomanes* subsp. *quadrivalens, Asplenium viride, Athyrium distentifolium, Cystopteris fragilis* subsp. *fragilis, Cystopteris fragilis* subsp. *alpina, Cystopteris fragilis* subsp. *dickieana, Gymnocarpium dryopteris, Hemionitis marantae, Heminiotis persica, Heminiotis pteridioides, Woodsia alpina*

5.7.4 SCOTOPHILUS (SHADE-LOVING) PTERIDOPHYTES: 18 (35%)

Asplenium adiantum-nigrum subsp. *adiantum-nigrum, Asplenium sagittatum, Asplenium scolopendrium, Athyrium filix-foemina, Dryopteris affinis, Dryopteris caucasica, Dryopteris dilatata, Dryopteris filix-mas, Dryopteris pallida, Onoclea struthiopteris, Polystichum aculeatum, Polystichum braunii, Polystichum lonchitis, Polystichum woronowii, Polypodium vulgare, Pteridium aquilinum, Pteris cretica, Pteris vittata*

Besides aquatic pteridophytes, many pteridophyte species share a habitat preference characterized by humid, shady, and rocky terrain. This habitat type, less favored by other vascular plant groups, enhances the competitiveness of pteridophytes in nature. Habitat diversity rates from Irano-Anatolian are given in Figure 5.4.

5.8 RARE AND ENDEMIC TAXA OF PTERIDOPHYTES IN THE IRANO-ANATOLIAN HOTSPOT

Rare and endemic species are genes formed under isolated conditions. Isolated conditions begin with geographical separation. If the special climate-dependent microclimate and different ecosystem or habitat diversity supports this situation, biological diversity will gradually increase. When isolation conditions occur, different genes influence each other and cause gene diversity, which forms the basis of biological diversity. Around the world, pteridophytes reproduce by spores and spread over wide areas, and they are less affected by isolation than other vascular plant groups. This situation causes rare and endemic taxa, especially pteridophytes, to be less common than other vascular plants. Below are some of the rare Pteridophyte species in the hotspot (Table 5.3).

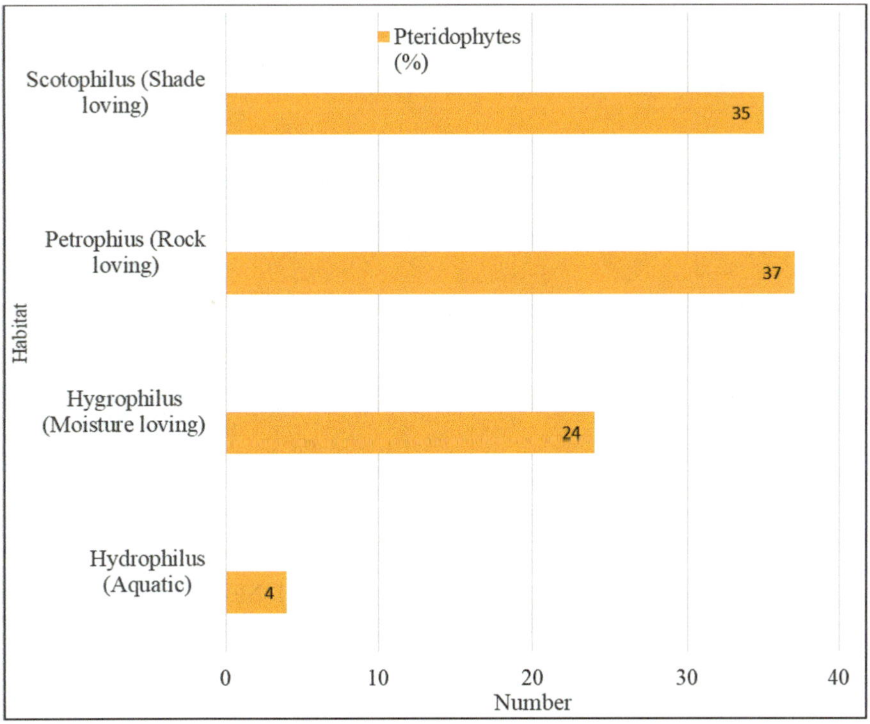

FIGURE 5.4 Habitat diversity rates of Irano-Anatolian hotspot pteridophytes.

1. ***Asplenium adiantum-nigrum*** L. subsp. ***woronowii*** (Christ) Fraser-Jenk. in New Sp. Syndr. Indian Pteridol.: 44 (1997).

   900–3000 m in Armenia (Gabrieljan and Greuter, 1984), calcareous rocks in alpine zone, 1500–2300 m N.E. Anatolia, Caucasus, Transcaucasus, N. Iran. Euro-Sib. Element (Davis, 1988).

2. ***Asplenium lepidum*** C. Presl subsp. ***haussknechtii*** (God. & Reut. ex Milde) Brownsey in Bot. J. Linn. Soc. 72: 261 (1976). Syn: *A. haussknechtii* God. & Reut. ex Milde, Fil. Eur. 78 (1867).

   In calcareous rock crevices, 1400–1500 m. Scattered Anatolia, Lebanon, Latakia, Transcaucasus, Iraq, Iran (Davis, 1965) (Figure 5.5).

FIGURE 5.5 *Asplenium lepidum* subsp. *haussknechtii* (Photo H.Yaşayacak).

3. ***Asplenium sagittatum*** (DC.) Bange in Bull. Mens. Soc. Linn. Lyon 21: 84 (1952). *P. sagittata* (DC.) Guinea & Heywood, Collect. Bot. Barcelona, 4(2): 246 (1954). Syn: *Scolopendrium sagittatum* DC. in Lam. & DC., Fl. Fr. ed. 3, 5: 238 (1815); *Scolopendrium hemionitis* Schwarz non L. Ie.: Guinea & Heywood, loc. cit., p. 245 f. 1.

   Shady limestone rocks, 450–1500 m. rare species. *S. Anatolia*, Mediterranean area. Mediterranean element (Davis, 1965) (Figure 5.6).
4. ***Dryopteris caucasica*** (A. Hr.) Fraser-Jenkins & Corley in Brit. Fern Gaz. 10:221 (1972). Syn: *Aspidiuum affine* Fisch. & Mey. ex Hohen. in Bull. Soc. Nat. Mosc. 11:240 (1838) non-Blume (1830!) nec Wall. (1828). *A. caucasicum* A. Br. in Flora 24:707 (1841).

FIGURE 5.6 *Asplenium sagittatum* (Photo H. Yaşayacak).

Humid forests. N. Türkiye, Caucasus, N. Iran. Hyrcano-Euxine element (Davis, 1988) (Figure 5.7).

5. ***Polystichum woronowii*** Fomin in Mon. Jard. Bot. Tiflis 18: 21, f. (1910).

Beech and other shady woods (Bobrov et al., 1934). Rarely NE Türkiye, Transcaucasus, N. Iran. Hyrcano-Euxine element (Davis, 1965).

Due to the dry climate characteristics of the hot region, pteridophytes are numerically low, and the rates of rare and endemic species are also low. The increase in the population of pteridophytes as they approach the warm regions of the Mediterranean and the Caucasus is a clear indicator of this situation.

Of the 51 pteridophyte taxa in the region, 5 taxa are rare and subendemic for the hot region. Although the 10% rare and subendemic rate in pteridophytes is relatively low, the 2500 other vascular endemic plant taxa in the Irano-Anatolian hot region show the biodiversity importance of the region.

FIGURE 5.7 *Dryopteris caucasica* (Photo H. Yaşayacak).

5.9 THE THREATS FACTORS AND CONSERVATION ACTIVITIES IN IRANO-ANATOLIAN HOTSPOT

As in numerous biodiversity hotspots, the Irano-Anatolian region continues to experience adverse effects on the forests, which harbor dense populations of pteridophytes, impacting the livelihoods and well-being of approximately 180 million people, the majority of whom are affected by poverty. A clear indicator of this is deforestation. Inevitably, the decrease in forest areas, intense human pressure, and the conversion of these areas to agricultural land will directly affect the number and distribution of pteridophytes. Forests and wetlands with the highest density of pteridophytes are under constant pressure around the world because they also provide important ecosystem services for human life, such as clean water supplies, shelter, and subsistence. The ravaging impact of humans on ecosystems is rather due to the permissibility of ever-increasing human activities on these fragile ecosystems.

Ensuring the protection of natural habitats in hotspots is closely linked to poverty alleviation efforts, which involve implementing sociological and economic measures and raising awareness through educational activities. Another critical concern with potentially devastating effects on pteridophytes is climate change, as many fern species are expected to remain vulnerable to its impacts. The survival of most pteridophytes is contingent upon soil and air moisture conditions, making prolonged periods without rainfall a significant source of stress for many species.

As a precautionary measure, efforts to combat climate change should be intensified, and the establishment of protected areas, particularly deciduous forests, should be expanded, with a focus on safeguarding pteridophyte habitats.

Nevertheless, prejudices regarding the perceived lack of interest in pteridophytes, their relatively low economic value, and insufficient information about them often lead regional policies and administrations to overlook the importance of pteridophytes. One anticipated consequence of climate change is the reduction in the range of mesic pteridophytes and the expansion of certain xerophytic pteridophytes' distributions. This expectation holds particular relevance for the Irano-Anatolian hotspot.

One expectation is that the potential effects of the *Subsaharo-Sindian* geographical region, which predominates in the southern part of the hotspot, will intensify within the Irano-Anatolian hotspot. This may lead to an increase in the numerical abundance and distribution of taxa specific to the *Subsaharo-Sindian* region within this hotspot.

The perception that pteridophytes are not visually striking and therefore unimportant often results in this group of plants being overlooked in botanical gardens and research institutes. Recognizing the landscape potential of pteridophytes and conducting research on them is crucial. Active conservation measures should include prioritizing rare pteridophytes in parks, arboretums, and botanical gardens, as well as raising awareness about these species.

Invasive species, which pose a major problem for native species all over the world, pose a danger to sensitive ecosystems. *Azolla filiculoides* (An., Ir.) from the Salviniaceae family (Davis, 2000), which has an invasive character in the region, is an invasive alien species and has spread to many wetlands in a very short time. It mainly harms the aquatic ecosystem (Figure 5.8).

The species was introduced into Northern Iran during the 1980s in order to enhance the nitrate content of rice-cultivated fields. It soon escaped to neighboring wetlands and became an aggressive invasive plant, apparently due to its capability of vegetative reproduction. Now, it has invaded almost all freshwater ecosystems and rice fields, and is even found in the saline waters of the Caspian Sea. It has taken possession of the natural habitats of many native aquatic plants of the northern wetlands, covering the water surface almost completely. It filters the sunlight required for photosynthesis by phytoplankton and the submerged aquatic flora, being a hazard for the survival of other aquatic organisms and a threat to fish production (Khoshravesh et al., 2009).

FIGURE 5.8 *Azolla filiculoides* (Photo H. Yaşayacak).

5.10 CONCLUSIONS AND RECOMMENDATIONS

The primary challenge for biodiversity is the destruction of habitats and climate change caused by human activities, particularly due to the increased mobility of people. Regarding pteridophytes, it is essential to raise awareness among individual people and government officials about the importance of endemic and rare species. It is crucial to establish policies that are focused on protecting these species and guaranteeing their long-term survival. Employing economic and agricultural measures to address human-nature conflict will significantly decrease the degree of risks. Expanding the quantity of protected areas, in addition to creating buffer zones, might provide advantages.

The most significant threat to wetlands will be the devastating effects of global warming caused by climate change. This phenomenon will particularly impact the aquatic pteridophytes in the region, such as *Equisetum hyemale*, *E. fluviatile*, *E. palustre*, *Salvinia natans*, and *Thelypteris palustris*. Furthermore, it is necessary to recognize another significant problem faced by these species due to the existence of invasive alien species, including *Azolla filiculoides*. Protection measures against existing or potential threats to pteridophytes in the region should be developed. Efforts should be made to increase the number of protected areas or expand their ranges specifically for pteridophytes. This would be beneficial for regional development, as it would help eliminate prejudices about pteridophytes and bring them to the forefront in botany and economy, along with conservation measures.

By increasing floristical studies on Pteridophytes in the Irano-Anatolian hotspot, we can better understand the region's potential in terms of the exact diversity of the group. The discovery of the local endemic *Polystichum asiae-minoris* taxon in Northwest Türkiye in 2020 indicates that the region needs to be more thoroughly investigated (Tunçkol et al., 2020). Floristic studies and concentrating research on the ferns in the area should be given priority. Determining the diversity of ferns in the Irano-Anatolian hotspot will also provide important data about Pteridophytes in the Caucasus and Mediterranean areas.

ACKNOWLEDGEMENTS

I would like to thank Mr. Barış BANİ and Mr. Avni YILDIZBAŞ, who contributed to the language and grammar editing of the article.

KEYWORDS

- **climate change threats**
- **habitat loss impacts**
- **Mediterranean–Caucasus influence**
- **moisture limitation**
- **pteridophyte diversity**
- **regional distribution patterns**

REFERENCES

Atlas for the End of the World. Hotspots: Irano-Anatolian. https://atlas-for-the-end-of-the-world.com (accessed April 21, 2026).

Bobrov, E. G.; Fedchenko, B. A.; Fomin, A. V.; Iljin, M. M.; Krishtofovich, A. N.; Komarov, V. L.; Yuzepchuk, S. V. In *Flora SSSR*; Komarov, V. L., Ed.; Izdatel'stvo Akademii Nauk SSSR: Leningrad, 1934; Vol. 1.

Christian, M. Biodiversity Hotspots: A Shortcut for a More Complicated Concept. In *Global Ecology and Conservation*; University Quebec: Canada, 2015; pp 297–309.

Davis, P. H. *Flora of Turkey and the East Aegean Islands (Supplement)*; Vol. 10; Edinburgh University Press: Edinburgh, England, 1988; pp 3–8.

Davis, P. H. *Flora of Turkey and the East Aegean Islands (Supplement)*; Vol. 11; Edinburgh University Press: Edinburgh, England, 2000; p 4.

Davis, P. H. *Flora of Turkey and the East Aegean Islands*; Vols. 1–9; Edinburgh University Press: Edinburgh, England, 1965–1985.

Davis, P. H. In *Flora of Turkey and the East Aegean Islands*; Edinburgh University Press: Edinburgh, 1965; Vol. 1, pp 31–63.

Eken, G.; Evans, M.; Karataş, A.; Balkız, Ö.; Karaçetin, E.; Kılıç, T.; Özbağdatlı, N.; Neumann-Denzau, G.; Gem, E.; Karataş, A. Irano-Anatolian. In *Hotspots Revisited: Earth's Biologically Richest and Most Endangered Terrestrial Ecoregions*; Mittermeier, R. A.; Robles Gil, P.; Hoffmann, M.; Pilgrim, J.; Brooks, T.; Mittermeier, C. G.; Lamoreaux, J.; da Fonseca, G. A. V., Eds.; Conservation International: 2004; pp 43, 46, 286–295.

Ekim, T.; Koyuncu, M.; Vural, M.; Duman, H.; Aytaç, Z.; Adıgüzel, N. *Turkish Plants Red Data Book*; Doğal Hayatı Koruma Derneği: Ankara, 2000.

Fraser-Jenkins, C. R.; Parris, B. S. Pteridophytes of Turkey—A revised check-list and their relationship to Europe and the West Himalaya. *Indian Journal of Forestry* **2020**, *43*, 263–284.

Gabrieljan, E. C.; Greuter, W. A revised catalogue of the Pteridophyta of the Armenian SSR. *Willdenowia* **1984**, *14,* 1.

Güner, A.; Özhatay, N.; Ekim, T.; Başer, K. H. C. *Flora of Turkey and the East Aegean Islands*; Edinburgh University Press: Edinburgh, 2000; Vol. 11.

Khoshravesh, R.; Akhani, H.; Eskandari, M.; Greuter, W. Ferns and Fern Allies of Iran. *Rostaniha* **2009**, *10* (Supplement 1), 1–130.

Myers, N.; Mittermeier, R. A.; Mittermeier, C. G.; Fonseca, G. A.; Kent, J. Biodiversity hotspots for conservation priorities. *Nature* **2000**, *403*, 853–858.

Sayyad, S. I.; Sohrab, A.; Afshin, A. S. Modeling Habitat Suitability of the Red-backed Shrike (*Lanius collurio*) in the Irano-Anatolian Biodiversity Hotspot. *Journal of Zoological Research* **2021**, *3* (2), 2. http://ojs.bilpublishing.com/index.php/jzr (accessed April 21, 2026).

Takhtajan, A. L. *Floristic Regions of the World*; University of California Press: Berkeley, London, 1986; p 522.

Tunçkol, B.; Yaşayacak, H.; Liang, Z.; Aksoy, N.; Zhang, L. *Polystichum asiae-minoris* (Dryopteridaceae), a New Fern from Kastamonu, Turkey. *Phytotaxa* **2020**, *447*, 296–300.

Zohary, M. *Geobotanical Foundations of the Middle East*; Gustav Fischer Verlag: Stuttgart, 1973.

CHAPTER 6

FISH DIVERSITY IN THE IRANO-ANATOLIAN HOTSPOT REGION

HOSSEIN MOSTAFAVI[1*], AZAD TEIMORI[2*], and NAJMEH TABASINEZHAD[3]

[1]*Department of Biodiversity and Ecosystem Management, Environmental Sciences Research Institute, Shahid Beheshti University, Tehran, Iran*

[2]*Department of Biology, Faculty of Sciences, Shahid Bahonar University of Kerman, Iran*

[3]*Department of Biology, Faculty of Sciences, Shahid Bahonar University of Kerman, Kerman, Iran*

*Corresponding authors

ABSTRACT

The Irano-Anatolian biodiversity hotspot region is a critical area of biodiversity, home to various endemic and migratory fish species. The hotspot is home to around 90 freshwater fish species, nearly 30% of which are found nowhere else, particularly in isolated lakes and rivers with closed basins. Many of these endemic species are at risk worldwide, such as the flathead trout (*Salmo platycephalus*), which is classified as Critically Endangered. Despite its importance, the region's aquatic ecosystems are increasingly imperiled by habitat destruction, water pollution, climate change, and invasive species. Dam construction, irrigation-related water diversion, and urbanization have fragmented habitats and disrupted migratory pathways. Concurrently, agricultural and industrial pollution has degraded water quality, while climate change exacerbates stressors by

Biodiversity Hotspot of the Irano-Anatolian Region. T. Pullaiah (Ed.)

DOI: 10.1201/9781779645760-6

altering hydrological regimes and thermal conditions. Effective conservation strategies—such as protected area establishment, ecosystem restoration, and sustainable fisheries management—are urgently needed. Success hinges on international cooperation and local community engagement to ensure the long-term persistence of the region's fish diversity. Addressing these challenges is critical to safeguarding freshwater ecosystems and their ecological functions.

6.1 INTRODUCTION TO THE IRANO-ANATOLIAN HOTSPOT REGION

The Irano-Anatolian region spans seven nations, including parts of Turkmenistan, Iraq, Turkey, Georgia, Iran, Azerbaijan, and Armenia and encompasses multiple ecoregions (Figure 6.1). This biodiversity hotspot is home to a wide variety of unique plant and animal species. A wide range of plant species are found within the area, including 400 plant species whose distribution is mainly limited to the Anatolian Diagonal. The region has some of the world's unique plant species, such as the salt plants found in Anatolia and Iran. The section of the region located in Turkey has a variety of orchids endemic to the country, most of which are classified as threatened. Individuals collect the orchids in vast quantities and utilize them to make traditional drinks.

In addition to its rich flora, the Irano-Anatolian region supports a diverse array of wildlife. It serves as a critical breeding ground for threatened bird species such as the white-headed duck and the great bustard. The area is also home to around 140 mammal species, 10 of which are endemic, including the critically endangered Asiatic cheetah. Several freshwater fish species are also unique to the region.

However, deforestation poses a major threat to this biodiversity, leading to habitat loss and endangering many of the region's unique species. Conservation efforts are crucial to preserving this ecologically significant area.

The Irano-Anatolian hotspot region is a unique and ecologically significant area located at the convergence of three major biodiversity regions: the Irano-Turanian, the Mediterranean, and the Euro-Siberian regions (Mittermeier et al., 2004). This area's intersection creates a zone of immense ecological complexity, as it serves as both a meeting point and a transition area between distinct biomes, each with its unique environmental and climatic conditions. Consequently, the landscape harbors extraordinary biological diversity, with a high proportion of species exclusive to the region.

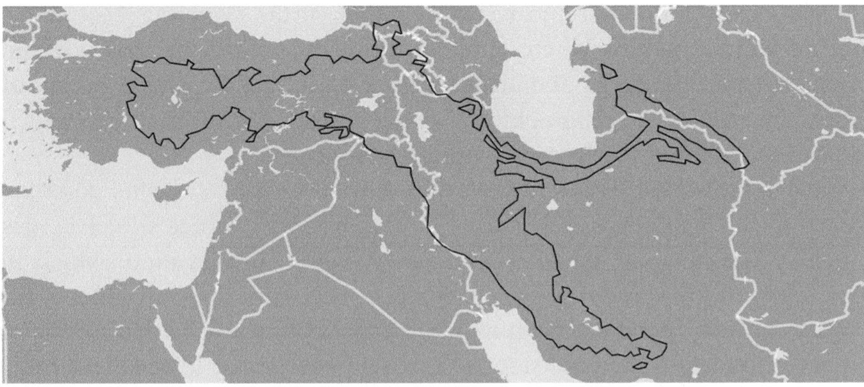

FIGURE 6.1 The geographic overview of the Irano-Anatolian region. It stretches across seven nations, which include parts of Turkmenistan, Iraq, Turkey, Georgia, Iran, Azerbaijan, and Armenia.

Source: https://www.cepf.net/our-work/biodiversity-hotspots/irano-anatolian.

According to Myers et al. (2000), this area ranks as one of Earth's most ecologically important biodiversity hotspots. Therefore, it is considered as a focal point for conservation efforts and scientific study. Spanning parts of the Middle East—including Iran, Anatolia (modern-day Turkey), the Caucasus, and parts of neighboring countries like Iraq, Armenia, Azerbaijan, and Georgia—the Irano-Anatolian hotspot is not only ecologically significant but also culturally and historically important. The variety of ecosystems here is extraordinary, ranging from dry desert expanses to lush, temperate forests and alpine meadows. Each ecosystem supports a unique set of species, many of which are endemic to the region (Brooks et al., 2006). This high level of endemism, along with the presence of many rare and endangered species, places the Irano-Anatolian hotspot among the top priorities for biodiversity conservation globally.

At the core of the hotspot lies the convergence of three biogeographic realms—the Irano-Turanian, the Mediterranean, and the Euro-Siberian regions—which influence the climate, vegetation, and wildlife of the area. The Irano-Turanian region, located in the central part of the hotspot, is characterized by its arid climate and is home to a variety of drought-tolerant plant species and specialized fauna adapted to harsh, dry conditions.

To the west, the Mediterranean influence brings more temperate and humid conditions, supporting diverse plant and animal communities, including broad-leaved forests and scrublands. The Euro-Siberian region to the north contributes to the cooler and wetter climates of the Caucasus Mountains and northern Anatolia, where temperate forests, grasslands, and wetland ecosystems are found in abundance (Vural et al., 2014). The interplay of these contrasting biogeographic influences results in the rich ecological mosaic that defines the Irano-Anatolian hotspot.

The region's ecological diversity is further enhanced by its complex topography. The Zagros and Taurus Mountain ranges, which stretch across much of the hotspot, create a dramatic variation in elevation, which in turn generates a variety of microclimates. This varied terrain allows for a wide range of habitats, from alpine meadows and coniferous forests at higher altitudes to lowland steppes and semi-deserts at lower elevations. Such variation enables the development of numerous ecological niches, providing opportunities for a high level of species diversification and endemism (Sichieri et al., 2011). For example, the Iranian Plateau is known for its arid steppe ecosystems, whereas the Anatolian Peninsula supports more temperate forests and shrublands. These distinctive habitats play a crucial role in fostering species adapted to different climatic conditions and physical landscapes.

Beyond the natural world, the Irano-Anatolian region holds significant historical and cultural importance. Over the millennia, the area has been home to various civilizations and empires, including the ancient Persians, the Hittites, and the Ottoman Empire. These civilizations thrived in part because of the region's ecological wealth, which provided fertile lands for agriculture and access to critical resources such as water and timber. Additionally, the region's location at the crossroads of Europe, Asia, and Africa made it a hub for trade and cultural exchange, influencing both the development of human societies and the movement of species across continents (Sassoon, 2003).

Despite its immense ecological value, the Irano-Anatolian hotspot faces numerous conservation challenges. Rapid urbanization, agricultural expansion, and infrastructure development have resulted in widespread habitat destruction, fragmentation, and degradation. Climate change is also exacerbating these threats, leading to shifts in ecosystems and the distribution of species. Invasive species, introduced both intentionally and unintentionally, further threaten native biodiversity by outcompeting local species and disrupting ecological balance. As a result, the Irano-Anatolian region is now experiencing significant biodiversity loss, with many species facing the risk of extinction (Heath, 2013).

The ecological importance of the Irano-Anatolian hotspot, combined with the urgency of its conservation challenges, makes it an area of critical focus for both local and international conservation efforts. Global conservation organizations, governmental agencies, and local stakeholders have recognized the need to protect this unique region and its biodiversity. Strategies such as the establishment of protected areas, habitat restoration, sustainable land management practices, and increased research efforts have been implemented to address these challenges (Tearfund, 2010). The region has also been included in international biodiversity conservation initiatives, such as the Critical Ecosystem Partnership Fund (CEPF), which supports conservation efforts in biodiversity hotspots around the world (CEPF, 2010).

In conclusion, the Irano-Anatolian hotspot is an area of exceptional ecological and cultural significance, home to a vast array of endemic species and unique habitats. Its rich biodiversity, coupled with the challenges it faces due to human activity and climate change, underscores the importance of protecting this critical region. Ongoing research, conservation, and sustainable development efforts will be essential in safeguarding the ecological treasures of the Irano-Anatolian hotspot for future generations.

6.2 GEOGRAPHICAL EXTENT AND ECOLOGICAL FEATURES

The Irano-Anatolian hotspot region is located at the crossroads of several large biogeographic realms, making it a transitional zone between European, Asian, and Mediterranean ecosystems (Eken et al., 2004). It is bordered by the Caspian Sea to the north, the Zagros Mountains to the east, and the Mediterranean Sea to the southwest. The region encompasses parts of modern-day Turkey, Iran, Iraq, Armenia, Azerbaijan, and Georgia, making it a diverse and complex landscape in terms of both topography and climate (Güner, 2017).

The hotspot is predominantly characterized by mountainous terrain, with the Zagros and Taurus Mountain ranges being two of the most prominent features (Teimori et al., 2016). These mountain ranges serve as important ecological barriers and contribute to the formation of distinct biomes, such as temperate forests, steppe grasslands, and arid deserts

(Yazdani and Parsa, 2015). Due to the varied topography, the Irano-Anatolian region includes a range of microhabitats, allowing for high species diversity (Sichieri et al., 2011).

The climatic conditions within the region are equally varied, ranging from the harsh, dry conditions of the Iranian Plateau to the more temperate and Mediterranean climates along the coasts of Turkey and the Caspian Sea (Karami et al., 2014). This variation in climate has contributed to the development of diverse ecosystems, from lush forests and wetlands in the north to arid deserts and scrublands in the south and east. Such environmental heterogeneity is a key factor driving the high levels of endemism and biodiversity observed in the region (Zohary, 2004).

6.3 BIODIVERSITY SIGNIFICANCE

The Irano-Anatolian biodiversity hotspot harbors exceptional levels of species endemism, supporting numerous plant and animal taxa that are globally restricted to this region (Mittermeier et al., 2004). Over 6,000 plant species have been documented in the region, of which approximately one-third are endemic (Ekim et al., 2000). The plant life in the Irano-Anatolian region includes a mix of Mediterranean, temperate, and desert species, making it an area of particular interest for botanists and ecologists (Vural et al., 2014). The region is known for its diverse tree species, including oaks, pines, and junipers, as well as a variety of wildflowers, shrubs, and herbs that are of both ecological and medicinal importance (Khachikian, 2011).

Beyond its renowned plant endemism, the Irano-Anatolian ecoregion sustains critical populations of specialized fauna, many exhibiting unique adaptations to local environmental conditions. Mammals such as the Persian leopard, the wild goat, and the Caspian seal, along with numerous bird species, amphibians, and reptiles, populate the region's varied ecosystems (Sillero-Zubiri et al., 2004). The region serves as a migratory route for birds traveling between Europe, Asia, and Africa (Salewski et al., 2009). The diversity of species in the Irano-Anatolian hotspot is not limited to well-known animals, but also includes numerous smaller, lesser-known species that play important roles in maintaining ecosystem balance (Arlettaz et al., 2016).

The hotspot is particularly significant for the study of evolutionary biology, as it has been identified as a key center of speciation. The varied habitats, along with the physical barriers provided by the mountain ranges and seas, have led to the development of distinct populations and species that have adapted to specific microenvironments (Harrison, 2008). The Irano-Anatolian region is therefore a natural laboratory for studying processes of diversification and adaptation, and it has been a focus of research into biogeography and evolutionary theory (Bittkau et al., 2011).

6.4 HISTORICAL AND CULTURAL CONTEXT

The Irano-Anatolian region has a rich historical and cultural legacy that spans thousands of years. It has been home to many ancient civilizations, such as the Hittites, Urartians, and Persians, as well as several powerful empires, including the Achaemenid and Seljuk Empires (Baines and Malek, 1999). The area's diverse ecological conditions provided not only sustenance and shelter for these civilizations but also influenced the development of agriculture, trade, and settlements (Lindner, 2004).

Archaeological evidence suggests that the region has been inhabited for millennia, with significant developments in agriculture, metallurgy, and architecture. The establishment of early agricultural practices in this region is thought to have contributed to the spread of farming technologies to other parts of the world, influencing the development of human societies. Additionally, the Irano-Anatolian region's location at the crossroads of Asia, Europe, and Africa made it a major hub for trade, culture, and the exchange of ideas (Sassoon, 2003).

In terms of biodiversity, the historical relationship between humans and the environment in the Irano-Anatolian region is complex. While many traditional farming and pastoral practices have coexisted with the region's ecosystems, human activities have also had significant impacts on the environment. Deforestation, land conversion for agriculture, and the introduction of non-native species have led to habitat loss and fragmentation, threatening the region's biodiversity (McNeely et al., 2005).

6.5 FISH DIVERSITY IN THE IRANO-ANATOLIAN HOTSPOT REGION

The Irano-Anatolian hotspot region is not only ecologically significant for its terrestrial biodiversity but also for its remarkable aquatic ecosystems. The region's fish diversity is particularly notable, given the wide range of freshwater habitats, from large rivers and lakes to smaller streams and wetlands, which support the biological diversification of the species. These aquatic ecosystems are located in the Mediterranean and the Irano-Turanian zones, providing unique habitats for native and migratory fish species. The fish diversity of the region is shaped by several factors, including its diverse climates, varied topography, and complex hydrological systems. This section explores the key fish species found in the Irano-Anatolian hotspot, highlighting the ecological importance of these species and the challenges they face due to habitat loss and environmental change.

6.5.1 FRESHWATER ECOSYSTEMS AND FISH HABITATS

The Irano-Anatolian region is home to several major river systems, each supporting a distinct aquatic environment for fish. Some of the most notable rivers in the region include the Euphrates, Tigris, Kura, and Aras rivers, along with several smaller tributaries. These rivers and their associated catchment areas are rich in biodiversity and provide important habitats for numerous fish species. In addition to the large rivers, the region also contains numerous freshwater lakes, reservoirs, and wetlands, which further contribute to the overall richness of the aquatic ecosystem.

The Mediterranean and Irano-Turanian influences play a significant role in shaping the fish populations of the region. The Mediterranean climate, particularly in the western parts of Anatolia and along the coastlines, supports temperate aquatic ecosystems with a high diversity of fish species. In contrast, the central and eastern parts of the region, influenced by the arid conditions of the Iranian Plateau, are home to more specialized, drought-tolerant species adapted to variable and often harsh environmental conditions (Güner, 2017).

Fish species in the region occupy a range of habitats, from fast-flowing rivers and streams to slower-moving lakes and marshes. The variation in water quality, temperature, and oxygen levels, combined with differing altitudes and latitudes, creates a diverse array of ecological niches. For example, the Zagros and Taurus Mountain ranges create barriers that affect the distribution of freshwater species, leading to the development of distinct populations in isolated catchments (Ramsar Convention, 2008). These varying ecological conditions have fostered the evolution of a wide range of species, many of which are endemic to the region.

6.5.2 *KEY FISHES OF THE IRANO-ANATOLIAN REGION*

The fish fauna of the Irano-Anatolian hotspot is composed of several unique and diverse species, many of which are of high conservation importance. The region supports native species and migratory species that use its river systems as breeding grounds or stopover points along migratory routes. Some of the most prominent species include members of the family Cyprinidae, which are widely represented across the region, as well as a variety of other families, such as Salmonidae, Siluridae, and Esocidae.

6.5.2.1 CYPRINIDAE (CARP AND MINNOWS)

The Cyprinidae family, which includes carp, minnows, and barbs, is the most diverse group of fish in the Irano-Anatolian hotspot. Several species of freshwater carp, such as the goldfish (*Cyprinus auratus*) (Figure 6.2) and the Iranian barb (*Barbus grypus*), are important both ecologically and economically. These species thrive in rivers, lakes, and reservoirs, where they play essential roles in nutrient cycling and maintaining the health of aquatic ecosystems. The common carp, in particular, has been introduced into many freshwater habitats outside its native range but remains one of the most abundant species in the region (Ekim et al., 2000).

FIGURE 6.2 The photo of *Carassius auratus* (Cyprinidae) from Caspian Sea basin, SL = 17.4cm.

Additionally, several minnow species and small cyprinids, such as *Capoeta* species, are also endemic to the region and found in its freshwater systems. These species are particularly important as prey for larger fish and aquatic birds, thus contributing to the region's complex food webs (Brooks et al., 2006). The diversity of cyprinid species reflects the variety of aquatic environments, from fast-flowing rivers to slow-moving marshes.

6.5.2.2 SALMONIDAE (SALMON AND TROUT)

Trout species, particularly *Salmo trutta* (brown trout) (Figure 6.3), are significant to the freshwater systems of the Irano-Anatolian region, particularly in the cooler, high-altitude rivers of the Caucasus and the Anatolian plateau. Brown trout are well-adapted to the cold waters of mountain streams and lakes and are important species for biodiversity and local fisheries. The fish is an indicator species, meaning its population dynamics can reflect the health of freshwater ecosystems (Ramsar Convention, 2008). Moreover, in the upper reaches of rivers like the Tigris and Euphrates, salmon and trout populations historically supported local human populations through fishing.

FIGURE 6.3 The photo of *Salmo trutta* (Salmonidae) from norh of Iran, SL = 25.9cm.

6.5.2.3 SILURIDAE (CATFISH)

Catfish, particularly those from the *Silurus* genus, are present in many of the region's larger rivers, such as the Euphrates and Tigris, and play important ecological roles as both predators and scavengers. The Tigris catfish (*Silurus triostegus*) is one of the largest fish species in the region and is found in slow-moving waters where it feeds on smaller fish and detritus (Vidal et al., 2007). These species are an important part of the food web and are often of economic importance to local fisheries.

6.5.2.4 ENDEMIC SPECIES

In addition to more common species, the Irano-Anatolian hotspot is home to numerous endemic fish species that have adapted to the unique conditions of isolated river systems or lakes. For example, the genus *Alburnus*, which includes small cyprinids, has several species found only in particular rivers or lakes within the region. Other important examples are high diversification of endemic fishes, including loaches and killifishes (Figure 6.3), and the members of minnows or family Leuciscidae (Cyprinoformes). These species are vulnerable to habitat loss and environmental changes, making them a key focus of conservation efforts (Güner, 2017). The presence of such endemic species is a direct result of the region's diverse geography, where many isolated basins and waterways have allowed populations to evolve in relative isolation. As a result, these species are often highly specialized and vulnerable to environmental disturbances, including water pollution, overfishing, and habitat fragmentation. Below are some of the endemic and endangered fish species from this region:

1. ***Cobitis linea* (Figure 6.4 a)**
   - **Habitat and Ecology:** Prefers muddy bottoms, an uncommon habitat in this part of Iran where seasonal rain scours rivers.
   - **Reproduction:** Spawns June–July.
   - **Threats:** Water diversions severely reduce water level, especially pumping water from the aquifer which feeds the systems; drainage rehabilitation; water pollution from domestic and agricultural sources; drought in recent years; an intensive aquaculture industry which added a high nutrient load and seven exotic fish species. Some of these species are as follows: *Cyprinus carpio*, *Hypophthalmichthys molitrix*, *Ctenopharyngodon idella*, *Carassius auratus*, *Pseudorasbora parva* (Family Cyprinidae), *Gambusia holbrooki* (Family Poeciliidae), *Oncorhynchus mykiss* (Family Salmonidae).
   - **Conservation Actions:** No conservation actions have been taken.
2. ***Oxynoemacheilus tongiorgii* (Figure 6.4 b)**
   - **Habitat:** Lives in a natural spring-stream system in Southwestern Iran, Zagros Mountain. The benthic substrate typically consists of fine particulate mud or small-diameter gravel (2–10mm).
   - **Reproduction:** Not yet characterized.

- **Threats:** The aquatic system faces multiple threats, including severe water-level reduction from pump well extraction, structural alterations due to drainage rehabilitation, pollution from domestic and agricultural sources, recurrent drought conditions, and the introduction of seven non-native species through intensive aquaculture development—all of which jeopardize spring ecosystem integrity.
- **Conservation Actions:** Remains to be empirically documented.

3. ***Esmaeilius sophiae* (Figure 6.4 c–d)**
   - **Habitat and Ecology:** Found in freshwater (rivers, springs), prefers benthic zones (bottom-dwelling) in slow-moving or stagnant waters, mainly herbivorous, feeding on algae and plant matter, nonmigratory, with populations influenced by environmental stressors (temperature, salinity, flooding).
   - **Reproduction:** Life-history traits (growth rate, age at maturity, fecundity) vary depending on habitat conditions; environmental fluctuations (e.g., temperature changes) affect reproductive success.
   - **Threats:** Habitat degradation due to water pollution, agricultural runoff, and dam construction; competition with invasive species (e.g., *Gambusia affinis*); climate change exacerbating salinity and temperature fluctuations.
   - **Conservation Actions:** Habitat protection (e.g., preserving the Kor River basin and Tigris River ecosystems), pollution control to reduce agricultural and industrial contaminants, and research on population dynamics, genetic diversity, and ecological requirements; and monitoring programs to assess threats from invasive species and climate change.

4. ***Garra typhlops*** (Iran Cave Barb) **(Figure 6.4 d)**
   - **Habitat and Ecology:** Endemic to caves in Iran (within the Zagros Mountains), troglobitic (obligate cave-dwelling), fully adapted to subterranean life, blind (anophthalmic) and lacks pigmentation (depigmented), typical of cave-adapted species, likely feeds on detritus, biofilm, and small invertebrates in cave streams.
   - **Reproduction:** Unknown (no specific studies cited), but likely has low fecundity due to energy-limited cave environments, may exhibit selected traits (slow growth, delayed maturation, fewer offspring) like other cavefish.

- **Threats:** Extremely limited range, making it vulnerable to extinction, habitat destruction from groundwater extraction, pollution, or cave disturbances, climate change altering cave hydrology (e.g., drying underground streams), potential competition/predation if invasive species enter caves.

FIGURE 6.4 Some endemic fishes from different families within the Irano-Anatolian hotspot region: (A) *Alburnoides qanati* (Leuciscidae); (B) *Acanthobrama persidis* (Leuciscidae); (C) *Alburnoides qanati* (Leuciscidae), (D) *Garra typhlops* (Cyprinidae); (E) *Oxynoemacheilus persa* (Nemacheilidae); (F) *Cobitis linea* (Cobitidae) (Teimori et al., 2016). Total length for “A–D” is minimum 12.5 cm to maximum 16.5 cm, and for “E–F” is minimum 10.5cm to maximum 11.5 cm.

The other cavefish in this region of the Irano-Anatolian hotspot are *Garra lorestanensis*, *G. tashanensis*, and *Eidinemacheilus smithi* (Zagroz blind loach). Regionally, Iraqi cavefish (e.g., *Typhlogarra widdowsoni*) face similar threats.

6.5.3 FISH DIVERSITY SPECIFIED IN EACH COUNTRY IN THE REGION

6.5.3.1 IRAN

Iran is home to a diverse array of freshwater fish species, with a total of 300 recorded species. The country exhibits a rich ichthyofauna, characterized by a significant proportion of native and endemic species, reflecting its unique freshwater ecosystems.

6.5.3.1.1 Species diversity

Iran's freshwater fish diversity is influenced by a range of aquatic habitats, including rivers, lakes, and wetlands. Of the 300 species, 170 are classified as native, while 30 are endemic to Iran. Additionally, the country hosts 12 exotic species, introduced either intentionally or accidentally.

6.5.3.1.2 Conservation status

According to the IUCN Red List, the conservation status of Iran's freshwater fishes varies considerably:

- **Least Concern (LC):** 152 species.
- **Endangered (EN):** 19 species.
- **Vulnerable (VU):** 15 species.
- **Critically Endangered (CR):** 14 species.
- **Not Evaluated (NE):** 74 species.

The presence of 14 critically endangered and 19 endangered species highlights conservation concerns. Many species face threats from habitat destruction, pollution, overfishing, and climate change. Furthermore, the 74 species that remain unevaluated indicate gaps in conservation knowledge and the need for further research.

6.5.3.1.3 Threats and conservation challenges

Iran's freshwater fish populations are under increasing pressure due to several factors:

- **Water Scarcity and Habitat Degradation:** Climate change and excessive water extraction for agriculture are reducing river flow and impacting fish populations.
- **Pollution:** Industrial discharge and agricultural runoff contribute to water pollution, affecting fish health and habitats.
- **Overfishing and Illegal Fishing Practices:** Unregulated fishing leads to declining populations of commercially valuable species.
- **Invasive Species:** The introduction of exotic fish can disrupt native ecosystems by outcompeting local species.

Given these challenges, conservation efforts should focus on habitat restoration, stricter fishing regulations, and improved monitoring of fish populations.

6.5.3.2 GEORGIA

Georgia hosts 75 freshwater fish species, with a mix of native, endemic, and exotic species. The country's diverse river systems, including the Kura and Rioni rivers, provide critical habitats for many species.

6.5.3.2.1 Species diversity

Of the 75 species, a majority (55 species) are native, 6 are endemic, and 8 are classified as exotic. The country's freshwater ecosystems are influenced by both the Caspian and Black Sea drainage basins, creating a unique fish assemblage.

6.5.3.2.2 Conservation status

Georgia's fish populations are categorized as follows:

- **Least Concern (LC):** 50 species.
- **Endangered (EN):** 5 species.
- **Vulnerable (VU):** 4 species.
- **Critically Endangered (CR):** 3 species.
- **Not Evaluated (NE):** 10 species.

While most species are of least concern, the presence of critically endangered and endangered species suggests conservation challenges. Many of these species are sensitive to habitat degradation and climate change.

6.5.3.2.3 Threats and conservation challenges

- **Hydropower Development:** Dam construction alters river flow, fragmenting fish habitats and disrupting migration patterns.
- **Pollution and Deforestation:** Industrial and agricultural pollution impact water quality, while deforestation increases sedimentation in rivers.
- **Climate Change:** Altered precipitation patterns affect water availability and fish populations.

Georgia needs enhanced conservation programs, including habitat protection and sustainable water management strategies, to ensure the long-term survival of its freshwater fish species.

6.5.3.3 ARMENIA

Armenia's freshwater fish diversity is relatively low compared to its neighboring countries, with 54 recorded species. However, the country hosts some unique and ecologically significant species.

6.5.3.3.1 Species diversity

Among Armenia's 54 freshwater fish species:

- 40 species are native.
- 3 species are endemic, indicating a limited but crucial biodiversity component.
- 5 species are exotic, potentially impacting native fish populations.

6.5.3.3.2 Conservation status

The conservation status of Armenia's freshwater fish is as follows:

- **Least Concern (LC):** 35 species Endangered (EN): 4 species.
- **Vulnerable (VU):** 3 species.
- **Critically Endangered (CR):** 2 species.
- **Not Evaluated (NE):** 7 species.

Despite the relatively small number of species, the presence of endangered and critically endangered fish raises conservation concerns. Many of these species are highly sensitive to environmental changes and habitat degradation.

6.5.3.3.3 Threats and conservation challenges

- **Water Pollution:** Industrial discharge and agricultural runoff pose significant threats to Armenia's freshwater ecosystems.
- **Habitat Modification:** Dams and water diversions impact natural river flow and fish migration.
- **Overfishing:** Unsustainable fishing practices deplete fish stocks, especially in Lake Sevan.

Armenia urgently needs to implement focused conservation strategies addressing these threats through (1) enhanced water quality monitoring and enforcement, and (2) expansion of protected habitats with integrated management plans.

6.5.3.4 AZERBAIJAN

Azerbaijan has a moderate level of freshwater fish diversity, with 79 species recorded in its aquatic ecosystems. The country's rivers, lakes, and wetlands support a variety of fish species that contribute to regional biodiversity.

6.5.3.4.1 Species diversity

- 55 species are native to Azerbaijan's waters.
- Five species are endemic, highlighting unique biodiversity elements.
- Nine species are exotic, some of which may pose ecological risks.

6.5.3.4.2 Conservation status

- **Least Concern (LC):** 48 species.
- **Endangered (EN):** 7 species.
- **Vulnerable (VU):** 6 species.
- **Critically Endangered (CR):** 5 species.
- **Not Evaluated (NE):** 13 species.

The high number of endangered and vulnerable species indicates that conservation measures need to be strengthened in Azerbaijan.

6.5.3.4.3 Threats and conservation challenges

- **Wetland Degradation:** Draining of wetlands for agriculture affects fish habitats.
- **Water Pollution:** Industrial and urban waste contribute to declining water quality.
- **Illegal Fishing:** Unregulated fishing practices pose a significant risk to fish populations.

Conservation strategies should focus on habitat preservation, pollution control, and strict enforcement of fishing regulations.

6.5.3.5 TÜRKIYE

Türkiye exhibits the highest freshwater fish diversity in the region, with 424 species recorded. Its extensive river systems and lakes provide essential habitats for a wide range of species.

6.5.3.5.1 Species diversity

- 230 species are native.
- 31 species are endemic, indicating a rich evolutionary history.
- 13 species are exotic, with potential ecological consequences.

6.5.3.5.2 Conservation status

- **Least Concern (LC):** 200 species.
- **Endangered (EN):** 35 species.
- **Vulnerable (VU):** 26 species.
- **Critically Endangered (CR):** 20 species.
- **Not Evaluated (NE):** 100 species.

Türkiye has the highest number of endangered and critically endangered species, signaling the need for urgent conservation action.

6.5.3.5.3 Threats and conservation challenges

- Overfishing and habitat loss threaten fish populations.
- Pollution from agriculture and industry impacts water quality.
- Climate change alters aquatic ecosystems, affecting fish survival.

Türkiye's conservation policies should focus on sustainable fisheries management, habitat restoration, and stricter pollution control measures.

6.5.4 COMPARISON OF FISH DIVERSITY AMONG COUNTRIES

6.5.4.1 TOTAL SPECIES AND ENDEMISM

Türkiye has the highest fish diversity, with 424 species, followed by Iran with 300. The other countries—Azerbaijan, Georgia, and Armenia—have significantly fewer species, with Armenia having the lowest at 54 (Table 6.1). Endemism is highest in Türkiye (31 species), while Iran also has a notable number (30). Smaller countries like Georgia and Armenia have fewer endemic species.

6.5.4.2 CONSERVATION STATUS

Iran and Türkiye have the highest number of species in conservation concern categories, including endangered, vulnerable, and critically endangered species. Türkiye has 35 endangered species, the most among the studied countries, while Iran follows with 19. Countries like Armenia and Georgia have fewer endangered species, likely due to their smaller total species counts.

6.5.4.3 THREATS ACROSS THE REGION

- **Habitat Degradation:** All countries face threats from habitat degradation due to water scarcity, pollution, and development projects.
- **Overfishing:** More prevalent in Türkiye and Iran due to high fish diversity and economic reliance on fisheries.
- **Hydropower and Climate Change:** Particularly severe in Georgia and Armenia, affecting river ecosystems.
- **Invasive Species:** Notable in Iran and Azerbaijan, were exotic species impact native biodiversity.

TABLE 6.1 Fish Diversity and Conservation Comparison Table

Country	Total species	Native	Endemic	Exotic	LC	EN	VU	CE	NE
Iran	300	170	30	12	152	19	15	14	74
Georgia	75	55	6	8	50	5	4	3	10
Armenia	54	40	3	5	35	4	3	2	7
Azerbaijan	79	55	5	9	48	7	6	5	13
Türkiye	424	230	31	13	200	35	26	20	100

6.5.5 CONSERVATION STATUS AND THREATS

- Türkiye and Iran have the highest fish diversity and the largest number of species in conservation concern.
- Georgia and Armenia have fewer total species but still face threats from habitat destruction and climate change.
- Azerbaijan has a moderate number of species but faces challenges with wetland degradation and pollution.

6.5.6 COMPARISON OF FISH FAMILY DIVERSITY

Fish family diversity varies across Iran, Türkiye, Georgia, Armenia, and Azerbaijan, with notable differences in the number of families and the dominant taxonomic groups. Iran has the highest fish family diversity, with 43 families, followed closely by Türkiye (37 families) (Figure 6.5). In contrast, Armenia has the lowest diversity, with only 14 families recorded.

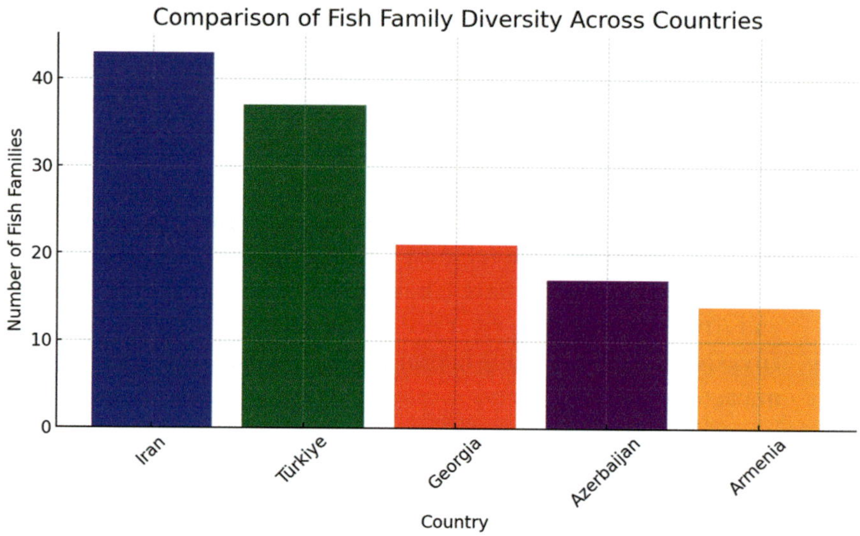

FIGURE 6.5 Comparison of fish family diversity among the countries.

The Leuciscidae family dominates in Türkiye, Georgia, Azerbaijan, and Armenia, with the highest species count in Türkiye (124 species). However, in Iran, the Cyprinidae family is the most diverse, comprising 74 species. This variation suggests that Iran's aquatic ecosystems support a different dominant group, possibly due to environmental or ecological factors (Table 6.2). Table 6.2 provides an overview of the fish family diversity, highlighting the most abundant families and those represented by a single species. All five countries have multiple fish families represented by only a single species. These least-represented families include Petromyzontidae, Carcharhinidae, Clariidae, and others, which may indicate limited distribution, specific habitat requirements, or recent introductions.

Overall, Iran and Türkiye emerge as key regions for fish family diversity, while Armenia and Azerbaijan have lower family richness. Understanding these patterns is essential for conservation efforts, particularly in managing vulnerable and endemic species.

TABLE 6.2 A Summary Table of the Fish Family Diversity Across the Five Countries

Country	Total Fish Families	Most Diverse Family (Highest Species Count)	Least Represented Families (Single Species)
Iran	43	Cyprinidae (74 Species)	Multiple (e.g., Petromyzontidae, Carcharhinidae, etc.)
Türkiye	37	Leuciscidae (124 Species)	Multiple (e.g., Adrianichthyidae, Pristigasteridae, etc.)
Georgia	21	Leuciscidae (20 Species)	Multiple (e.g., Petromyzontidae, Anguillidae, etc.)
Azerbaijan	17	Leuciscidae (22 Species)	Multiple (e.g., Petromyzontidae, Anguillidae, etc.)
Armenia	14	Leuciscidae (19 Species)	Multiple (e.g., Petromyzontidae, Clariidae, etc.)

Fish diversity varies significantly among the five studied countries—Iran, Türkiye, Georgia, Armenia, and Azerbaijan—both in terms of species richness and composition. Türkiye and Iran exhibit the highest diversity, with Türkiye hosting 424 species across 39 families and Iran having 298 species across 44 families. This suggests that these two countries provide a wide range of aquatic habitats supporting diverse fish populations (Table 6.3).

Endemism is a key factor distinguishing these regions. Türkiye has the highest number of endemic species (53), followed by Iran (21), while Georgia and Azerbaijan show no endemism, and Armenia has only one endemic species. This highlights Türkiye and Iran as crucial regions for fish conservation due to their unique species.

When considering native and exotic species, Iran (247 native species) and Türkiye (335 native species) again stand out, demonstrating a rich native biodiversity. However, both countries also have the highest number of exotic species, with 30 in Iran and 33 in Türkiye, indicating a higher level of species introduction, which could impact native ecosystems. In contrast, Azerbaijan has the lowest count of exotic species (5), suggesting minimal human-induced fish introductions.

Overall, the data underscores the importance of Türkiye and Iran as biodiversity hotspots, while Georgia, Armenia, and Azerbaijan have comparatively lower fish diversity. Conservation efforts should focus on protecting endemic species and managing exotic species introductions to maintain ecological balance.

TABLE 6.3 Fish Diversity Varies Significantly Among the Five Studied Countries

Country	Families	Total Species	Endemic Species	Native Species	Exotic Species
Iran	44	298	21	247	30
Georgia	22	75	0	64	9
Armenia	15	54	1	38	13
Azerbaijan	18	79	0	74	5
Türkiye	39	424	53	335	33

6.5.7 CONSERVATION OF FISH DIVERSITY IN THE IRANO-ANATOLIAN HOTSPOT REGION

The Irano-Anatolian hotspot region is home to a diverse and unique range of freshwater fish species, many of which are endemic to the area. However, the region's aquatic ecosystems are under significant threat from human activities, climate change, and other environmental pressures. These challenges have serious implications not only for the region's fish populations but also for the broader ecological balance. Addressing these issues is crucial for ensuring the survival of both common and endemic species, as well as the overall health of freshwater ecosystems in the region. This section examines the conservation concerns, challenges, and efforts related to fish diversity in the Irano-Anatolian hotspot. Major conservation challenges are given below.

6.5.7.1 HABITAT DESTRUCTION AND FRAGMENTATION

One of the most pressing threats to the fish populations in the Irano-Anatolian hotspot is habitat destruction, which is primarily driven by human development activities such as dam construction, water diversion for agriculture, and urban expansion. The region's freshwater habitats, including rivers, lakes, and wetlands, have been dramatically altered by the construction of dams and reservoirs. These human-made structures disrupt the natural flow of rivers, change water temperatures, and create barriers that hinder the movement and migration of fish species.

Migratory species, such as the Acipenser sturgeon, have been particularly impacted by dam construction in the Tigris and Euphrates River systems. Historically, these sturgeon species migrated upstream to spawn in the region's freshwater rivers, but the construction of large dams has blocked their migratory routes, severely affecting their population and reproductive success (Heath, 2013). The disruption of natural river flow also impacts species that rely on specific flow regimes for breeding and feeding, which can lead to population declines and the loss of biodiversity in aquatic ecosystems.

In addition to damming, water diversion for irrigation, a common practice in many parts of the region, further exacerbates the situation. These activities reduce the water volume available in rivers and lakes, leading to the drying up of vital habitats. The alteration of these ecosystems diminishes the area available for fish species to thrive, putting pressure on their survival.

6.5.7.2 WATER POLLUTION

Water pollution is another significant concern for the fish populations of the Irano-Anatolian hotspot. Agricultural runoff, industrial discharge, and untreated sewage often contaminate freshwater ecosystems, introducing harmful chemicals and toxins into the water. Pesticides and fertilizers used in agriculture contribute to nutrient pollution, leading to eutrophication—an excess of nutrients that causes oxygen depletion and harms aquatic life. Additionally, heavy metals and industrial chemicals released into the water pose direct threats to fish health and reproductive success.

Pollution also weakens the resilience of fish species to environmental changes. Many native fish species, particularly those adapted to pristine and unpolluted habitats, struggle to survive in increasingly degraded waters. For example, species of *Barbus* and *Alburnus*, which are prevalent in the freshwater systems of the Irano-Anatolian region, are especially sensitive to changes in water quality and can experience population declines due to pollution (Brooks et al., 2006).

6.5.7.3 CLIMATE CHANGE

Climate change presents a long-term and pervasive threat to the aquatic ecosystems of the Irano-Anatolian hotspot. As global temperatures rise, freshwater habitats are experiencing shifts in precipitation patterns, altered water temperatures, and changes in seasonal flows. These changes can have profound effects on fish species, which are highly sensitive to temperature and water flow conditions. For example, higher temperatures may limit the range of cold-water fish species like trout, pushing them to higher elevations or causing local extinctions in certain river systems (Güner, 2017).

Changes in precipitation patterns are also altering the seasonal flooding and water availability in rivers and lakes, which in turn impacts the reproductive cycles of many fish species. Furthermore, altered flow patterns can exacerbate the effects of habitat fragmentation, as altered river courses may prevent fish from accessing critical breeding and feeding areas.

6.5.7.4 INVASIVE SPECIES

The spread of invasive species, both aquatic and terrestrial, has become an increasing problem for the region's freshwater ecosystems. Non-native species, often introduced through human activity, can outcompete native species for food, space, and other resources. Invasive fish species may prey on native species, disrupt food webs, or carry diseases that can harm native populations. For example, the introduction of non-native carp species into freshwater systems has altered the balance of fish populations, leading to declines in native species that are less adapted to the new competition (Heath, 2013). Invasive species can also affect the physical environment by altering habitats. For example, invasive aquatic plants can change the structure of riverbeds and alter water flow, which negatively impacts native fish species that rely on specific habitat conditions for breeding and shelter.

6.5.8 CONSERVATION EFFORTS IN THE IRANO-ANATOLIAN HOTSPOT

Efforts to conserve the fish diversity of the Irano-Anatolian hotspot are critical to mitigating the impacts of these threats and preserving the region's unique aquatic ecosystems. Several conservation strategies have been implemented to address the pressing issues of habitat loss, pollution, and invasive species, although much work remains to be done.

6.5.8.1 PROTECTED AREAS AND ECOSYSTEM RESTORATION

The establishment of protected areas, including national parks, wildlife reserves, and Ramsar sites (wetlands of international importance), has been one of the key strategies to safeguard aquatic habitats and the fish species that depend on them. These protected areas help to ensure that critical habitats remain intact and free from direct human disturbance. For example, the establishment of protected wetlands and riverine areas within the Irano-Anatolian hotspot has helped conserve important fish breeding grounds and migration routes, providing a safe haven for vulnerable species (Ramsar Convention, 2008).

Ecosystem restoration efforts are also underway to rehabilitate degraded river systems and improve water quality. River restoration projects, including the removal of dams and the restoration of natural flow patterns, are being implemented in some areas to help restore fish migration routes and improve habitat conditions (Tearfund, 2010). These efforts not only benefit fish populations but also contribute to the overall health of the region's aquatic ecosystems.

6.5.8.2 SUSTAINABLE FISHERIES MANAGEMENT

Sustainable fisheries management is an essential component of fish conservation in the Irano-Anatolian region. Overfishing has long been a concern in many parts of the region, and implementing sustainable fishing practices is critical for maintaining healthy fish populations. Regulations on fishing quotas, fishing gear, and seasonal closures have been introduced in some areas to prevent overfishing and allow fish populations to recover. Local communities have also been involved in fisheries management, ensuring that conservation efforts are tailored to the specific needs of the region's ecosystems and communities (WWF, 2017).

6.5.8.3 INTERNATIONAL COLLABORATION

The importance of international collaboration in the conservation of the Irano-Anatolian region's fish diversity cannot be overstated. The region spans multiple countries, each with its own environmental policies and practices, making coordinated efforts crucial for effective conservation. Global initiatives such as the CEPF have been instrumental in supporting conservation projects in the region (CEPF, 2010). These efforts involve partnerships between governmental agencies, non-governmental organizations, local communities, and international bodies, working together to address the region's environmental challenges and ensure the protection of its unique biodiversity.

6.6 CONCLUSION

The fish diversity of the Irano-Anatolian hotspot is a valuable ecological and economic resource, yet it faces numerous threats from habitat destruction, pollution, climate change, and invasive species. Effective conservation efforts, including the establishment of protected areas, sustainable fisheries management, and international collaboration, are essential for ensuring the long-term survival of these species and maintaining the health of freshwater ecosystems. By addressing these challenges and continuing to invest in conservation strategies, it is possible to safeguard the fish diversity of the Irano-Anatolian region for future generations.

KEYWORDS

- **Caspian Sea basin**
- **conservation**
- **freshwater ecosystem**
- **habitat degradation**
- **Irano-Anatolian hotspot**
- **Zagros mountains**

REFERENCES

Arlettaz, R.; Schaub, M.; Fournier, J.; Reichlin, T. S.; Sierro, A.; Watson, J. E. M.; Braunisch, V. The Role of Biodiversity in Sustainable Development. *Ecology* **2016**, *97* (8), 1812–1821.

Baines, J.; Málek, J. *Cultural Atlas of Ancient Egypt*, Revised ed.; Oxford University Press, 1999.

Bittkau, C.; Comes, H. P.; Kadereit, J. W. Patterns of Molecular Differentiation in the Irano-Anatolian Region: Insights from a Phylogeographic Study of the Alpine Plant *Campanula rotundifolia* (Campanulaceae). *Biol. J. Linn. Soc.* **2011**, *104* (3), 518–532. https://doi.org/10.1111/j.1095-8312.2011.01742.x.

Brooks, T. M.; Mittermeier, R. A.; da Fonseca, G. A. B.; Gerlach, J.; Hoffmann, M.; Lamoreux, J. F.; Mittermeier, C. G.; Pilgrim, J. D.; Rodrigues, A. S. L. Global Biodiversity Conservation Priorities. *Science* **2006**, *313* (5783), 58–61. https://doi.org/10.1126/science.1127609.

CEPF. *Critical Ecosystem Partnership Fund: Biodiversity Hotspots*; CEPF, 2010.

Eken, G.; Evans, M.; Karataş, A.; Balkız, Ö.; Karaçetin, E.; Kılıç, T.; Özbağdatlı, N.; Neumann-Denzau, G.; Gem, E.; Karataş, A. Irano-Anatolian. In *Hotspots Revisited: Earth's Biologically Richest and Most Endangered Terrestrial Ecoregions*; Mittermeier, R. A.; Robles Gil, P.; Hoffmann, M.; Pilgrim, J.; Brooks, T.; Mittermeier, C. G.; Lamoreux, J.; da Fonseca, G. A. B., Eds.; Cemex, 2004; pp 43, 46, 286–295.

Ekim, T.; Koyuncu, M.; Dönmez, M. *Flora of Turkey and the East Aegean Islands*; Edinburgh University Press, 2000.

Güner, A. *Flora of Turkey and the East Aegean Islands*; Edinburgh University Press, 2017.

Heath, D. *Global Challenges in Biodiversity and Conservation*; Cambridge University Press, 2013.

Karami, V.; Keshavarz, M.; Vanclay, F. Climate Change and Agriculture: Impacts and Adaptive Responses in Iran. *J. Integrative Agric.* **2014**, *13* (4), 1–15. https://doi.org/10.1016/S2095-3119(13)60649-4.

Khachikian, M. *The Flora of Armenia: Diversity, Ecology, and Conservation*; Armenian Academy of Sciences: Yerevan, 2011.

Lindner, R. *The Environmental History of the Near East: Agriculture, Trade, and Settlements*; Cambridge University Press, 2004.

McNeely, J. A.; Scherr, S. J. *Ecoagriculture: Strategies to Feed the World and Save Wild Biodiversity*; Island Press, 2003.

Mittermeier, R. A.; Turner, W. R.; Larsen, F. W.; Brooks, T. M.; Gascon, C. *Hotspots Revisited: Earth's Biologically Richest and Most Endangered Terrestrial Ecoregions*; CEMEX, 2004.

Myers, N.; Mittermeier, R. A.; Mittermeier, C. G.; da Fonseca, G. A. B., Kent, J. Biodiversity Hotspots for Conservation Priorities. *Nature* **2000**, *403* (6772), 853–858.

Noroozi, J.; Khalvati, S.; Nafisi, H.; Mahmoodi, M.; Schneeweiss, G. M.; Schönswetter, P. Endemics Determine Bioregionalization in The Alpine Zone of the Irano-Anatolian Biodiversity Hotspot (South-West Asia). *Alp. Botany* **2021**, *131*, 177–186. https://doi.org/10.1007/s00035-021-00266-7.

Ramsar Convention. *Ramsar Sites: Protected Wetlands of the World*; Ramsar Secretariat, 2008.

Salewski, V.; Bruderer, B.; Schaub, M. Passerine Migration Strategies and Body Mass Variation along Geographic Sectors across East Africa, the Middle East and the Arabian Peninsula. *J. Ornithol.* **2009**, *150* (2), 369–381. https://doi.org/10.1007/s10336-008-0357-z.

Sassoon, J. *The Ottoman Empire: The Classical Age 1300–1600*; History Book House, 2003.

Sichieri, R.; Seemungal, C. Ecological Adaptations in the Irano-Anatolian Hotspot. *J. Biogeogr.* **2011**, *38* (8), 1376–1388.

Sillero-Zubiri, C.; Hoffmann, M.; Macdonald, D. W. *Canids: Foxes, Wolves, Jackals, and Dogs*; IUCN, 2004.

Sobrevila, C.; Hickey, V. *The Role of Biodiversity and Ecosystems in Sustainable Development (Environmental Strategy Analytical Background Papers)*; World Bank: Washington, DC, 2010.
Teimori, A.; Mostafavi, H.; Esmaeili, H. R. An Update Note on Diversity and Conservation of The Endemic Fishes in Iranian Inland Waters. *Turkish J. Zool.* **2016**, *40* (1), Article 11. https://doi.org/10.3906/zoo-1407-2.
Vural, M.; Dönmez, M.; Çalışkan, M. Floristic Diversity of the Mediterranean Region of Turkey: An Analysis of Patterns. *Ecology* **2014**, *21* (3), 215–223.
WWF. *The Conservation Status of the Irano-Anatolian Region*; World Wildlife Fund, 2017.
Yazdani, R.; Parsa, A. *Vegetation of the Iranian Plateau*; Springer, 2015.
Zohary, M. *The Making of the Modern Landscape: The Mediterranean Region*; Oxford University Press, 2004.

CHAPTER 7

THE PHYTOPLANKTON DIVERSITY OF FRESHWATER SYSTEMS IN THE IRANO-ANATOLIAN HOTSPOT: A REVIEW

NESE YILMAZ[1*], ALI THORABI HAGHIGHI[2], and IBRAHIM TAMER EMECAN[3]

[1]*Department of Freshwater Resources and Management, Faculty of Aquatic Sciences, Istanbul University, Fatih, Istanbul, Türkiye*

[2]*Water, Energy and Environmental Engineering Research Unit, Faculty of Technology, University of Oulu, Oulu, Finland*

[3]*Vocational School of Technical Sciences, Istanbul University–Cerrahpaşa, Istanbul, Türkiye*

*Corresponding authors

ABSTRACT

In recent years, due to excessive population growth, uncontrolled urbanization, intensive industrialization, and the wastes brought by industrialization, and in addition to all these, we have to face the problems of depletion and pollution in our water resources as a result of climatic changes that have become unstoppable. As is known, water is the primary substance needed for the continuity of life of all living beings on Earth. The food chain in water begins with phytoplankton communities that serve as primary producers. The existence of a healthy food pyramid that ends with humans depends on phytoplanktonic organisms. Therefore, studies on phytoplankton identification, especially in important freshwaters in the Irano-Anatolian hotspot, were compiled. The studies conducted draw

Biodiversity Hotspot of the Irano-Anatolian Region. T. Pullaiah (Ed.)

DOI: 10.1201/9781779645760-7

attention to the status of phytoplankton communities, changes that have occurred, possible future scenarios, and necessary precautions. Our review chapter focuses on studies on phytoplankton diversity in important water resources (rivers, dams, lakes, lagoons) in Türkiye, Iran, Armenia, and Azerbaijan located in the Irano-Anatolian hotspot region. In summary, more importance should be given to phytoplankton identification, density, and distribution in ecological studies, and sufficient resources and budgets should be allocated to support researchers so that biomonitoring studies can continue without slowing down.

7.1 INTRODUCTION

Water is the most essential substance necessary for the continuation of life on Earth. Approximately 97% of the earth's water is found in the ocean, 2% is frozen as ice in the poles, and the remaining 1% is available in freshwater, used for drinking and daily needs, industries, and irrigation activities. Water pollution problems in the world are mainly caused by climate changes, population growth, excessive urbanization and industrialization, and unconscious consumption of available natural resources. For all these reasons, the rapid consumption of freshwater resources has made access to clean, hygienic, and safe water more complex as the day passes (Yilmaz, 2013). The rapid pollution of water resources and the risk of extinction are at the top of environmental issues. Therefore, protecting our water resources and improving trophic conditions have been given greater importance recently (Yilmaz et al., 2018). The characteristics of water bodies, especially lakes and rivers, must be known in detail to use potable water resources more effectively (Yilmaz and Guleçal, 2021). Since phytoplankton are located at the bottom of the food pyramid, they are primary producers of aquatic systems. Furthermore, phytoplanktonic organisms, which are sensitive to changes, can be used as bioindicators of water quality since they respond very quickly (Yilmaz, 2013). The diversity and composition of phytoplankton species give good information about the trophic condition, productivity proportion, nutrient concentrations, water quality, and pollution in aquatic environments (Yilmaz, 2018). Also, their relations with other living organisms at upper levels of the food pyramid, such as zooplankton, fish, invertebrates, and humans, further show the critical role of phytoplanktonic organisms in waters (Yilmaz and Aykulu, 2010).

In addition, phytoplankton is one of the four biological elements recommended in the EU Water Framework Directive, which came into force in 2000 to assess the ecological status and potential of surface waters in Europe (Yilmaz, 2013). Determination of phytoplankton composition and diversity and monitoring of seasonal changes will allow for estimation of undesirable situations in waters, and timely precautions can be taken (Yilmaz, 2018). As in other aquatic environments, phytoplankters form the first ring of the food chain in stream systems. Pollution is also carried to seas and lakes by streams along with water. These carried wastes and their derivatives can pose a serious threat to the chemical, physical, and biological characteristics of water, especially in lakes. In comparison, there is less research conducted on algae in streams than in lakes. More ecologic and taxonomic studies must be carried out on the phytoplankton of streams, due to their significant effects on lakes (Yilmaz and Guleçal, 2021).

Lagoon systems are special aquatic ecosystems and host various species that constitute a habitat for freshwater and marine life forms. They also provide essential areas for fish and other aquatic life forms. Due to their high nutrient concentrations, lagoons are among the most efficient coastal ecosystems. They are also important because they provide suitable conditions for agriculture, aquaculture, fishing, tourism, and recreation. Since lagoons are of great ecological importance, they must be kept under constant protection.

Therefore, periodically monitoring the algal flora as the natural biological source of the world in lagoon systems is critical. As highly sensitive ecosystems, lagoons play important roles in protecting and sustaining biodiversity. Coastal lagoons are under pressure from marine and physical, chemical, and biological changes in terrestrial environments. Therefore, changes in all these parameters can be used as an indicator of pollution in lagoons (Yilmaz et al., 2021a). The goal of the study is to compile studies on phytoplankton communities in freshwaters (including streams, lagoons, and lakes) in the Irano-Anatolian hotspot, one of the important regions in the world, and to reveal the phytoplankton diversity of this region with its similarities and differences.

7.2 INFORMATION ABOUT THE IRANO-ANATOLIAN HOTSPOT

The Irano-Anatolian hotspot has been considered the east-west transit area of the historical Silk Road for centuries. It has formed the main trade route and cultural exchange between these two regions by connecting the Mediterranean Basin with Western Asia. The Irano-Anatolian hotspot is a topologically highly diverse region consisting of mountains and closed basins extending from Türkiye to Iran and Turkmenistan. The Irano-Anatolian hotspot includes most of central and eastern Türkiye, a small part of southern Georgia and parts of Azerbaijan and Armenia, northeastern Iraq, northwestern Iran from the Zagros Mountains region to the GNU Biosphere Reserve, from the Alborz Mountains region to the Kopet Dag Mountains in Turkmenistan. The lowest point of this region is the Kopet Dag peak, with an altitude of 300 m and the west of the Zagros Mountains region, and the highest point of this region is the Ararat volcano in Türkiye and Damavand in Elburz with an altitude of 5165 and 5671 m, respectively. The Anatolian Plateau extends into Armenia and western Iran, and its altitude changes from 800 to 2,000 meters. The area covered by the Irano-Anatoliann Hotspot is 899,773 km^2, of which 134,966 km^2 is vegetation. Its acceptance as a new hotspot is due to its high endemism and rich flora and fauna. This hotspot is represented by 2,500 endemic plant species, 10 mammal, 12 reptile, 30 fish, and 2 amphibian species which are endemic. The total number of plant species in the region is reported as 6000 species, with mammals 140 species, birds 362 species, reptiles 116 species, amphibians 18 species, and fish 90 species. The population density of this area is 58 human beings per km^2, and the protected area is 56,193 thousand km^2.

Since 1970, many species in the hotspot have experienced significant declines due to the negative effects of deforestation and overgrazing, which the doubling of the human population has exacerbated. The greatest threat, especially to the Turkish part of the Irano-Anatolian Hotspot, is the development of irrigation schemes for agriculture and related infrastructures such as dams. In addition, overgrazing, excessive harvesting of woody plants for firewood, and mining activities are among the factors that lead to habitat destruction in this hotspot. Additionally, it is known that military operations, especially in Iraq, Iran, and Türkiye, have caused severe damage to forests and wetlands. Climatologically, this region experiences hot summers and very cold winters, with annual rainfall ranging from 100 to 1000 mm (Ilanloo et al., 2021).

7.3 THE IMPORTANCE OF INVESTIGATIONS ON PHYTOPLANKTON

Phytoplankton is one of the most critical biological indicators used in water pollution studies in aquatic systems. Since they respond quickly to variations in water, species belonging to phytoplankton assemblages are used as the main bioindicators. They are considered primary producers in aquatic environments and constitute the first ring of the food chain. Day by day, the improvement and protection of existing water resources are becoming more and more critical with the reorganization of usage of the areas. Therefore, it is necessary to determine the water quality and pollution level by evaluating biological indicators, physicochemical parameters, and heavy metal concentrations collectively in water resources (Yilmaz et al., 2021b).

In order to use existing water resources more efficiently in all sectors, the limnological traits of the water bodies need to be defined all over the world. Water, and additionally pollutants, are carried to seas and lakes by watercourses. Waste substances and their derivatives carried via waterways might also negatively affect the chemical, physical, and organic composition of water bodies. Phytoplankton are the number one producer in the food chain in streams, as well as in different aquatic environments. Phytoplankton are one of the important communities that respond fast to pollutants in aquatic structures. Phytoplankton composition may be applied to become aware of the trophic state, productivity, nutrient content, water quality, and pollutants in aquatic structures. When looking at the studies conducted, it is seen that algae in streams have been studied less than in lakes. Nevertheless, ecological and taxonomical studies on the phytoplankton of streams, which have significant effects on the structure of lakes, are of great importance (Yilmaz and Aykulu, 2010; Yilmaz and Guleçal, 2021).

In aquatic ecosystems, primary producers are phytoplankton species, which are used as indicator organisms for water pollution detection because of their sensitivity towards rapid aquatic changes. Phytoplankton species are indicated by their functional groups in terms of their tolerance and sensitivity level to different combinations of physicochemical and biological properties of aquatic systems (Yilmaz, 2015). Phytoplankton reproduction, as biomass increases, is important because it is in the first step of the food chain and could be used as a bioindicator in aquatic systems because it reflects changes immediately. Therefore, periodically monitoring the algal flora as the natural biological source of the world in lagoon systems is a critical issue in determining the quantitative levels. Being highly fragile ecosystems, lagoons play important roles in protection and maintaining biological diversity. In conjunction with this, as an essential issue for monitoring the pollution level in lagoon systems, the determinations of the composition, trophic structure, and productivity of phytoplankton community, as well as heavy metal concentrations, are required (Yilmaz et al., 2021a). Numerous scientific investigations have indicated that changes in climate have brought various environmental issues. The swift contamination of water resources and the threat of extinction are prioritized among these environmental concerns. The escalating population, unregulated industrialization, and urbanization are progressively intensifying this phenomenon daily. Consequently, the safeguarding of our water supplies and the enhancement of trophic conditions have recently been prioritized.

The ability of lagoon systems to support a wide variety of living species makes them extremely unique aquatic habitats. They serve as a home for both marine and freshwater organisms. They serve as crucial spawning grounds for fish and other water life, as well as places for birds to take refuge. Lagoons are among the most fruitful ecosystems on the coast because of their high nutrient concentrations. They also offer favorable circumstances for fishing, aquaculture, agriculture, and tourism. Lagoons need to be continuously protected for all of these causes. Physical, chemical, and biological changes in both terrestrial and marine settings are putting stress on coastal lagoons. Changes in these characteristics can therefore be thought of as a sign of contamination.

In aquatic systems, phytoplanktonic organisms form the first step of the food chain by using light-assisted photosynthesis to produce high-energy organic molecules from carbon dioxide and inorganic substances.

Standing at the bottom level of the food chain, phytoplankton bear significant importance with respect to their role in aquatic ecosystems and their relations with other living organisms at upper levels (Yilmaz and Aykulu, 2010). The phytoplankton is at the top of the food chain, followed by zooplankton, tiny and large fish, and humans. Phytoplanktonic organisms serve as the foundation of primary production in aquatic environments and are essential for feeding animals that live in freshwater and oceans. Additionally, they are among the biological indicator species that are widely employed in water contamination investigations across the globe (Yilmaz et al., 2018).

Phytoplankton are the primary producers in the food chain, and they are very important organisms in aquatic systems. Phytoplankton species can be used as indicators of water quality, given their sensitivity and dynamic responses to changes in the surrounding environment. The EU Water Framework Directive, which was introduced in 2000, states phytoplankton are one of the four biological elements suggested for assessing the ecological status and potential of European surface waters. Taxonomic studies of phytoplankton are very practical in re-evaluation of the uses and stability of lakes. In particular, data on the algal flora of water sources are commonly used in the preservation, improvement, and control of water quality (Yilmaz, 2013).

In recent years, due to extreme population growth, unplanned urbanization, unconscious industrialization, and global warming, our available water resources have been expeditiously consumed. For this reason, it is necessary to do more research on water quality and pollution each passing day. It is essential to study the physical, chemical, and biological properties of waters and evaluate them together to determine their trophic level. Phytoplankters are at the forefront of bioindicator organisms used in aquatic ecosystems because they respond quickly to the changes occurring in the aquatic environments. The composition of phytoplankton gives information about the primary production, nutrient level, and water pollution of aquatic systems. Determination of the phytoplankton composition and monitoring of seasonal changes will allow for estimation of undesirable situations that may occur in the aquatic systems and timely precautions to be taken (Yilmaz, 2018).

7.4 PHYTOPLANKTON DIVERSITY OF IRANO-ANATOLIAN HOTSPOT

The studies conducted on phytoplankton diversity in freshwater resources (rivers, lakes, dams, and lagoons) located in the Irano-Anatolian hot spot region in Türkiye, Armenia, Iran, and Turkmenistan are compiled below:

7.4.1 LAKE KARAMIK (AFYON, TÜRKİYE)

An investigation by Gönülöl and Obalı claims that (1986), the phytoplankton of Karamık Lake (Afyon, Türkiye) consists of 175 species, belonging to Bacillariophyta, Chlorophyta, Chrysophyta, Cryptophyta, Cyanophyta, Dinophyta, and Euglenophyta. A significant part of the lake is covered with rushes, which has increased the number of species of *epipelic* and *epiphytic* origin. In phytoplankton, Chlorococcales and Desmidiales members have been recorded as widespread and abundant, but Cyanophyta members have been of second importance. Bacillariophyta is considered the richest group, with 76 species in terms of species richness. Species of *Achnanthes*, *Cymbella*, *Epithemia*, *Navicula*, *Rhopalodia gibba* have been recorded as the most common diatoms at the lake.

7.4.2 LAKE MANYAS (BALIKESIR, TÜRKİYE)

Çelik and Ongun (2008) carried out the spatial and temporal dynamics of the steady-state phytoplankton assemblages in Lake Manyas, a temperate shallow hypertrophic lake, between 2003 and 2004. Totally, 165 phytoplankton species were identified at the lake. During the study, *Achnanthes microcephala* (Kützing) Cleve of diatoms and Euglenozoa division member *Phacus pusillus* Lemmermann were detected as the dominant species. Also, Cyanobacteria division members *Anabaena spiroides* Klebahn and *Microcystis aeruginosa* (Kützing) Kützing were recorded as important species in Lake Manyas.

7.4.3 LAKE ULUABAT (BURSA, TÜRKİYE)

In an investigation by Karacaoğlu et al. (2004, 2006), a total of 331 phytoplanktonic taxa were identified in Lake Uluabat. The Bacillariophyta division was dominant, 46% of the total taxa (152). The other groups that constituted the phytoplankton were Chlorophyta 27% (89 taxa), Cyanophyta 13% (42 taxa), Euglenophyta 9% (31 taxa), Dinophyta 3% (11 taxa), Cryptophyta 1% (4 taxa), and Chrysophyta 1% (2 taxa). Cyanobacteria members and pennate diatoms were primarily recorded during the summer period, while centric diatoms were recorded in substantial amounts almost all year. When *Aulacoseira subarctica* (O.Müll.) E.Y.H, *Cyclotella* spp., and *Stephanodiscus dubius* (Fricke) Hust. were the most common centric diatoms, *Asterionella formosa* Hassall, *Cymatopleura solea* (Bréb.) W.Sm., *Synedra delicatissima* var. *angustissima* Grunow, *Synedra ulna* (Nitzsch) Ehrenb., *Nitzschia acicularis* (Kütz.) W.Sm., *N. holsatica* Hust., *N. palea* (Kütz.) W.Sm., *N. lorenziana* var. *subtilis* Grunow, and *Surirella biseriata* Bréb. were the most common pennate diatoms in the phytoplankton. The second dominant group was Chlorophyta, and Chlorococcales order was recorded as dominant with 59 taxa. *Actinastrum*, *Ankistrodesmus*, *Chlamydomonas*, *Coelastrum*, *Eudorina*, *Micractinium*, *Monoraphidium*, *Pandorina*, *Pediastrum*, *Scenedesmus*, *Sphaerocystis*, and *Tetraedron* were recorded as the most common species (Karacaoğlu et al., 2004). The authors noted that cyanobacteria were at the highest density at the beginning of the survey in July 1998, accounting for 52% of the total phytoplankton density, and were largely represented by *Microcystis aeruginosa*, *Anabaena affinis*, and *Anabaena spiroides* (Karacaoğlu et al., 2006).

In another study, the composition, density, and changes in Lake Uluabat's phytoplankton throughout time and space were investigated by Dalkıran et al. (2016), and a total of 169 taxa were identified belonging to Bacillariophyta (56), Chlorophyta (51), Cyanobacteria (25), Euglenophyta (19), Miozoa (7), Charophyta (6), and Ochrophyta (5). As a result, when compared with the previous studies, the authors indicated that the phytoplankton taxa richness was approximately declined by half at the lake. According to Dalkıran et al. (2016), it was determined that the phytoplankton of Lake Uluabat changed over the years. It was also revealed that phytoplankton species diversity and composition are an important group of organisms that can be used to determine trophic level and ecological quality. They pointed out that the presence of some Cyanobacteria taxa that can produce toxins indicates that the lake is at risk.

7.4.4 LAKE İZNIK (BURSA, TÜRKİYE)

Driving factors affecting the phytoplankton functional groups were investigated in a deep alkaline lake, Lake İznik, by Oğuz et al. (2020). In this study, 103 taxa were found to have an abundance of Bacillariophyta (36 taxa) and Chlorophyta (42 taxa). On the other hand, Charophyta, Cryptophyta, Cyanobacteria, Miozoa, and Euglenozoa contributed to the phytoplankton community with low species numbers. During the study, a heavy cyanobacterial bloom (*C. ovalisporum*) was determined in the lake. It has been reported that high biomass was reached, showing signals of eutrophication by the species *D. mendotae*, *S. aphanizomenoides*, *R. raciborskii*, and *P. rubescens* of the Cyanobacteria division.

7.4.5 ÇAYGÖREN RESERVOIR (BALIKESIR, TÜRKİYE)

Sevindik (2010) investigated the phytoplankton of the Çaygören Reservoir, and a total of 192 taxa belonging to 8 divisions have been identified, including Chlorophyta (75 taxa), Bacillariophyta (60 taxa), Cyanobacteria (19 taxa), Euglenophyta (19 taxa), Charophyta (8 taxa), Myzozoa (6 taxa), Cryptophyta (3 taxa), and Heterokontophyta (2 taxa). While most species have been replaced by their widespread occurrence, some taxa are rarely found in Türkiye. Also, 29 of these taxa were reported as new records for Turkish freshwater algae. Chlorophyta was detected as the dominant phytoplankton group with the most incredible species richness (75 taxa) with large contribution of Chlorococcales (45 taxa), mostly consisted of *Pediastrum, Scenedesmus,* and *Tetrastrum* species. Bacillariophyta was the second dominant group with 60 taxa. *Navicula* and *Nitzschia* were dominant genus. Species richness of Cyanobacteria increased in summer and represented by the species of *Planktothrix* sp., and *Aphanocapsa holsatica* were significant at summer and autumn months. *Aphanothece clathrata, Aphanizomenon flos-aquae, Anabaena planctonica, Pseudanabaena catenata,* and *Gomphosphaeria aponina* at the reservoir. *Trachelomonas volvocina, T. hispida,* and *T. volzii* var. *intermedia* were the most common taxa found at the reservoir of Euglenophyta division. Charophyta was represented with 8 taxa of the order Zygnematales, and *Cosmarium* was the most dominant genus.

7.4.6 LAKE GÖLBAŞI (HATAY, TÜRKİYE)

The flora of Gölbaşı Lake consisted of 41 taxa belonging to Bacillariophyta (24), Chlorophyta (12), Pyrrophyta (2), Cyanophyta (2), and Chrysophyta (1). The dominant group of the lake was recorded as Chrysophyta, followed by Bacillariophyta, Chlorophyta, Cyanophyta, and Pyrrophyta. *Pediastrum bonyanum, Pediastrum clathratum, Pediastrum simplex, Pediastrum duplex, Pediastrum tetras,* and *Pediastrum* sp. of Chlorophyta were noted as frequently observed species. *Dinobryon sertularia,* of Chrysophyta, *Closterium* of Desmidiales, *Peridinium* and *Ceratium* of Pyrrophyta were observed. *Merismopedia* sp. and *Gamphospheria* sp. of Cyanophyta, were rarely found (Naz and Türkmen, 2005).

7.4.7 LAKE CERNEK (SAMSUN, TÜRKİYE)

In a study carried out by Taş et al. (2002) in Lake Cernek, in spring and autumn, *Cyclotella ocellata* and *Nitzschia acicularis* of Bacillariophyta produced blooms in the phytoplankton. The phytoplankton's most common and prominent algae group was detected as the Chlorophyta division, and *Monoraphidium caribeum, M. irregulare, Scenedesmus quadricauda, Oocystis borgei,* and *Pediastrum boryanum* have been the most prevalent species when they bloom in the spring and fall. In summer, *Microcystis aeruginosa, Anabaena affinis,* and *A. spiroides* of

Cyanophyta members produced blooms. *Euglena gracilis, E. polymorpha* of Euglenophyta were generally found. The division of Cryptophyta was represented by *Cryptomonas ovata*, Dinophyta division by *Gymnodinium* sp., and *Xanthophyta* division by *Characiopsis cylindrica*.

7.4.8 LAKE SIMENIT (SAMSUN, TÜRKİYE)

A total of 167 taxa belonging to the Cyanophyta, Bacillariophyta, Chlorophyta, Cryptophyta, Dinophyta, Euglenophyta, and Xanthophyta groups were recorded at Lake Simenit (Samsun, Türkiye). Bacillariophyta and Euglenophyta division members were determined to be most common according to the number of species. It was detected that species of the genera *Chaetoceros*, *Cyclotella*, *Cocconeis*, *Scenedesmus, Anabaena*, *Euglena,* and *Trachelomonas* caused water blooms (Ersanlı and Gönülol, 2006).

7.4.9 LAKE LADIK (SAMSUN, TÜRKİYE)

Members of Cyanoprokaryota (8), Bacillariophyta (46), Chlorophyta (54), Euglenophyta (30), Cryptophyta (2), Dinophyta (1), and Xanthophyta (I) were among the taxa found in Ladik Lake by Maraşlıoğlu et al. (2005). Common organisms in the phytoplankton included *Cryptomonas ovata*, *Aulacoseria distans*, *T. volvocina* var. *punctata*, *Trachelomonas hispida*, *Cyclotella ocellata*, *Pediastrum* spp., and *Scenedesmus* spp. The common recorded species were *Cyclotella ocellata*, *Aulacoseria distans*, *Trachelomonas hispida*, *T. volvocina* var. *punctata*, *Cryptomonas ovata*, *Pediastrum* spp., and *Scenedesmus* spp.

7.4.10 HASAN UĞURLU DAM LAKE (SAMSUN, TÜRKİYE)

Using samples taken from two sites between July 1992 and December 1993, the phytoplankton composition and seasonal fluctuations of Hasan Uğurlu Dam Lake were investigated. About 57 taxa were found to be members of the divisions of phytoplankton known as Bacillariophyta, Chlorophyta, Cyanophyta, Dinophyta, and Euglenophyta. In the phytoplankton, *Asterionella formosa*, *Cyclotella planctonica*, *Pediastrum simplex,* and *Ceratium hirundinella* produced water blooms during specific months (Gönülol and Obalı, 1998).

7.4.11 LAKE GAGA (ORDU, TÜRKİYE)

Taş (2012) examined the phytoplankton composition of Gaga Lake on a monthly basis for a year, from February 2005 to January 2006. About 99 algal taxa were identified and categorized into seven taxonomic groupings. The taxonomic categories of Chlorophyta (35 taxa), Ochrophyta (22 taxa), Charophyta (16 taxa), Cyanobacteria (14), Myzozoa (5), Euglenozoa (5), and Heterokontophyta (2) ranged from the most species-rich to the least species-rich. Out of all the species, green algae predominated. The study's main taxa were *Ceratium hirundinella*, a member of the Dinophyceae, and *Dinobryon divergens* and *D. sertularia*, members of the Chrysophyceae.

7.4.12 UZUNCAYIR DAM LAKE (TUNCELI, TÜRKİYE)

Of the 46 phytoplankton taxa that were identified, Bacillariophyta had the highest species number, accounting for 24 taxa. When it came to both abundance and regularity of occurrence, diatoms were the most noticeable phytoplankton. The least frequent were Euglenozoa, Miozoa, Ochrophyta, Cyanobacteria, Chlorophyta, and Charophyta. Nonetheless, a number of diatoms, including *Gyrosigma attenuatum, Hippodonta hungarica, Meridion circulare, Nitzschia clausii, Nitzschia frustulum, Pantocsekiella ocellata,* and *Rhoicosphenia abbreviata;* charophytes, including *Cosmarium granatum* and *Mougeotiopsis calospora;* Cyanobacteria, including *Leiblenia epiphytica* and *Planktothrix agardhii;* dinoflagellates, including *Ceratium hirundinella, Parvodinium africanum, Parvodinium inconspicuum,* and *Peridinium cinctum;* and Ochrophyta, including *Dinobryon pediforme*, were also discernible, occurring throughout the year (Kutlu et al., 2020).

7.4.13 LAKE SAPANCA (SAKARYA, TÜRKİYE)

Bacillariophyta 30, Chlorophyta 12, Cyanophyta 5, Euglenophyta 3, Dinophyta 2, Chrysophyta 1, and Cryptophyta 1 were among the 54 phytoplankton species that were identified. In terms of species numbers and density, Bacillariophyta, Chlorophyta, and Cyanophyta composed the majority of the phytoplankton (Yilmaz and Aykulu, 2010).

7.4.14 HIRFANLI DAM LAKE (KIRŞEHIR, TÜRKİYE)

A total of 189 taxa, comprising 174 phytoplanktonic organisms (11) and 15 zooplanktonic organisms, were identified in Hirfanlı Dam Lake.

Cyclostephanos dubious was the dominant and widespread diatom in Hirfanlı Dam Lake. *Peridinium cinctum* (Muell.) Ehr., *Ceratium hirundinella* (O.F. Muell.) Dujardin, *Asterionella formosa* Hassall, *Microcystis aeruginosa* Kuetz., and some species of *Chlamydomonas* were also identified. These species sometimes show increases in density, especially in early summer and autumn in Hirfanlı Dam Lake. However, *Nitzschia palea* (Kützing) W. Smith, *Navicula pupula* Kützing, and *Navicula cryptocephala* Kützing, characteristic species of domestic sewage contaminated waters (18), were rare in Hirfanlı Dam Lake phytoplankton but frequent and abundant in the benthic flora (Baykal et al., 2006).

7.4.15 LAKE GALA (EDIRNE, TÜRKİYE)

A total of 112 taxa from 5 divisions were identified during the study period. Chlorophyta was the most diverse group in the lake with 47 taxa, and diatoms were found to have the highest cell counts with a mean value of 670011 cell / L. The general pattern of seasonal succession in phytoplankton of the lake was represented with Chlorophyta in June and by Cyanophyta in September, and diatoms were the dominant group of the lake in all other months. Spatial heterogeneity was observed in the lake where a slight *Microsystis* spp. increase occurred in the early autumn months. The qualitative and quantitative composition of the phytoplankton community of Lake Gala differed significantly from previous phytoplankton data, indicating the degradation of the lake (Öterler et al., 2015).

7.4.16 SAZLIDERE DAM LAKE (ISTANBUL, TÜRKİYE)

Bacillariophyta, Charophyta, Chlorophyta, Cryptophyta, Cyanobacteria, Euglenozoa, Miozoa, and Ochrophyta were the members of the phytoplankton. The 19 groups (B, D, N, P, T, S2, X1, Y, E, F, J, H1, LO, LM, M, MP, R, W1, and W2) that made up the phytoplankton community of the dam lake were determined via the functional group approach. These groups were most prevalent in mesotrophic and eutrophic waters. In the study, a total of 67 taxa were identified. The phytoplankton community of the dam lake consists of Bacillariophyta (46%), Charophyta (7%), Chlorophyta (19%), Cryptophyta (2%), Cyanobacteria (13%), Euglenozoa (6%), Miozoa (5%), and Ochrophyta (2%) members. In terms of species diversity and phytoplankton density, the Bacillariophyta division was found to be dominant. It was determined that *Cyclotella atomus*

Hustedt, *C. ocellata* Pantocsek, and *Ulnaria ulna* (Nitzsch) Compere of Bacillariophyta, *Anabaena flos-aquae* Brebisson ex Bornet & Flauhault and *Planktothrix rubescens* (De Candolle ex Gomont) Anagnostidis & Komarek of Cyanobacteria were increased in spring and summer; *Sphaerocystis planctonica* (Korshikov) Bourrelly, *Scenedesmus* spp., and *Mougeotia* species of Chlorophyta were increased in autumn and winter. While numbers of Chlorophyta members were decreased, numbers of Cryptophyta and Miozoa members were increased (Yilmaz, 2018).

7.4.17 LAKE KÜÇÜKÇEKMECE (İSTANBUL, TÜRKİYE)

Yilmaz et al. (2021a) conducted research on seasonal fluctuations in phytoplankton composition, heavy metal contamination, and certain water quality metrics across the channel, which serves as a waterway between Lake Küçükçekmece and the Sea of Marmara. The lake, which is regarded as a Class B wetland, has remained inside the Istanbul metropolitan region over time. Throughout March through November, samples were collected every month from five different sampling locations. About 46 species from the divisions of Bacillariophyta (20), Charophyta (4), Chlorophyta (6), Cryptophyta (1), Cyanobacteria (4), Euglenozoa (3), and Miozoa (8) were identified following experimental procedures. Throughout the channel, the phytoplankton densities ranged from 76 individual cm^{-3} to 3283 individual cm^{-3}, while the chlorophyll-a concentrations varied from 0.99 to 55.32 mg m^{-3}. According to the data obtained in this study, the area where research was conducted was determined as having eutrophic characteristics.

7.4.18 LAKE TERKOS (İSTANBUL, TÜRKİYE)

Bacillariophyta (17), Charophyta (2), Chlorophyta (5), Cryptophyta (1), Cyanobacteria (5), Euglenozoa (2), Miozoa (1), and Ochrophyta (1) were the classifications assigned to 34 taxa from 8 divisions. It was discovered that the blue-green algae *Anabaena spiroides* (Klebahn) dominated the ecosystem in Lake (Yilmaz et al., 2021b).

7.4.19 BÜYÜKÇEKMECE DAM LAKE (İSTANBUL, TÜRKİYE)

Yilmaz (2019) examined the contamination and water quality of Buyukcekmece Dam Lake, the second-largest supply of drinking water in Istanbul, Türkiye, and its influent streams, Karasu, Izzettin, Eskice,

Ahlat, Beylikçayı, Çekmece, Çakmaklı, and Tahtaköprü. The lake and its streams were examined for certain physicochemical characteristics, nitrogen concentrations, and phytoplankton composition. Seasonally, water samples were taken from nine sampling locations in May, August, November, and February of 2017. Bacillariophyta (22), Charophyta (6), Chlorophyta (14), Cryptophyta (2), Cyanobacteria (8), Euglenozoa (8), Miozoa (2), and Ochrophyta (1) division members make up the phytoplankton composition of the lake and its feeding streams. The findings showed that the Buyukcekmece Dam Lake basin has characteristics that are similar to those of eutrophic conditions. Relevant authorities require that Büyükçekmece Dam Lake and its creeks be continuously protected to improve its water quality.

7.4.20 RIVA STREAM (İSTANBUL, TÜRKİYE)

The water quality and heavy metal pollution status of Riva Stream, which is related to Ömerli Dam Lake, one of the water resources of Istanbul (Türkiye), was determined by Yilmaz et al. (2014). Some physicochemical parameters, phytoplankton compositions, some nutrients, and heavy metal concentrations in Riva Stream's surface water were investigated. Samples were collected at six sampling sites in the course of the stream in February 2012 and May 2012. A total of 19 taxa, belonging to five divisions, were identified. The phytoplankton density varied from 47 to 264 ind./cm^3; chlorophyll-*a* contents varied between 1.52 and 1.83 mg/m^3. Measured concentrations of heavy metals and nutrients showed differences related to sampling points. It was observed that some heavy metals highly pollute Riva Stream, and a number of heavy metal resistant/tolerant phytoplankton species are present in the stream. The physicochemical characteristics of Riva Stream, as well as its high concentrations of nutrients, make it similar to eutrophic conditions. It is required that Riva Stream should be taken under protection as soon as possible by the appropriate authorities for enhancing the quality of the water.

7.4.21 KAMIL ABDUŞ LAGOON (TUZLA LAKE) (İSTANBUL, TÜRKİYE)

Yilmaz et al. (2018) investigated the phytoplankton composition, several physicochemical parameters, and nutrient concentrations in order to find out the water quality and pollution condition of Kamil Abduş Lagoon

(Tuzla Lake), which is situated in the Istanbul Metropolitan area. Monthly water samples were taken at three sampling locations between February 2016 and January 2017. About 32 taxa were identified, including those from the divisions of Bacillariophyta (12), Charophyta (1), Chlorophyta (4), Cryptophyta (1), Cyanobacteria (5), Euglenozoa (5), and Miozoa (4). In the research area, both freshwater and marine species were observed. The lake's trophic level is close to eutrophic, as evidenced by the presence of mesotrophic and eutrophic phytoplankton species, as well as high levels of nutrients and chlorophyll-*a*. In this study, water quality characteristics were examined in order to assess the importance of the input data through the use of artificial neural network analysis, particularly to guarantee that regression analysis overfitting was limited. The lagoon has to be preserved immediately since pollution is endangering it. Therefore, in order to choose the best management strategies for enhancing the lagoon's water quality, thorough research on phytoplankton, including physicochemical parameters and nutrients, must be conducted.

7.4.22 ALIBEYKÖY DAM LAKE (İSTANBUL, TÜRKİYE)

Alibeyköy Dam is a vital source of drinking water for the Istanbul metropolitan area. The pollution levels of Alibeykoy Dam Lake were evaluated by looking at the distribution of phytoplankton, physico-chemical parameters, nutrient content, and heavy metal concentrations. In January, February, May, and September of 2021, water samples were taken from the lake and six influent streams: Cebeci, Pirincci, Sidan, Bolluca, Gulgen, and Kocaman. Bacillariophyta (14), Charophyta (2), Chlorophyta (9), Cryptophyta (1), Cyanobacteria (3), Euglenozoa (4), and Miozoa (3) were among the seven divisions that comprised the 36 taxa that were identified. The group with the highest species richness and abundance was determined to be diatoms. *Microcystis aeruginosa* in Alibeyköy Dam and Sidan Creek, *Cyclotella ocellata* in Cebeci and Gulgen Creeks, *Scenedesmus* sp. in Pirincci Creek, *Navicula cryptocephala* in Bolluca Creek, and *Sphaerocystis planctonica* in Kocaman Creek were the most common species found in each location. The lake may be approaching eutrophic conditions, according to mesotrophic and eutrophic phytoplankton species. High levels of fertilizers and heavy metals found in the water tests further corroborated this result. The ecological condition of Alibeyköy Dam Lake, a vital source of drinking water in Istanbul, is thoroughly understood, thanks to this study, which makes it noteworthy buildup by determining the prevalent species and pollution levels. These results are essential for putting into practice efficient water resource management plans, guaranteeing the sustainability of the lakes, and safeguarding public health (Yilmaz et al., 2025).

7.4.23 LAKE HAZAR (ELAZIĞ, TÜRKİYE)

During the sampling period, the phytoplankton community of the pelagic zone in Lake Hazar included 50 species of Bacillariophyta, which were reported by Koçer and Şen (2014), who described the qualitative changes of the diatom community in the surface water. In addition to the diatoms reported by Koçer and Şen (2014), *Chlorella vulgaris*, *Closteriopsis acicularis*, *Oocystis apiculata*, *Oocystis pusilla*, *Planctonema* sp., *Chroococcus dispersus*, *Chroococcus minor*, *Chroococcus minutus*, *Chroococcus turgidus*, *Planktolyngbya contorta*, and *Peridinium cinctum* were identified as species belonging to divisions Chlorophyta, Cyanophyta, and Pyrrophyta in the present study. The seasonal succession of the phytoplankton in the pelagic zone of Lake Hazar was mainly governed by several dominant and subdominant taxa. *Gomphonema* species in December and *Cyclotella meneghiniana* and *Melosira* sp. in March were the dominant taxa. When total algal abundance decreased highly in June, *Meridion circulare* was dominant in surface waters. The dominance of *Planktolyngbya contorta* and *Cyclotella meneghiniana* at the other depths during June held in September. Only diatoms and *Planktolyngbya contorta* were dominant in all seasons and depths. *Fragilaria* and *Nitzschia* species, *Chlorella vulgaris*, *Closteriopsis acicularis*, and *Planctonema* sp. were subdominant taxa in different seasons and depths.

7.4.24 MELEN RIVER (AKÇAKOCA, TÜRKİYE)

A total of 135 taxa belonging to Cyanophyta, Bacillariophyta, Chlorophyta, Dinophyta, Euglenophyta, Heterokontophyta, Streptophyta, and Xanthophyta were identified. The total biomass of stations 1 and 2 was mainly characterized by Dinophyta, and at other stations by Bacillariophyta. Five major genera (*Peridinium, Ceratium, Phacus, Cyclostephanos*, and *Melosira*) accounted for over 70% of the total phytoplankton abundance. The community consisted of 136 taxa, distributed among 50 genera. Most species belong to Cyanophyta (10), Bacillariophyta (83), Chlorophyta (20), Dinophyta (4), Euglenophyta (9), Heterokontophyta (1), Streptophyta (8), and Xanthophyta (1). The results indicated that Bacillariophyta was qualitatively dominant, while Chlorophyta was subdominant in the river. The preponderance among different groups was quite common. Dinophyta was dominant at stations 1 and 2 in late spring, summer and early autumn. Except at station 2, Euglenophyta was intermittently dominant at the other stations. Throughout the river mouth, the planktonic algae ratio within the biomass was between 0.69% and 0.78%, which was mainly contributed by organisms such as *Navicula cryptocephala, Nitzschia palea, N. acicularis N. fonticola,* and *N. amphibia* (Baykal et al., 2011).

7.4.25 KARKAMIŞ DAM LAKE (ŞANLIURFA, TÜRKİYE)

During the study carried out by Sönmez et al. (2017), the phytoplankton of Karkamış Dam Lake were from 85 taxa including 56 in Bacillariophyta, 15 in Chlorophyta, 1 in Chrysophyta, 10 in Cyanophyta, 2 in Dinophyta and 1 in Euglenophyta. About 44 of these taxa were significant because they were found in all stations. The most frequently represented algae from these groups were *Gomphonema* (8 taxa), *Epithemia*, *Fragilaria*, *Navicula* and *Ulnaria* (4 taxa each), *Cymbella*, *Gyrosigma* and *Nitzschia* (3 taxa each) from Bacillariophyta, and *Phormidium* (3 taxa) from Cyanophyta. Based on the data, the highest number of taxa was found in the 2nd station (59 taxa), while the lowest taxa were found in the 5th station (50 taxa). Among all algae, 30 taxa from Bacillariophyta and 5 taxa from Cholorophyta were recorded in all stations. These algae are; *Asterionella formosa*, *Aulacoseira granulata* var. *angustissima, Campylodiscus noricus*, *Cocconeis pediculus*, *Cocconeis placentula*, *Cymatopleura solea*, *Cymbella affinis*, *Cymbella cistula*, *Cymbella cymbiformis*, *Diatoma vulgaris*, *Ellerbeckia arenaria*, *Epithemia argus*, *Epithemia turgida*, *Fragilaria acus*, *Fragilaria brevistriata*, *Fragilaria crotonensis*, *Gomphonema acuminatum*, *Gomphonema olivaceum*, *Gomphonema truncatum*, *Gyrosigma acuminatum*, *Gyrosigma balticum*, *Halamphora veneta*, *Navicula rhynchocephala*, *Nitzschia sigma*, *Rhoicosphenia abbreviata*, *Tryblionella levidensis*, *Ulnaria amphirhynchus*, *Ulnaria capitata*, *Ulnaria delicatissima var. angustissima* and *Ulnaria ulna* (30 species) from Bacillariophyta, *Microspora tumidula*, *Pediastrum duplex*, *Spirogyra affinis*, *Volvox aureus* and *Zygnema circumcarinatum* (5 species) from Chlorophyta. *Dinobryon divergens* (1 species) from Chrysophyta, *Geitlerinema amphibium*, *Lyngbya major*, *Oscillatoria limosa*, *Oscillatoria tenuis*, *Phormidium breve* and *Phormidium lucidum* (6 species) from Cyanophyta, and *Ceratium hirundinella* and *Peridinium cinctum* (2 species) from Dinophyta.

7.4.26 ANAMUR (DRAGON) STREAM (MERSIN, TÜRKİYE)

Yilmaz et al. (2017) assessed the level of water contamination in Anamur Creek, one of Mersin's (Türkiye) water sources. They looked into the phytoplankton composition and a few physicochemical parameters in Anamur Creek's surface water. Between April and June 2010, samples were taken at five different locations along the stream. About 15 taxa from the divisions of Euglenozoa (1), Miozoa (2), Bacillariophyta (11), and Cryptophyta (1) were found. Chlorophyll-a values in the stream range from 4.04 to 26.23 mg/m3, indicating eutrophic conditions. River sports, fish farms, and agriculture all use Anamur Creek. The project, which started in 2011 and was finished in 2015, has started to supply water from the Anamur Creek to the Geçitköy Dam in the Turkish Republic of Northern Cyprus. Because of this, the trophic status of this creek is significantly influenced by the identification of its usage regions and water levels. The appropriate authorities must place Anamur Creek under protection in order to improve the quality of its water. The relevance of input data sets utilized to examine and track the water quality in the defined study area is successfully imagined using artificial neural network analysis.

7.4.27 SEVAN LAKE (ARMENIA)

In a study carried out by Asatryan et al. (2022) in Sevan Lake, the main groups of phytoplankton indicated as Bacillariophyta (diatoms), Cyanophyta (blue-green), Chlorophyta (green), Xanthophyta (yellow-green), and Euglenophyta (*Euglena*). The study concludes on some dangerous transformations in the phytoplankton community of the biggest freshwater reservoir of the Caucasus. Particularly, both the intensity of algal "blooms" and the role of Cyanobacteria in that have increased. Additionally, it was mentioned that the dominating species of the possibly hazardous genus *Dolichospermum*/*Anabaena* led HABs. When compared to "blooms" dominated by Bacillariophyta species, spatial interpolations showed that HABs had much smaller extents. HABs were mostly found along the estuaries of Lake Sevan's main tributaries in the littoral zone. Considering temperature preferences of different groups of algae, such transformation indicates the future perspectives of Lake Sevan ecosystem under the increasing threat of global warming. Spatial analysis of algal "blooms" allows to conclude that HABs in Lake Sevan are still more local and probably connected with the load from tributaries as spatially they are close to estuaries.

In another study carried out by Hovhannisyan et al. (2010) in Lake Sevan, the successional development of phytoplankton demonstrated seasonal features typical for hydro ecosystems. In particular, there was a dominance of green and blue-green algae in summer and of diatom algae in autumn. In summer, the dominant taxa were the green algal genus *Oocystis* sp. and the cyanobacterium *Aphanothece clathrata*. In autumn, the dominant position was occupied by the diatom *Melosira granulata*. According to Chl-*a* quantity, Lake Sevan can be classified as a mesotrophic lake. During the investigated period, any clear correlations between phytoplankton abundance, vertical distribution, and abiotic factors were observed. Hambaryan et al. (2020) noted that the main groups of phytoplankton of Lake Sevan belong to divisions Bacillariophyta, Cyanophyta, Chlorophyta, and, in recent years, Euglenophyta.

7.4.28 LAKE YEREVAN (ARMENIA)

In total, 72 species (2012), 85 species (2013), and 136 species (2014) were identified in Lake Yerevan waters. Three groups of planktonic algae in Lake Yerevan waters were dominating in proportional share: diatoms (Bacillariophyta), greens (Chlorophyta), and cyanobacteria (Cyanophyta). Certain genera were marked with algal representatives such as *Peridinium* sp., *Euglena* sp. and so forth, and some others–with an insignificant share. The seasonal successions of Lake Yerevan phytoplankton community correspond to the stage of meso–to eutrophy, with peaks registered for hot season and frequent dominance by certain genus (Minesyan, 2016).

7.4.29 ARPA RIVER (ARMENIA)

The mass species belonged to the following groups: Xantophyta (1), Euglenophyta (4), Chlorophyta (15), Cyanophyta (17), and Bacillariophyta (Diatoms) (47). According to their quantitative and qualitative indices, diatoms accounted for 57 and 60% of the total number and biomass of the Arpa River's phytoplankton, as well as those of its principal tributaries. *Rhoicosphenia curvata*, *Ceratoneis arcus*, *Fragilaria crotonensis*, *Cocconeis placentula*, *Diatoma vulgare*, *Cymbella ventricosa*, and *Melosira granulata* were among the species with high rates. Cyanobacteria, which are signs of pollution, have yielded more than 20 species. The genera *Oscillatoria* (5) and *Spirulina* (4) showed variety. *Pleurocapsa minor*, *Aphanizomenon flosaquae*, *Anabaena spiroides*, and *Coelosphaerium kuetzingianum* were identified at different observation stations (Hambaryan and Gevorgyan, 2019).

7.4.30 URMIA LAKE (IRAN)

In general, 14 phytoplankton genera were identified, among which 10 genera belonged to Bacillariophyta (diatoms), 2 genera to Chlorophyceae (green algae), and 2 genera to Cyanophyceae (bluegreen algae). The most abundant microalgae identified in the lake were genera *Dunaliella* (92.1%), *Navicula* (2.4%), *Synedra* (1.9%), and *Nitzschia* (1.1%) (Dahesht et al., 2010).

7.4.31 ARAS DAM RESERVOIR (IRAN)

In total, 72 species belonging to 5 divisions were determined. Cyanobacteria contained the highest density (74%) during the study period, with *Pseudanabaena limnetica* as the most abundant species. This group retained its dominance the whole year round, which indicated the poor quality and high nutrient load of the Aras reservoir, mainly due to human activities. Cyanobacteria, regardless of its lower species numbers, was the most abundant group in all seasons. Some species, such as *Cyclotella* sp., *Microcystis botrys*, and *Oscillatoria* sp., were present in samples in all four seasons. This trend was observed in all seasons. Euglenophyta (14%) formed the second dominant phytoplankton division, with a low species diversity (Mohebbi et al., 2016).

7.4.32 ZAYENDEH-ROOD DAM LAKE (IRAN)

A total of 112 species and 53 genera belonging to 6 divisions were identified. The species belonging to Bacillariophyta, Chlorophyta, Cyanophyta, Euglenophyta, and Dinophyta were found frequently. Chrysophyta species were found in low numbers. *Cyclotella meneghinina* Kütz., a centric diatom, was the dominant species at all stations. Regarding physicochemical analysis and phytoplankton composition, it seems that Zayandeh-Rood Dam Lake is oligo-mesotrophic, and some species of phytoplankton can be used as indicators for evaluating water quality (Shams et al., 2012).

7.4.33 MIGHAN LAKE (IRAN)

In this study, 12 algal species were identified in Mighan Lake. Bacillariophyta were the most conspicuous taxa (5 species), while Cyanobacteria and Chlorophyta included 3 and 4 species, respectively. *D. salina,* a halophilic green alga, composed 87.36% of phytoplankton composition. Finally, phytoplankton assemblages adapt to habitat properties at different salinity levels; these levels are the main limiting factor determining phytoplankton associations. However, the presence of *Artemia* species, as well as nutrient levels, should be considered (Hafezieh et al., 2020).

7.4.34 BUKAN DAM RESERVOIR (IRAN)

Azari et al. (2011) declared that, in total, 55 phytoplankton taxa were recorded, and Chlorophyta was the richest algal group, with 28 taxa, followed by Diatoms, with 10 taxa. Other algal groups were Dinophyta (6 taxa), Cyanobacteria (5 taxa), Euglenophyta (3 taxa), Chrysophyta (2 taxa), and Cryptophyta (1 taxon).

7.4.35 SHAHID RAJAEE RESERVOIR (IRAN)

During the sampling period, 107 phytoplankton species were identified from 8 major divisions, namely Bacillariophyta, Pyrrophyta, Chlorophyta, Cyanophyta, Euglenophyta, Chrysophyta, Xantophyta, and Cryptophyta. The relative abundance of Bacillariophyta was slightly higher than Pyrrophyta. Bacillariophyta and Pyrrophyta were the most dominant divisions, and they formed more than 95% of phytoplankton abundance. The number of species in Chlorophyta (28) was almost similar to the number of species in Bacillariophyta (27) and Pyrrophyta (23), but the contribution of Chlorophyta in phytoplankton abundance was only 2.3%. In this study, Bacillariophyta was the dominant phylum in the phytoplankton community (Makhlough et al., 2017).

7.4.36 AZAD DAM LAKE (IRAN)

According to Fazli et al. (2022), a total of 51 taxa of phytoplankton belonging to six phyla were identified. Bacillariophyta (diatoms) were most diverse with 22 taxa, comprised more than 90% of the total density. Chlorophyta was the second prominent phytoplankton phylum in the mentioned study. The results of this study showed that Azad Dam Lake was oligotrophic, with its phytoplankton communities exhibiting a seasonal pattern and being dominated by Bacillariophyta, which resulted in low diversity. In Azad Dam Lake, *Cyclotella menenghiniana* (Bacillariophyta) has the highest density and spread all over the area. The other important species are *Chlorella* sp. and *Scenedesmus quadricaudata* (Chlorophyta).

7.4.37 DENIZKOL RESERVOIR (TÜRKMENISTAN)

149 samples were taken from the Dengizkol Reservoir, and research work was carried out. In the Dengizkol Reservoir, microscopic algae include 123 species belonging to 4 divisions, 9 classes, 19 orders, 25 families, and 32 genera. *Chlorella vulgaris* from green phytoplankton, rich in physiologically active substances, and *Scenedesmus obliquus* (Turp.) Kütz. isolated algologically pure cells of the species, propagated in laboratory and semi-industrial conditions, and *Hypophthalmichthys molitrix* Val. used as supplementary food for fish (Shodmonov et al., 2023).

7.4.38 CASPIAN SEA (TURKMENISTAN, IRAN)

In accordance with the survey carried out by Tahami (2008) in the Southern Caspian Sea, five Phyla of phytoplankton were identified, of which the Crysophyta (diatom) has the maximum number and the biomass in all seasons. After the Crysophyta phylum, the Pyrophyta phylum has the highest biomass. Three Phyla, such as Cyanophyta, Chlorophyta, and Euglenophyta, had very low biomass.

During the study which was carried out by Nasrollahzadeh et al. (2008) in the Iranian Caspian Sea, it was observed that the dominant species such as *Rhizosolenia calcaravis* (Chrysophyta) and *Exuviaella cordata* (Pyrrophyta) disappeared or were found in low abundance in summer. The dominant species was mainly replaced by *Oscillatoria* sp. (Cyanophyta). However, some of the Chrysophyta like *Chaetoceros subtilis* and *Tribonema vulgare* were found to have increased in abundance, even more than before the introduction of an alien species. During autumn, a new succession between species occurred: *S. laxissma* showed a sharp increase in number. At the same time, *Chaetoceros mulleri* appeared in high abundance and persisted throughout the winter. Finally, as recovery begins in winter, the dominant species such as *T. nitzschioides* and *Nitzshia* sp. (Chrysophyta) and *Exuviaella cordata* (Pyrrophyta) reappeared. During the same period, the Cyanophyta species decreased in abundance and diversity.

According to a study by Kurochkina et al. (2023) in the Eastern Middle Caspian Sea, the phytoplankton composition included 50 species from four systematic groups that were registered in the eastern region of the Middle Caspian: Chlorophyta (6), Bacillariophyta (24), Dinoflagellata (13), and Cyanophyta (7). *Rhizosolenia calcar-avis*, *Anabaena bergii*,

Exuviaella cordata, and *Binuclearia lauterbornii* were the most important phytoplankton species.

7.5 CONCLUSION AND SUGGESTIONS

As seen in the studies on phytoplankton diversity in lakes, rivers, dam lakes, and lagoons located in the Irano-Anatolian hot spot region, phytoplankton species compositions are identified. In addition, chlorophyll-*a*, a good indicator of phytoplankton biomass, is analyzed along with nutrient salts (nitrite, nitrate, phosphate, etc.) that enable phytoplankton to grow. Usually, monthly sampling is done to reveal the seasonal changes in the water source studied and species distribution. In this way, information is obtained about the trophic status of the aquatic environment, primary production, water quality, and pollution loads. In particular, certain algae groups are used as bioindicators, and in this way, it is possible to improve water resources and prevent pollution. In summary, more importance should be given to phytoplankton identification, density, and distribution in ecological studies conducted in aquatic systems, and sufficient funds and budgets should be allocated to researchers by the necessary authorities so that biomonitoring studies can continue without slowing down.

KEYWORDS

- **conservation strategies**
- **habitat degradation**
- **Irano-Anatolian hotspot**
- **phytoplankton**
- **water pollution**
- **water quality**

REFERENCES

Asatryan, V.; Stepanyan, L.; Hovsepyan, A.; Khachikyan, T.; Mamyan, A.; Hambaryan, L. The Dynamics of Phytoplankton Seasonal Development and Its Horizontal Distribution in Lake Sevan (Armenia). *Environ. Monit. Assess.* **2022**, *194* (10), 757, 1–17. https://doi.org/10.1007/s10661-022-10446-5.

Azari, A. M.; Mohebbi, F.; Asem, A. Seasonal Changes in Phytoplankton Community Structure in Relation to Physico-Chemical Factors in Bukan Dam Reservoir (Northwest Iran). *Turkish J. Bot.* **2011**, *35* (1), 77–84.

Baykal, T.; Açıkgöz, İ.; Udoh, A. U.; Yıldız, K. Seasonal Variations in Phytoplankton Composition and Biomass in a Small Lowland River-Lake System (Melen River, Turkey). *Turkish J. Biol.* **2011**, *35* (4), 485–501.

Baykal, T.; Salman, S.; Açıkgöz, İ. The Relationship between Seasonal Variation in Phytoplankton and Zooplankton Densities in Hirfanlı Dam Lake (Kırşehir, Turkey). *Turkish J. Biol.* **2006**, *30* (4), 217–226.

Çelik, K.; Ongun, T. Spatial and Temporal Dynamics of the Steady-State Phytoplankton Assemblages in a Temperate Shallow Hypertrophic Lake (Lake Manyas, Turkey). *Limnology* **2008**, *9*, 115–123.

Dahesht, L. E.; Negarestan, H.; Eimanifar, A.; Mohebbi, F.; Ahmadi, R. The Fluctuations of Physicochemical Factors and Phytoplankton Populations of Urmia Lake, Iran. *Iran. J. Fish. Sci.* **2010**, *9* (3), 368–381.

Dalkıran, N.; Karacaoğlu, D.; Dere, Ş.; Çınar, Ş.; Bulut, C.; Savaşer, S. Uluabat Gölü Fitoplankton'unun Tür Kompozisyonu ve Zamansal-Mekansal Değişimi. *LimnoFish* **2016**, *2* (3), 121–135. https://doi.org/10.17216/LimnoFish.27972.

Ersanlı, E.; Gönülol, A. A Study on the Phytoplankton of Lake Simenit, Turkey. *Cryptogamie, Algologie* **2006**, *27* (3), 289–305. Adac. Tous Droits Réservés.

Fazli, H.; Tahami, F. S.; Saravi, H. N.; Roohi, A.; Eker-Develi, E.; Daryanabard, G. R.; Vahedi, F. The Response of Phytoplankton Assemblages to Environmental Parameters in the Azad Dam Lake in the West of Iran. *Int. J. Limnol.* **2022**, *58*, 15.

Gönülol, A.; Obalı, O. A Study on the Phytoplankton of Hasan Uğurlu Dam Lake (Samsun, Turkey). *Turkish J. Biol.* **1998**, *22* (4), 447–462.

Gönülol, A.; Obalı, O. Phytoplankton of Karâmık Lake (Afyon), Turkey. *Commun. Fac. Sci. Univ. Ank.* **1986**, *4*, 105–128.

Hafezieh, M., Mohebbi, F., Seidgar, M., Ahmadi, A. Phytoplankton Population Structure in Mighan Salt Lake (Arak, Markazi Province). *J. Phycol. Res.* **2020**, *4* (2), 534–543.

Hambaryan, L. R.; Gevorgyan, G. A. Investigations of Formation of Phytoplankton Community in the Arpa River (Armenia) and Its Main Tributaries. *Chem. Biol.* **2019**, *53* (1), 46–52.

Hambaryan, L.; Khachikyan, T.; Ghukasyan, E. Changes in the Horizontal Development of Phytoplankton of the Littoral of Lake Sevan (Armenia) in Conditions of Water Level Fluctuations. *Limnol. Freshwater Biol.* **2020**, *4*, 662–664.

Hovhannisyan, R. H.; Hovsepyan, A. A.; Hambaryan, L. R. Phytoplankton Distribution and Abundance in Relation to Physico-Chemical Factors in Lake Sevan, Armenia. *Natl. Acad. Sci. RA Electron. J. Nat. Sci.* **2010**, *15* (2), 28–37.

Ilanloo, S. S.; Ashrafi, S.; Shabani, A. A. Modeling Habitat Suitability of the Red-Backed Shrike (*Lanius collurio*) in the Irano-Anatolian Biodiversity Hotspot. *J. Zool. Res.* **2021**, *3* (2), 1–8.

Karacaoğlu, D.; Dalkıran, N.; Dere, Ş. Factors Affecting the Phytoplankton Diversity and Richness in a Shallow Eutrophic Lake in Turkey. *J. Freshwater Ecol.* **2006**, *21* (4), 575–581. https://doi.org/10.1080/02705060.2006.9664118.

Karacaoğlu, D.; Dere, Ş.; Dalkıran, N. A Taxonomic Study on the Phytoplankton of Lake Uluabat (Bursa). *Turkish J. Bot.* **2004**, *28* (5), Article 2.

Koçer, M. A. T.; Şen, B. Some Factors Affecting the Abundance of Phytoplankton in an Unproductive Alkaline Lake (Lake Hazar, Turkey). *Turkish J. Bot.* **2014**, *38* (4), Article 16.

Kurochkina, T. F.; Nasibulina, B. M.; Bakhshalizadeh, S.; Popov, N.; Kuanysheva, G.; Fazio, F.; Ali, A. M. Plankton Community Structure and Biomass in the Eastern Middle Caspian Sea. *Water* **2023**, *15*, 138. https://doi.org/10.3390/w15010138.

Kutlu, B.; Aydın, R.; Danabas, D.; Serdar, O. Temporal and Seasonal Variations in Phytoplankton Community Structure in Uzunçayır Dam Lake (Tunceli, Turkey). *Environ. Monit. Assess.* **2020**, *192*, 105.

Makhlough, A.; Saravi, H. N.; Ebrahimzadeh, M. The Water Quality of the Shahid Rajaee Reservoir (Mazandaran, Iran): Based on Phytoplankton Community. *Iran J. Sci. Technol. Trans. Sci.* **2017**, *41*, 627–635.

Maraşlioğlu, F.; Soylu, E. N.; Gönülol, A. Seasonal Variation of the Phytoplankton of Lake Ladik (Samsun, Turkey). *J. Freshwater Ecol.* **2005**, *20* (3), 549–553. https://doi.org/10.1080/02705060.2005.9664770.

Minasyan, A. Taxonomic Richness and Quantitative Characteristics of the Phytoplankton Community in the Epilimnion of Lake Yerevan (Armenia). *Lakes Reservoirs Ponds* **2016**, *10* (2), 77–95.

Mohebbi, F.; Riahi, H.; Sheidaei, M.; Shariatmadari, Z. Phytoplankton of Aras Dam Reservoir (Iran): An Attempt to Assess Water Quality. *Iran. J. Fish. Sci.* **2016**, *15* (4), 1318–1336.

Nasrollahzadeh, H. S.; Din, Z. B.; Foong, S. Y.; Makhlough, A. Trophic Status of the Iranian Caspian Sea Based on Water Quality Parameters and Phytoplankton Diversity. *Cont. Shelf Res.* **2008**, *28* (9), 1153–1165. https://doi.org/10.1016/j.csr.2008.02.015.

Naz, M.; Türkmen, M. Phytoplankton Biomass and Species Composition of Lake Gölbaşı (Hatay, Turkey). *Turkish J. Biol.* **2005**, *29* (1), 49–56. https://journals.tubitak.gov.tr/biology/vol29/iss1/8 (accessed April 21, 2026).

Oğuz, A.; Akçaalan Albay, R.; Köker, L.; Gürevin, C.; Dorak, Z.; Albay, M. Driving Factors Affecting the Phytoplankton Functional Groups in a Deep Alkaline Lake. *Turkish J. Bot.* **2020**, *44* (6), 633–646. https://doi.org/10.3906/bot-2004-33.

Öterler, B.; Albay, M.; Çamur-Elipek, B.; Güher, H.; Kırgız, T. Spatial and Temporal Distribution of Phytoplankton in Lake Gala (Edirne, Turkey). *Trakya Univ. J. Nat. Sci.* **2015**, *16* (2), 71–80.

Sevindik, T. O. Phytoplankton Composition of Çaygören Reservoir, Balıkesir, Turkey. *Turkish J. Fish. Aquat. Sci.* **2010**, *10*, 295–304.

Shams, M.; Afsharzadeh, S.; Atıcı, T. Seasonal Variations in Phytoplankton Communities in Zayandeh-Rood Dam Lake (Isfahan, Iran). *Turkish J. Bot.* **2012**, *36* (6), Article 12.

Shodmonov, F.; Ruzibaeva, D.; Allamuratova, G.; Abdurayimov, T.; Abdurasulov, S. Determining the Algoflora of Dengizkol Lake and Breeding Promising Species in Laboratory Conditions. *E3S Web Conf.* **2023**, *421*, 01003. SERBEMA.

Sonmez, F.; Kutlu, B.; Sesli, A. Spatial and Temporal Distribution of Phytoplankton in Karkamış Dam Lake (Şanlıurfa, Turkey). *Fresenius Environ. Bull.* **2017**, *26* (10), 6234–6245.

Tahami, F. S. A. G. Study on the Distribution and Biomass of Phytoplankton in Southern Caspian Sea. *Health Environ. Res. Online* **2008**, *28*, 59, 60.

Taş, B. *Energy Education Science and Technology Part A: Energy Science and Research* **2012**, *30* (1), 33–44.
Taş, B. I.; Gönülol, A.; Taş, E. A Study on the Seasonal Variation of the Phytoplankton of Lake Cernek (Samsun, Turkey). *Turkish J. Fish. Aquat. Sci.* **2002**, *2*, 121–128.
Yilmaz, N. Assessment of Seasonal Variation of Phytoplankton and Related Water Quality Parameters of Sazlidere Dam Lake (Istanbul, Turkey). *Desalin. Water Treat.* **2018**, *131*, 107–113.
Yilmaz, N. Diversity of Phytoplankton in Küçükçekmece Lagoon Channel, Turkey. *Maejo Int. J. Sci. Technol.* **2015**, *9*, 32–42.
Yilmaz, N. Phytoplankton Composition of Sazlidere Dam Lake, Istanbul, Turkey. *Maejo Int. J. Sci. Technol.* **2013**, *7*, 203–211.
Yilmaz, N. Water Quality Assessment Based on the Phytoplankton Composition of Büyükçekmece Dam Lake and Its Influent Streams (Istanbul, Turkey). *Desalin. Water Treat.* **2019**, *159*, 3–12.
Yilmaz, N.; Aykulu, G. An Investigation on the Seasonal Variation of the Phytoplankton Density on the Surface Water of Sapanca Lake, Turkey. *Pakistan J. Bot.* **2010**, *42*, 1213–1224.
Yilmaz, N.; Dogan, I.; Demir, G.; Yalcin, I. E.; Khan, S.; Ozyigit, I. I. Assessment of Pollution in Alibeyköy Dam Lake (Istanbul, Türkiye) and Its Influent Streams: Phytoplankton Composition and Heavy Metal Accumulation. *Water Environ. Res.* **2025**, *97* (2).
Yilmaz, N.; Elhag, M.; Yasar, U. Consideration of Phytoplankton Composition and Water Quality of Anamur (Dragon) Creek, Turkey. *Desalin. Water Treat.* **2017**, *91*, 386–394.
Yilmaz, N.; Gulecal, Y. Phytoplankton Community of Terkos Lake and Its Influent Streams, Istanbul, Turkey. *Pakistan J. Bot.* **2012**, *44*, 1135–1140.
Yilmaz, N.; Ozyigit, I. I.; Demir, G.; Yalcin, I. E. Determination of Phytoplankton Density and Study of the Variation of Nutrients and Heavy Metals in the Surface Water of Riva Stream, One of the Water Sources of Istanbul, Turkey. *Desalin. Water Treat.* **2014**, *55* (3), 810–820. https://doi.org/10.1080/19443994.2014.952674.
Yilmaz, N.; Özyiğit, İ. İ.; Demir, H. H.; Yalcin, I. E. Assessment of Phytoplankton Composition and Heavy Metal Pollution in a Drinking Water Resource: Lake Terkos (Istanbul, Turkey). *Desalin. Water Treat.* **2021b**, *225*, 265–274.
Yilmaz, N.; Özyiğit, İ. İ.; Dogan, I.; Demir, G.; Yalcin, I. E. A Case Study Performed in Küçükçekmece Lagoon Channel (Istanbul, Turkey): How Heavy Metal Contamination and Seasonal Variations on Phytoplankton Composition Influence Water Quality. *Desalin. Water Treat.* **2021a**, *239*, 126–136.
Yilmaz, N.; Yardimci, C. H.; Elhag, M.; Dumitrache, C. A. Phytoplankton Composition and Water Quality of Kamil Abduş Lagoon (Tuzla Lake), Istanbul, Turkey. *Water* **2018**, *10*, 1–13.

CHAPTER 8

ZOOPLANKTON DIVERSITY OF THE IRANO-ANATOLIAN HOTSPOT

PINAR GÜRBÜZER[1,2*]

[1]*Department of Hydrobiology, Faculty of Fisheries, Sinop University, Sinop, Türkiye*

[2]*Department of Astronautical Engineering, University of Turkish Aeronautical Association, Ankara, Türkiye*

*Corresponding author

ABSTRACT

Zooplanktonic organisms are essential for biodiversity in aquatic environments for their role of grazing on primary producers and serving as prey for higher predators. Aquatic habitat regions and climate have an effect on their distribution and adaptations to it. The Irano-Anatolian Hotspot region extends from high altitude mountainous regions to lowland lakes, so zooplankton diversity is expected to be high. In this study, the zooplankton community research, based on literature analysis from 37 lakes, reservoirs, and streams across 9 regions, was conducted in the hotspot region. A total of 373 zooplankton taxa have been documented, and 289 of these are from Rotifera, followed by 49 from Cladocera, and 35 from Copepoda. The most common species were identified as *Keratella cochlearis*, *Keratella quadrata, Lecane luna,* and *Polyarthra dolichoptera* among rotifers; *Bosmina longirostris* among cladocerans, each found in 28 and above aquatic habitats; and the copepod species *Cyclops vicinus* were recorded in 20 different locations. These species are cosmopolitan in the Palearctic region and important to predicting the impacts of future environmental threats.

Biodiversity Hotspot of the Irano-Anatolian Region. T. Pullaiah (Ed.)

DOI: 10.1201/9781779645760-8

8.1 INTRODUCTION

Microscopic organisms play a vital part in biogeochemical cycles and contribute significantly to biodiversity in most environments (Martinez-Garcia et al., 2025). Many aquatic environments, such as lakes, ponds, rivers, lagoons, sinkholes, marshes, cave waters, troughs, creeks, dams, and oceans, are home to a majority of microscopic organisms (Durmaz et al., 2022; Gürbüzer et al., 2023; Ustaoğlu et al., 2012). There is at least one group of zooplanktonic organisms in these ecosystems, especially Rotifera and crustaceans such as Cladocera and Copepoda. Small zooplankton feed herbivorously on algae and other primary producers while also serving as prey for carnivores. Through this dual role, they exert grazing and predation pressure on algae, thereby controlling eutrophication and phytoplankton growth, while also playing a key role in the dynamics of the pelagic food web (Gyllström et al., 2005; Jeppesen et al., 2000). The abundance and composition of zooplankton species may be impacted by a number of biotic and abiotic factors, including temperature, habitat variations, and the presence or absence of fish and macrophytes. Their sensitivity to environmental changes makes them valuable bioindicators for monitoring aquatic ecosystem health. However, anthropogenic activities such as pollution, habitat modification, and climate change are increasingly altering zooplankton populations, potentially disrupting food web dynamics, and biogeochemical processes (Kirillin, 2010; Wang et al., 2023). Also, the lake's region and climate have an effect on the distribution and adaptations of zooplankton. In temperate regions lakes, seasonal changes influence zooplankton vertical migration and life cycles, while in tropical lakes relatively stable temperatures and high richness create year-round competition shaping their feeding strategies, spatial distribution, etc. (Adrian et al., 2009; Hrycik et al., 2022; Sommer et al., 1986).

Globally, 36 regions are recognized as biodiversity hotspots, collectively encompassing only 2.5% of Earth's terrestrial surface with their remaining intact habitats (Conservation International, n.d.). The Irano-Anatolian hotspot, situated in Asia, retains approximately 3.6% of its area as undisturbed natural vegetation (Gür, 2022; Sloan et al., 2014). Aquatic ecosystems are affected by the environmental conditions of the regions where they are located, such as minimum and maximum temperature, annual precipitation, wind speed, and and other environmental drivers (Abbaspour et al., 2009). The Irano-Anatolian hotspot spans central and eastern Turkey, western Iran, northern Iraq, and some parts of Armenia,

Azerbaijan, Turkmenistan, and Georgia. It encompasses high-altitude plateaus and mountain ranges, including the Zagros and Alborz Mountains, and extends into adjacent lowlands and steppe regions.

Most of the borders of the Irano-Anatolian region are located in the Anatolian region of Türkiye and the northwestern parts of Iran. The region is mostly BSk (Dry, Semi-arid steppe, Cold), CSb (Temperate, Dry summer, Warm summer), and DSb (Continental, Dry summer, Warm summer) in the Köppen-Geiger Climate Classification (Kottek et al., 2006). The central and east part of Anatolia is located in the Irano-Anatolian Hotspot. The central part of the region is mainly lower than 1500m and occupied by shallow lakes (Gür, 2016; Kuzucuoğlu et al., 2019). The other part, which rises slowly towards the east, also called the high Anatolian Plateau (eastern part), is mostly above 1500m and generally characterized by fast-flowing rivers (Gür, 2016; Kuzucuoğlu et al., 2019). Türkiye's coasts (north, south, and west) have a more temperate climate, while the Anatolian region has an arid and cold climate. Iran is semiarid and arid, so evaporation is high, and although the country has mountainous regions, the climatic conditions bring salinization in aquatic ecosystems, which affects the composition and richness of aquatic organisms (Abbaspour et al., 2009; Malekzadeh-Viayeh, 2021).

The Irano-Anatolian region is predominantly mountainous but contains both basin lakes and lakes along alpine and sub-alpine belts. The highest peaks of the region include Mount Ararat (5,165 m a.s.l.) and Mount Damavand (5,671 m a.s.l.) in the Alborz range. However, due to the limited number of studies conducted outside Turkey, the current knowledge about the aquatic life in these lakes is insufficient.

Currently, there are 312 lakes and reservoirs with a surface area larger than 50 km^2, and 133 lakes and ponds with a surface area larger than 0.1 km^2 but smaller than 50 km^2 (Figure 8.1). Lake data were obtained from the HydroSheds database, with additional information on lake area, elevation, and related attributes sourced from the HydroLakes database and the Global Lakes and Wetlands Database (GLWD) layers (Messager et al., 2016). The highest altitude lake in the region is Tar Lake (2899 meters above sea level), located in Iran.

Zooplankton studies in Iran, especially on the Rotifera phylum, were detailed in Malekzadeh-Viayeh in 2021 (Malekzadeh-Viayeh, 2021). Eight provinces of Iran are part of the Irano-Anatolian Hotspot. According to Malekzadeh-Viayeh's research, a total of 236 taxa were identified in the regions. Among them, 53 taxa were reported in Fars Province, 35 in Merkezi Province, 149 in West Azarbaijan, 20 in East Azarbaijan, 15 in Chaharmahal and Bakhtiary, 18 in Isfehan, 18 in Ardabil Province, and 3 in Lorestan. Additionally, Khoei's (Khoei et al., 2011) research conducted a study on rotifer species in aquatic ecosystems in and around Tehran Province. While Tehran city is not entirely within the hotspot region, its northern, northeastern, and southern parts are included. Therefore, it has been added to the zooplankton list for the Irano-Anatolian Hotspot shown below.

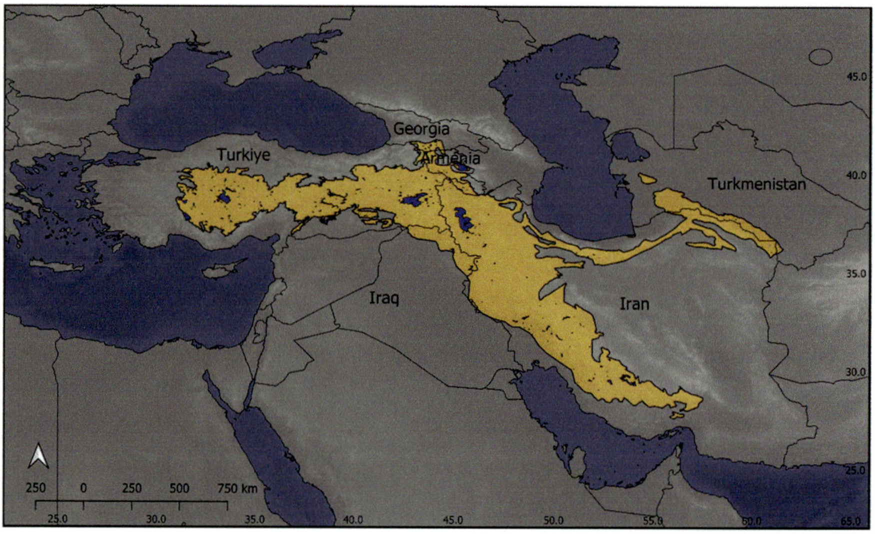

FIGURE 8.1 Map showing the location of the Irano-Anatolian Hotspot and the lakes in the region

A total of 373 zooplankton taxa have been documented in the literature based on studies conducted by various researchers in the region. Table 8.1 presents the species list compiled from studies carried out in 11 lakes, 1 pond, 19 dam lakes, 3 streams, 2 karstic sinkholes, and 1 marsh area within the region. For each country, literature searches were performed using the keywords "zooplankton," "rotifera," "cladocera," and "copepod." The list was compiled based on data from these articles focusing on lakes located within the hotspot area. Accordingly, data were obtained from 34 wetland studies in Turkiye, 10 in Iran, 1 in Iraq, and 1 in Armenia. It is important

TABLE 8.1 List of Aquatic Locations and Their Codes

Code	Location	Code	Location
T1	Sariyar Lake (Ergönül et al., 2016)	T24	Lake Van (Yildiz et al., 2010)
T2	Cubuk Dam Lake (Ergönül et al., 2016)	T25	Zernek Dam Lake (Yildiz, 2012)
T3	Kurtbogazi Dam Lake (Ergönül et al., 2016)	T26	Nazik Lake (Saler et al., 2019a)
T4	Ilgin Lake (Ergönül et al., 2016)	T27	Beysehir Lake (Altındağ and Yiğit, 2004)
T5	Mogan Lake (Altındağ et al., 2007; Ergönül et al , 2016)	T28	Akşehir Lake (Altindağ and Yiğit, 1999)
T6	Eymir Lake (Ergönül et al., 2016)	T29	Hirfanli Dam Lake (Yiğit and Altindağ, 2005)
T7	Asartepe Dam Lake (Buyurgan et al., 2010)	T30	Gelingullu Dam Lake (Kaya and Altindağ, 2007)
T8	Tercan Dam Lake (Saler and Selamoglu, 2020)	T31	Sultan Marshes (Kaya et al., 2012)
T9	Kuzgun Dam Lake (Saler and Selamoglu, 2020)	T32	Meyil Karstic Sinkhole (Durmaz et al., 2022)
T10	Demirdoven Dam Lake (Saler and Selamoglu, 2020)	T33	Kizoren Karstic Sinkhole (Durmaz et al., 2022)
T11	Kars River (Özbay and Altidag, 2009)	T34	Keban Dam Lake (Saler and Sen, 2010; Saler and İpek, 2020; Tellioğlu and Akman, 2007)
T12	Karasu River (Saler and Karakaya, 2023)	IR1	Lar Reservoir (Biria et al., 2011)
T13	Kigi Dam Lake (Bulut, 2018a)	IR2	Fars (Malekzadeh-Viayeh, 2021)
T14	Ozluce Dam Lake (İpek Alış and Saler, 2016)	IR3	Merkezi (Malekzadeh-Viayeh, 2021)
T15	Kapiacmaz Pond (Bulut, 2018b)	IR4	West Azarbaijan (Malekzadeh-Viayeh, 2021)
T16	Deriner Dam Lake (Bulut et al., 2024)	IR5	Lorestan (Malekzadeh-Viayeh, 2021)
T17	Uzuncayir Dam Lake (Saler et al., 2014)	IR6	East Azarbaijan (Malekzadeh-Viayeh, 2021)
T18	Cat Dam Lake (Saler et al., 2019b)	IR7	Chaharmahal and Bakhtiary (Malekzadeh-Viayeh, 2021)

TABLE 8.1 *(Continued)*

Code	Location	Code	Location
T19	Medik Reservoir (Saler, 2017)	IR8	Isfehan (Malekzadeh-Viayeh, 2021)
T20	Peri Stream (Saler et al., 2011)	IR9	Ardabil (Malekzadeh-Viayeh, 2021)
T21	Kalecik Dam Lake (Bulut and Saler, 2013)	IR10	Tehran (Hakimzadeh Khoei et al., 2011; Malekzadeh-Viayeh, 2021)
T22	Hazar Lake (Tellioğlu, 1998)	AR1	Lake Sevan (Hayrapetyan et al., 2022; Vezhnavets et al., 2024)
T23	Beyhan Dam Lake (Bulut and Saler, 2014)	IQ1	Lake Dukan (Dhahir and Ali, 2017)

T: Turkiye, IR:Iran, AR: Armenia, IQ: Iraq.

to note that many studies on aquatic areas within Iran are written in Farsi; therefore, only data from articles published in English were included. The names of the aquatic sites, their assigned codes, and the corresponding articles are provided in Table 8.1.

A total of 202 taxa were identified in studies conducted in the Irano-Anatolian Region of Turkiye, and the number of zooplanktonic organisms belonging to Rotifera, Cladocera, and Copepoda groups were 147, 40, and 15, respectively. In Iran, records indicate a total of 246 rotifer species. There are 34 taxa composed of 16 Rotifera, 8 Cladocera, and 10 Copepoda in Iraq, whereas 42 taxa comprising 21 Rotifera, 8 Cladocera, and 13 Copepoda in Armenia. The number of zooplankton groups listed for each country and relative percentages of the zooplanktonic organisms from the hotspot area were presented in Figure 8.2. Zooplankton species list and their locations are provided in Table 8.2.

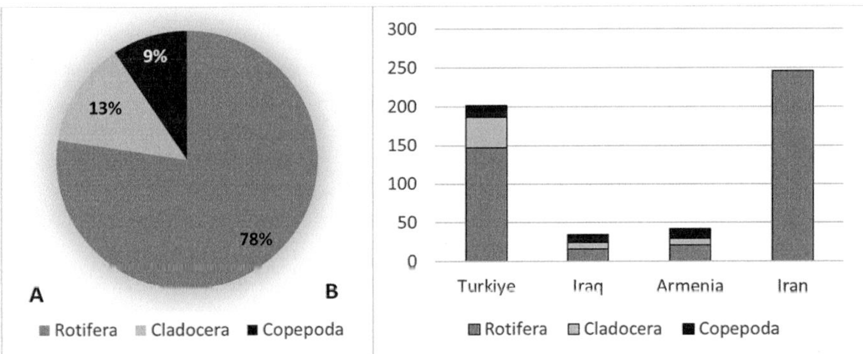

FIGURE 8.2 Zooplankton group percentages from whole Irano-Anatolian biodiversity hotspot region (A), species number and distribution from each country (B).

Following the data presented in Table 8.2, from studies carried out in 37 lakes, reservoirs, streams, and across 9 regions, *Keratella cochlearis* (65% frequency = F) was identified as the most common species, found in 30 different locations. This was followed by *Keratella quadrata, Lecane luna,* and *Polyarthra dolichoptera* among rotifers, and *Bosmina longirostris* among cladocerans, each found in 28 aquatic habitats (their frequencies are 61%). *Synchaeta pectinata* and *Daphnia longispina* were recorded in 25 different locations (54% F). *Notholca squamula* (23), *Asplanchna priodonta,* and *Euchlanis dilatata* (22), *Brachionus angularis, Cephalodella gibba,*

TABLE 8.2 List of Irano-Anatolian Hotspot Zooplanktonic Species and Their Distribution Throughout the Region*

Taxon	Locations
ROTIFERA	
Adineta sp.	IR4
Adineta vaga	IR3
Anuraeopsis fissa	T7, T30, IQ1, T5, IR4
Anuraeopsis sp.	IR6, IR7
Ascomorpha ecaudis	T30, T21, T12, T5, AR1
Ascomorpha ovalis	T16
Ascomorpha saltans	T30, T22, T5, T19, T26, T8, T9, T17, T14, T13, T16, T12, T15, T34
Ascomorpha sp.	IR1, IR6, IR7, IR8
Asplanchna brightwelli	IR1, IR4, IR2
Asplanchna girodi	T16, T30, AR1, T33, IR9
Asplanchna hyalina	T34
Asplanchna priodonta	T23, T18, T26, T1, T2, T3, T7, T16, T30, T29, T21, T12, T28, T5, T19, T20, T17, T25, T33, T34, IR4, IR2
Asplanchna sieboldi	T16, T30, T15, T12, T28, T5, T19, T26, T8, T9, T10, T17, T34, IR4
Asplanchna silvestris	T28
Asplanchna sp.	IR4, IR6, IR9, IR7, IR8
Bdelloida sp.	AR1, IR8, IR2
Beauchampiella eudactylota	IR2
Brachionus angularis	T7, T23, T30, T22, T29, T4, T1, T21, T12, T13, T28, T5, T24, T19, T32, T14, T34, IR4, IR3, IR2
Brachionus asplanchnoidis	IR4
Brachionus bennini	IR4
Brachionus bidentatus	T12, IR4, IR2
Brachionus calyciflorus	T30, T22, T29, T11, T28, T27, T5, AR1, T8, T9, T10, T25, T33, T34, IR4
Brachionus diversicornis	T1, IR2
Brachionus forficula	T4, T1
Brachionus havanensis	IR2
Brachionus leydigii	IR4
Brachionus nilsoni	IR4
Brachionus plicatilis	T22, T4, T28, T27, T24, T32, T34, IR4, IR8, IR2

TABLE 8.2 *(Continued)*

Taxon	Locations
Brachionus quadridentatus	T7, T30, T22, T29, T12, T28, T5, AR1, T19, T32, T33, T20, IR4, IR3, IR2
Brachionus rubens	IR10
Brachionus sp.	IR4, IR6, IR9, IR5, IR7, IR2
Brachionus urceolaris	T30, T29, T15, T28, T5, T32, T33, T8, T9, T25, T34, IR4, IR3, IR2
Brachionus variables	IR10
Cephalodella auriculata	T30, T23
Cephalodella belone	IR3
Cephalodella catellina	T7, T16, T30, T29, T12, T28, T5, AR1, T10, IR4
Cephalodella compressa	IR3
Cephalodella forficula	T21, T12, T11, T19, T26, T9, T10, T33, T34, IR4, IR3
Cephalodella gibba	T7, T23, T2, T30, T22, T21, T12, T11, T13, T28, T27, T5, T19, T14, T20, T8, T17, T17, T34, IR4
Cephalodella gibboides	IR4
Cephalodella gracilis	T7, IR3, IR4
Cephalodella inquilina	IR9, IR3
Cephalodella intuta	T7
Cephalodella lepida	IR4
Cephalodella maior	IR10
Cephalodella misgurnus	IR4, IR10
Cephalodella obvia	IR4
Cephalodella plicata	IR10
Cephalodella sp.	T32, IR4, IR6, IR9
Cephalodella stenroosi	IR4
Cephalodella sterea	IR4
Cephalodella tincaformis	IR4
Cephalodella ventripes	T7, T23, T11, IR4
Cephalodella vittata	IR4
Collotheca heptabrachiata	IR10
Collotheca mutabilis	T7, T30
Collotheca ornata	T7, T30, IR4
Collotheca pelagica	IR10
Collotheca sp.	AR1, IR1, IR6, IR7

TABLE 8.2 *(Continued)*

Taxon	Locations
Colurella adriatica	T7, T23, T30, T22, T21, T11, T28, T27, T5, T19, T33, T34, IR8, IR2
Colurella colurus	T7, T23, T15, T12, T11, T24, T19, T33, IR4, IR3
Colurella obtusa	T30, T12, T11, T8, T33, IR4, IR3
Colurella sinistra	IR4
Colurella sp.	IR4, IR6
Colurella uncinata	T11, T27, T5, T34, IR4, IR4
Conochiloides dossurarius	T18, IR10
Conochilus hippocrepis	T21, AR1, IR10
Conochilus sp.	T28, IR7
Conochilus unicornis	T30, T21, AR1
Dicranophoroides caudatus	IR2
Dicranophorus epicharis	T15, T5, T33, IR4
Dicranophorus forcipatus	T5, IR4, IR3
Dicranophorus grandis	T7, T30, T12, T11
Dicranophorus sp.	T16
Dipleuchlanis propatula	IR4
Diplois daviesiae	IR4
Dissotrocha aculeata	IR4
Dissotrocha macrostyla	IR10
Drilophaga judayi	IR3
Embata laticeps	IR10
Encentrum felis	T7
Encentrum putorius	T11
Encentrum saundersiae	T15, T12, T26, IR4
Encentrum sp.	IR4
Encentrum uncinatum	T11
Encentrum ussuriensis	IR3
Encentrum walterkostei	IR3
Eosphora anthadis	IR4
Eosphora ehrenbergi	IR4
Eosphora najas	IR4
Eothinia elongata	IR4
Epiphanes senta	T15, T19, T34, IR4, IR3

TABLE 8.2 *(Continued)*

Taxon	Locations
Euchlanis dilatata	T7, T23, T18, T29, T21, T15, T12, T13, T28, T27, IQ1, T5, AR1, AR1, T19, T26, T14, T8, T10, T33, IR4, IR8
Euchlanis incisa	T7, T11, IR10
Euchlanis lyra	IR4
Euchlanis meneta	T27
Euchlanis parva	IR2
Euchlanis sp.	IR7
Eusphora thoides	T28
Filinia brachiata	IR4
Filinia cornuta	T7, T30
Filinia limnetica	T7, T30, T28, T5, IR10
Filinia longiseta	T30, T22, T4, T21, T12, T28, T5, AR1, T32, T33, T26, T25, T34, IR4, IR2
Filinia maior	T24
Filinia minuta	T5
Filinia opoliensis	T19
Filinia pejleri	T5, T8, T10
Filinia sp.	IR4, IR6, IR9, IR7
Filinia terminalis	T7, T23, T18, T30, T22, T21, T15, T12, AR1, T26, T25, T34, IR4
Floscularia ringens	IR10
Gastropus minor	IR4
Gastropus stylifer	T16, T30, T21, T5
Habrotocha constricta	IR10
Harringia rousseleti	IR2
Hexarthra bulgarica	IR4
Hexarthra fennica	T30, T28, T27, T24, T34, IR4, IR2
Hexarthra intermedia	IR2
Hexarthra jenkinae	IR4
Hexarthra mira	T4, T5, AR1, T24, T19, T34, IR4, IR2
Hexarthra oxyuris	T5, IR10
Hexarthra polyodonta	T32, T33, IR10
Hexarthra sp.	IR4, IR7
Itura aurita	T7, IR4

TABLE 8.2 *(Continued)*

Taxon	Locations
Kellicottia longispina	T18, T30, T22, T4, T21, T12, T13, T19, T14, T20, T17, T17, T25, T9, T34, IR6
Keratella cochlearis	T7, T23, T18, T30, T29, T4, T2, T3, T21, T15, T12, T13, T27, IQ1, T5, IR1, T19, T14, T20, T8, T9, T10, T17, T17, T25, T34, IR4, IR3, IR2, IR9
Keratella edmondsoni	T12
Keratella irregularis	IR4
Keratella lenzi	IR4
Keratella quadrata	T7, T23, T18, T30, T29, T12, T13, T28, T27, T5, AR1, T24, T19, T32, T33, T6, T26, T14, T20, T8, T9, T10, T17, T17, T34, IR4, IR2, IR4
Keratella sp.	IR4, IR6, IR9, IR7, IR8
Keratella tecta	T23, T18, T16, T30, T4, T1, T2, T3, T2, T21, T15, T12, T5, T19, T8, T9, T10, IR4, IR3, IR4
Keratella tropica	IQ1, T5, T14, T33, IR4, IR2
Keratella valga	T19, IR2
Lapedella ovalis	T14
Lecane broaensis	IR3
Lecane bulla	T7, T23, T30, T12, T11, T27, IQ1, T5, T3, T33, IR4, IR8
Lecane closterocerca	T7, T30, T21, T11, T28, T27, T5, T3, T8, T9, T10, T33, IR4, IR3
Lecane depressa	IR4
Lecane elasma	IQ1
Lecane elegans	T27
Lecane elsa	IR10
Lecane flexilis	T27, IR4
Lecane furcata	T11, IR4
Lecane grandis	T27, T24, T33, IR10
Lecane hamata	T30, T11, T5, T33, IR4, IR2
Lecane hastata	IR4, IR3
Lecane inconspicua	IR9
Lecane inermis	IR3
Lecane intrasinuata	T18
Lecane lamellata	T27, T24, T32, IR4
Lecane leontina	IR10

TABLE 8.2 *(Continued)*

Taxon	Locations
Lecane ludwigi	T5
Lecane luna	T7, T23, T30, T22, T29, T15, T12, T11, T13, T28, T27, T5, AR1, T2, T14, T8, T9, T10, T17, T17, T25, T18, T21, T33, T34, IR4, IR8, IR2
Lecane lunaris	T7, T23, T30, T29, T11, T13, T27, T5, T8, T9, T10, T33, T18, T21, T34, IR4, IR3, IR8, IR2
Lecane nana	T33, IR10
Lecane obtusa	IR2
Lecane ohiensis	T28, T24, T26, T8, T9, T10, T33
Lecane opias	IR10
Lecane papuana	IR4
Lecane paradoxa	IR3
Lecane pumila	IR4
Lecane punctata	IQ1, IR4
Lecane pyriformis	T34, IR3
Lecane quadridentata	T5, IR4, IR8, IR2
Lecane rugosa	T27, T5
Lecane sp.	IR4, IR6, IR5, IR7
Lecane stenroosi	T30, T11, IR2
Lecane tenuiseta	IQ1
Lecane thalera	IR4, IR2
Lecane undulata	IQ1
Lepadella acuminata	T30, T27, IR4, IR3
Lepadella apsida	191IR4
Lepadella biloba	T11, IR3
Lepadella cornuta	T12, T34
Lepadella costata	T34, IR10
Lepadella discoidea	IR4
Lepadella heterostyla	IR2
Lepadella neglecta	IR3
Lepadella ovalis	T7, T23, T12, IQ1, T19, T8, T17, T17, T25, T34, IR4, IR3
Lepadella patella	T7, T23, T30, T29, T15, T12, T11, T28, T27, IQ1, T19, T8, T33, T34, IR4, IR3, IR8, IR2
Lepadella punctata	IR10

TABLE 8.2 *(Continued)*

Taxon	Locations
Lepadella quadricarinata	T7, T11, T5, IR10
Lepadella sp.	IR6, IR9, IR7
Lepadella triptera	T5, IR4, IR2
Liliferotrocha subtilis	T28
Limnias melicerta	IR2
Lindia janickii	IR4
Lindia truncata	IR4
Lophocharis salpina	T30, T28, T27, T5, IR4
Macrotrachela quadricornifera	IR10
Monommata actices	IR4
Monommata arndti	T7, T30
Monommata longiseta	IR10
Monommata sp.	T28
Mytilina mucronata	IR4, IR2
Mytilina spinigera	T24, T25
Mytilina ventralis	T28, T5, T19, IR4, IR2, IR8
Nitokra hibernica	T12, T19, T26
Notholca acuminata	T22, T29, T12, T28, T27, T5, AR1, IR1, T14, T17, T17, IR4
Notholca caudata	IR4
Notholca labis	IR10
Notholca salina	IR4
Notholca sp.	IR4, IR9, IR7
Notholca squamula	T7, T23, T18, T30, T22, T29, T21, T12, T13, T27, IQ1, T24, T19, T14, T20, T17, T17, T25, T34, IR4, IR8, IR2, IR10
Notholca striata	IR10
Notommata aurita	IR10
Notommata brachyota	IR4
Notommata copeus	T15, T5, IR2
Notommata cyrtopus	T7, T5
Notommata diasema	IR4
Notommata glyphura	T23, T8, T9, IR4
Notommata pygmaea	IR4
Notommata sp.	IR4

TABLE 8.2 *(Continued)*

Taxon	Locations
Notommata tripus	IR9
Pedalia sp.	IR1
Philodina citrina	IR4
Philodina megalotrocha	T7, T30, T27, T5, T33, IR4
Philodina paradoxus	IR10
Philodina roseola	T12, IQ1, T26, T34, IR4, IR3, IR2
Philodina sp.	IR1, IR4, IR6, IR5
Platyias quadricornis	IR4, IR2
Pleurotrocha petromyzon	T11, IR10
Polarthra vulgaris	T28
Polyarthra dolichoptera	T23, T18, T16, T30, T22, T29, T4, T1, T2, T3, T21, T15, T12, T11, T13, T5, AR1, IR1, T19, T26, T14, T8, T9, T10, T17, T17, T34, IR4
Polyarthra euryptera	IR4
Polyarthra major	T4
Polyarthra minor	IR4
Polyarthra remata	T7, T16, T30, T4, T1, T26, T20, T34, IR4, IR3, IR10
Polyarthra sp.	AR1, IR4, IR6, IR9, IR7, IR8
Polyarthra vulgaris	T7, T16, T30, T29, T4, T3, T27, IQ1, T5, AR1, T32, T33, T20, T25, 3a, T34, IR4, IR2
Pompholyx sp.	IR6, IR9, IR7
Pompholyx sulcata	T30, T19
Proales fallaciosa	T23, IR3
Proales theodora	IR10
Proalides sp.	IR6, IR9
Proalides tentaculatus	IR4
Ptygura furcillata	IR10
Ptygura melicerta	T5, IR4
Ptygura sp.	T27
Ptygura tridorsicornis	IR4
Resticula melandocus	T5, IR4
Resticula nyssa	IR4
Rhinoglena frontalis	T28
Rotaria citrina	IQ1, IR3

TABLE 8.2 *(Continued)*

Taxon	Locations
Rotaria mucrora	IR10
Rotaria neptunia	T30, T11, T5, T14, IR4, IR4
Rotaria rotatoria	T7, T13, T5, T19, IR4
Rotaria socialis	IR10
Rotaria sp.	IR6, IR9
Rotaria tardigrada	IQ1, T5
Scaridium longicaudum	IR4, IR8, IR2
Scaridium sp.	IR6
Sinantherina semibullata	IR10
Squatinella rostrum	IR4, IR2
Stephanoceros fimbriatus	IR10
Synchaeta littoralis	T30, IR4, T29
Synchaeta oblonga	T7, T23, T30, T21, T12, T13, T5, T14, IR4
Synchaeta pectinata	T7, T23, T16, T30, T29, T21, T15, T12, T13, T28, T5, AR1, IR1, T19, T14, T20, T17, T17, T25, T9, T10, T33, T34, IR4, IR2
Synchaeta	IR4, IR6, IR9
Synchaeta stylata	T30, IR10
Synchaeta tremula	IR4
Synchaeta verrucosa	T22
Synchaeta vorax	IR10
Testudinella incisa	IR10
Testudinella mucronata	IR10
Testudinella parva	T27, IR10
Testudinella patina	T27, T5, T19, T34, IR4, IR8, IR2
Trichocerca agnatha	IR4
Trichocerca bidens	T7, IR4
Trichocerca capucina	T18, T21, T12, T13, T14, T17, T17
Trichocerca cavia	IR4, IR8, IR2
Trichocerca cylindrica	T30, T24, T34, IR10
Trichocerca dixonnuttalli	IR2
Trichocerca elongata	T30, T5, IR4
Trichocerca longiseta	T4, T28, T19, IR4, IR2
Trichocerca myersi	IR10

TABLE 8.2 *(Continued)*

Taxon	Locations
Trichocerca parvula	T5
Trichocerca porcellus	IR4
Trichocerca pusilla	T7, T28, T5, T33, IR4, IR2
Trichocerca rattus	T5, IR4, IR2
Trichocerca similis	T7, T18, T2, T3, T30, T21, T15, T8, T10, IR4, IR3
Trichocerca sp.	IR4, IR6, IR9
Trichocerca stylata	T28, IR10
Trichocerca taurocephala	T24
Trichocerca tenuior	T30
Trichocerca weberi	IR4, IR2
Trichotria pocillum	T7, T27, T5, AR1, T33, IR4
Trichotria tetractis	T23, T21, T12, T28, T27, T5, T26, T14, T8, T10, T17, T17, T33, T34, IR4, IR2
Trihocerca sp.	T27
CLADOCERA	
Acroperus harpae	T27
Alona affinis	T31
Alona guttata	T11, T27, T33
Alona quadrangularis	T30, T31
Alona rectangula	T30, T21, T11, T27, T5, T19, T26, T31
Alona sp.	AR1
Alonella excisa	T31
Bosmina coregoni	T12, IQ1, IR1, T26
Bosmina longirostris	T7, T23, T18, T16, T30, T29, T4, T1, T2, T21, T15, T12, T13, IQ1, T5, AR1, T24, IR1, T19, T26, T14, T20, T31, T8, T9, T17, T17, T34
Ceriodaphnia pulchella	T19
Ceriodaphnia quadrangula	T30, T4, T2, T5
Ceriodaphnia reticulata	T16, T22, T12, T19, T14, T20, T17, T17, T34
Chydorus ovalis	T33
Chydorus sphaericus	T7, T23, T30, T29, T21, T15, T12, T13, T27, T5, AR1, T19, T14, T31, T8, T9, T10, T17, T17, T33
Cornigerius lacustris	T22
Coronatella rectangula	T16, T33, T34

TABLE 8.2 *(Continued)*

Taxon	Locations
Daphnia cucullata	T23, T16, T16, T21, T12, T13, T26, T8, T9, T10, T34
Daphnia curvirostris	T31
Daphnia galeata	T30, T29, T26
Daphnia hyalina	AR1
Daphnia laevis	IQ1
Daphnia longiremis	IQ1
Daphnia longispina	T23, T18, T16, T16, T30, T29, T21, T15, T12, T13, T27, T5, AR1, IR1, T19, T26, T14, T31, T8, T9, T10, T17, T17, T25, T34
Daphnia magna	T16, T16, T29, T15, T12, AR1, T26, T8, T9, T10, T25, T34
Daphnia obtusa	T16
Daphnia pulex	T29, T5, IR1
Daphnia rosea	IQ1
Daphnia similis	IQ1
Diaphanosoma birgei	T9, T10, T26
Diaphanosoma brachyurum	T27, IQ1, T9
Diaphanosoma lacustris	T22, T29, T21, T5, AR1, T14, T34
Diaphanosoma sp.	AR1
Disparalona rostrata	T23, T21, T27, T31
Eubosmina tubicen	IQ1
Eurycercus lamellatus	T29, T27
Graptoleberis testudinaria	T12, T27
Leptodora kindtii	T16, T16, T21, T12, T14, T20, T17, T17, T34
Leydigia leydigi	T30, T21, T11, T19, T14, T31
Macrothrix laticornis	T7, T29, T11, T5, T31
Macrothrix sp.	IR1
Moina brachiata	T4, T2
Moina macrocarpa	T24
Moina micrura	T12, T19, T8, T9, T10
Oxyurella tenuicaudis	T33
Pleuroxus aduncus	T30, T21, T5, T31
Pleuroxus truncatus	T12
Polyphemus pediculus	T16, T4

TABLE 8.2 *(Continued)*

Taxon	Locations
Sida crystallina	T16, T16, T34
Simocephalus vetulus	T3, T5, T31, T17, T17
COPEPODA	
Acanthodiaptomus denticornis	T23, T22, T29, T21, AR1, T24, T19, T26, T20, T8, T9, T10, T17, T12, T13, T17, T15, T16, T34
Acanthocyclops robustus	T23, T21, T10, T15
Arctodiaptomus acutilobatus	T29, T5, IR1
Arctodiaptomus bacillifer	T5, AR1
Arctodiaptomus osmanus	AR1
Arctodiaptomus sp.	T7, T30, T32, T25
Arctodiaptomus spinosus	T24
Cyclops vicinus	T18
Cyclops abyssorum	T29, AR1, T10
Cyclops gigas	IQ1
Cyclops lacustris	AR1
Cyclops nanus	IQ1
Cyclops scutifer	IQ1
Cyclops smirnovi	IR1
Cyclops sp.	T7, T27, T11, AR1
Cyclops strenuus	AR1, T24
Cyclops vernalis	IQ1
Cyclops vicinus	T23, T16, T30, T22, T21, T15, T12, T13, AR1, T19, T26, T14, T20, T8, T9, T10, T17, T17, T25, T34
Cyclops vulgaris	IR1
Diacyclops bicuspidatus	AR1
Diacyclops languididus	IQ1
Diaptomus dilobatus	IQ1
Diaptomus saltillinus	IQ1
Diaptomus theeli	IQ1
Eucyclops macruroides	T5
Eucyclops macrurus	IQ1
Eucyclops serrulatus	AR1
Eucyclops sp.	IR1

TABLE 8.2 *(Continued)*

Taxon	Locations
Eudiaptomus drieschi	T27
Macrocyclops albidus	T21, T19
Macrocyclops sp.	AR1
Megacyclops gigas	AR1
Megacyclops viridis	T29, T9, T24
Thermocyclops crassus	AR1
Tropocyclops parsinus	IQ1

**Location codes were given in the Table 8.1.*

Keratella tecta, a cladoceran species *Chydorus sphaericus,* and copepod species *Cyclops vicinus* (20) were other commonly observed species from different locations, and their frequencies are found to be 50%, 48%, and 43%, respectively.

An examination of zooplanktonic organisms in the studied lakes reveals that the species composition in Türkiye's aquatic habitats, 34 different sites, aligns with that of the Irano-Anatolian region. The frequency of these species in Türkiye varies between 88% and 59%. In studies conducted within Iran's boundaries, 10 different sites, the most common rotifer species were *Brachionus angularis* and *Brachionus* sp., recorded in six sites with a frequency of 60%. These were followed by *Asplanchna* sp., *Keratella cochlearis, Lecane lunaris, Lepadella patella,* and *Polyarthra* sp., which were recorded in five sites with a frequency of 50%.

Common species of rotifers listed in this study are cosmopolitan for the Palearctic region. It has been reported that the Palearctic region hosts 1350 rotifer taxa, including both Monogononta and Bdelloidae (Segers, 2008). In this study, 289 taxa were listed for the Irano-Anatolian Hotspot region, accounting for approximately 21.40% of the total. Among the cladocerans, following *B. longirostris*, *D. longispina,* and *C. sphaericus*, *D. magna* (12 locations) and *D. cucullata* (11 locations) were noted to exhibit a more restricted distribution. Among copepods, after *C. vicinus, Acanthodiaptomus denticornis* was the most common calanoid copepod, present in 19 different locations. Other copepods were found to be significantly less common in the region and detected in four or fewer aquatic habitats.

In conclusion, the Irano-Anatolian Hotspot region is known to be under various threats such as climate change, which affects the whole world, and anthropogenic impacts such as dam construction, draining of wetlands, unsustainable agricultural activities, excessive irrigation, habitat destruction, etc. (Ambarlı et al., 2016; Mittermeier et al., 2011; Moradi et al., 2025; Şekercioĝlu et al., 2011). In aquatic ecosystems, these pressures are expected to lead to increased salinity, resulting in a decline in both zooplankton richness and evenness (Beklioğlu et al., 2020). Moreover, dried wetlands may increase. A review of the literature reveals that both the northern and southern parts of the region contain unstudied areas. Since the zooplankton diversity in these regions is not fully known, predicting the impacts of future environmental threats is challenging. However, it is essential to recognize that hotspot areas are a harbor for biodiversity, and their diversity will be crucial in the future. Ecosystems function as interconnected networks, with all trophic levels influencing one another. Therefore, identifying primary consumers such as zooplankton in the region's still unexplored aquatic ecosystems is gaining importance in this regard.

KEYWORDS

- **aquatic ecosystems**
- **biodiversity hotspots**
- **environmental adaptation**
- **freshwater ecosystems**
- **habitat degradation**
- **Irano-Anatolian region**

REFERENCES

Abbaspour, K. C.; Faramarzi, M.; Ghasemi, S. S.; Yang, H. Assessing the Impact of Climate Change on Water Resources in Iran. *Water Resour. Res.* **2009**, *45*, 1–16.

Adrian, R.; O'Reilly, C. M.; Zagarese, H.; Baines, S. B.; Hessen, D. O.; Keller, W.; et al. Lakes as Sentinels of Climate Change. *Limnol. Oceanogr.* **2009**, *54*, 2283–2297.

Altındağ, A.; Yiğit, S. A Taxonomical Study on the Rotifer Fauna of Akşehir Lake. *Turkish J. Zool.* **1999**, *23*, 1–6.

Altındağ, A.; Yiğit, S. Beyşehir Gölü Zooplankton Faunası ve Mevsimsel Değişimi. *Gazi Univ. Gazi Educ. Fac. J.* **2004**, *24*, 217–225.

Altındağ, A.; Yiğit, S.; Ergönül, M. B. The Zooplankton Community of Lake Mogan, Turkey. *J. Freshwater Ecol.* **2007**, *22*, 709–711.

Ambarlı, D.; Zeydanlı, U. S.; Balkız, Ö.; Aslan, S.; Karaçetin, E.; Sözen, M.; et al. An Overview of Biodiversity and Conservation Status of Steppes of the Anatolian Biogeographical Region. *Biodivers. Conserv.* **2016**, *25*, 2491–2519.

Beklioğlu, M.; Bucak, T.; Levi, E. E.; Erdoğan, Ş.; Özen, A.; Filiz, N.; et al. Influences of Climate and Nutrient Enrichment on Multiple Trophic Levels of Turkish Shallow Lakes. *Inland Waters* **2020**, *10*, 173–185.

Biria, H. A.; Salavatian, S. M.; Vahedi, A. Investigation on Identification, Density, and Distribution of Zooplankton in Lar Reservoir in Iran. *Biosci. Biotechnol. Res. Asia* **2011**, *8*, 557–564.

Bulut, H. A Taxonomic Study on Zooplankton Fauna of Kiğı Dam Lake (Bingöl, Turkey). *Süleyman Demirel Univ. J. Fish. Fac.* **2018a**, *14*, 74–79.

Bulut, H. Seasonal Variation of Kapıaçmaz Pond Zooplankton in Elazığ, Turkey. *Turkish J. Agric. Food Sci. Technol.* **2018b**, *6*, 1617–1621.

Bulut, H.; Fidan, D.; Saler, S. Zooplankton of Deriner Dam Lake (Artvin, Türkiye). *Acta Aquat. Turc.* **2024**, *20*, 14–22.

Bulut, H.; Saler, S. Zooplankton of Beyhan Dam Lake (Elazığ, Turkey). *Turkish J. Sci. Technol.* **2014**, *9*, 23–28.

Bulut, H.; Saler, S. Zooplankton of Kalecik Dam Lake (Elazığ, Turkey). *Fırat Univ. J. Sci.* **2013**, *25*, 99–103.

Buyurgan, Ö.; Altındağ, A.; Kaya, M. Zooplankton Community Structure of Asartepe Dam Lake (Ankara, Turkey). *Turkish J. Fish. Aquat. Sci.* **2010**, *10*, 135–138.

Conservation International. Biodiversity Hotspot. https://www.cepf.net/our-work/biodiversity-hotspots/irano-anatolian/species (accessed April 21, 2026).

Dhahir, S. F.; Ali, L. A. Q. Zooplankton Community in Dukan Lake (Kurdistan Region, Iraq), with a New Record of *Craspedacusta sowerbii* Lankester (1880) (Cnidaria: Hydrozoa). *Baghdad Sci. J.* **2017**, *14*, 735–741.

Durmaz, O.; Altındağ, A.; Gürgen, G.; Berdi, D. Zooplankton Fauna and Seasonal Changes of Two Karstic Sinkhole Lakes: Meyil and Kızören (Konya, Türkiye). *J. Limnol. Freshw. Fish.* **2022**, *8*, 251–257.

Ergönül, M. B.; Erdoğan, S.; Altındağ, A.; Atasağun, S. Rotifera and cladocera fauna of several lakes from the central Anatolia, Marmara, and western Black Sea regions of Turkey. *Turkish Journal of Zoology* **2016**, *40*, 141–146.

Gür, H. The Anatolian Diagonal Revisited: Testing the Ecological Basis of a Biogeographic Boundary. *Zoology in the Middle East* **2016**, *62*, 189–199.

Gür, H. The Future Impact of Climate and Land-Use Changes on Anatolian Ground Squirrels Under Different Scenarios. *Ecological Informatics* **2022**, *70*, 101693.

Gürbüzer, P.; Altındağ, A.; Tekatlı, G.; Tekatlı, Ç. Zooplankton Fauna of High Mountain Lake: Sarıncof (Çamlıhemşin, Rize, Turkey). *KSU Journal of Agriculture and Nature* **2023**, *26*, 673–679.

Gyllström, M.; Hansson, L. A.; Jeppesen, E.; García-Criado, F.; Gross, E.; Irvine, K.; et al. The Role of Climate in Shaping Zooplankton Communities of Shallow Lakes. *Limnol. Oceanogr.* **2005**, *50*, 2008–2021.

Hakimzadeh Khoei, M.; Kaya, M.; Altındağ, A. New Records of Rotifers from Iran with Biogeographic Considerations. *Turk. J. Zool.* **2011**, *35*, 395–402.

Hayrapetyan, A.; Gevorgyan, G.; Schultze, M.; Shikhani, M.; Khachikyan, T.; Krylov, A.; et al. Contemporary Community Composition, Spatial Distribution Patterns, and

Biodiversity Characteristics of Zooplankton in Large Alpine Lake Sevan, Armenia. *J. Limnol.* **2022**, *81* (s1), 63–81.

Hrycik, A. R.; McFarland, S.; Morales-Williams, A.; Stockwell, J. D. Winter Severity Shapes Spring Plankton Succession in a Small, Eutrophic Lake. *Hydrobiologia* **2022**, *849*, 2127–2144.

İpek Alış, N.; Saler, S. Zooplankton Fauna of Özlüce Dam Lake (Bingöl, Turkey). *Bitlis Eren Univ. Fen Bilim. Derg.* **2016**, *5*, 86–90.

Jeppesen, E.; Jensen, J. P.; Sondergaard, M.; Lauridsen, T.; Landkildehus, F. Trophic Structure, Species Richness and Biodiversity in Danish Lakes: Changes along a Phosphorus Gradient. *Freshw. Biol.* **2000**, *45*, 201–218.

Kaya, M.; Altındağ, A. Zooplankton Fauna and Seasonal Changes of Gelingüllü Dam Lake (Yozgat, Turkey). *Turk. J. Zool.* **2007**, *31*, 347–351.

Kaya, M.; Altındağ, A.; Sezen, G. Seasonal Variation of Cladocera Fauna of Sultan Marshes and Its Environment (Central Anatolia, Türkiye). *Süleyman Demirel Univ. Fac. Arts Sci. J. Sci.* **2012**, *7*, 86–91.

Kirillin, G. Modeling the Impact of Global Warming on Water Temperature and Seasonal Mixing Regimes in Small Temperate Lakes. *Boreal Environ. Res.* **2010**, *15*, 279–293.

Kottek, M.; Greiser, J.; Beck, B.; Rudolf, B.; Rubel, F. World Map of the Köppen–Geiger Climate Classification Updated. *Meteorol. Z.* **2006**, *15*, 259–263.

Kuzucuoğlu, C.; Çiner, A.; Kazancı, N. The Geomorphological Regions of Turkey. *World Geomorphological Landscapes* **2019**, 41–178.

Malekzadeh-Viayeh, R. Rotifer Diversity in Iranian Waters: A Review. In *Tigris Euphrates Rivers Their Environ. from Headwaters to Mouth*; Jawad, L. A., Ed.; Springer: Cham, 2021; pp 705–741.

Martínez, A.; Bonaglia, S.; Di Domenico, M.; Fonseca, G.; Ingels, J.; Jörger, K.; et al. Fundamental Questions in Meiofauna—How Small but Ubiquitous Animals Can Help to Better Understand Nature. *EcoEvoRxiv* **2023**.

Messager, M. L.; Lehner, B.; Grill, G.; Nedeva, I.; Schmitt, O. Estimating the Volume and Age of Water Stored in Global Lakes Using a Geo-Statistical Approach. *Nat. Commun.* **2016**, *7*, 1–11.

Mittermeier, R. A.; Turner, W. R.; Larsen, F. W.; Brooks, T. M.; Gascon, C. Global Biodiversity Conservation: The Critical Role of Hotspots. In *Biodiversity Hotspots*; Zachos, F. E.; Habel, J. C., Eds.; Springer-Verlag: Berlin Heidelberg, 2011; pp 3–22.

Moradi, H.; Noroozi, J.; Fourcade, Y. Plant Endemic Diversity in the Irano-Anatolian Global Biodiversity Hotspot is Dramatically Threatened by Future Climate Change. *Biol. Conserv.* **2025**, *302*, 110963.

Özbay, H.; Altındağ, A. Zooplankton Abundance in the River Kars, Northeast Turkey: Impact of Environmental Variables. *Afr. J. Biotechnol.* **2009**, *8*, 5814–5818.

Saler, S. Diversity and Abundance of Zooplankton in Medik Reservoir of Turkey. *Maejo Int. J. Sci. Technol.* **2017**, *11*, 126–132.

Saler, S.; Bulut, H.; Güneş, S.; Alpaslan, K.; Karakaya, G. Seasonal Variations of Zooplankton in Nazik Lake (Turkey). *Turk. J. Sci. Technol.* **2019a**, *14*, 79–84.

Saler, S.; Bulut, H.; Karakaya, G. Zooplankton of Çat Dam Lake (Malatya, Turkey) with a New Record for Turkish Rotifers *Lecane intrasinuata* (Olofsson, 1917). *Iran. J. Fish. Sci.* **2019b**, *18*, 199–204.

Saler, S.; Eroğlu, M.; Haykır, H. Zooplankton of Peri Stream (Tunceli, Türkiye). *E-J. New World Sci. Acad.* **2011**, *6*, 14–20.

Saler, S.; Haykır, H.; Baysal, N. Zooplankton of Uzunçayır Dam Lake (Tunceli, Turkey). *J. Fish.* **2014**, *8*.

Saler, S.; İpek, N. İ. Zooplankton of Keban Dam Lake (Yurtbaşı Locality) (Elazığ, Turkey). *NWSA Acad. J. Ecol. Life Sci.* **2020**, *15*, 15–23.

Saler, S.; Karakaya, G. Zooplankton of Karasu River (Türkiye). *Dünya Sağlık ve Tabiat Bilim. Derg.* **2023**, *6*, 40–51.

Saler, S.; Selamoğlu, Z. Zooplankton Diversity of Three Dam Lakes in Turkey. *Iran. J. Fish. Sci.* **2020**, *19*, 2720–2729.

Saler, S.; Sen, B. Long-Term Changes in Rotifera Fauna of Gülüşkür Bay (Keban Dam Lake, Elazığ, Turkey). *J. Anim. Vet. Adv.* **2010**, *9*, 1909–1912.

Segers, H. Global Diversity of Rotifers (Rotifera) in Freshwater. In *Freshw. Anim. Divers. Assess.*; Balian, E. V.; Lévêque, C.; Segers, H.; Martens, K., Eds.; Springer: Netherlands, 2008; pp 49–59.

Şekercioğlu, Ç. H.; Anderson, S.; Akçay, E.; Bilgin, R.; Can, Ö. E.; Semiz, G.; et al. Turkey's Globally Important Biodiversity in Crisis. *Biol. Conserv.* **2011**, *144*, 2752–2769.

Sloan, S.; Jenkins, C. N.; Joppa, L. N.; Gaveau, D. L. A.; Laurance, W. F. Remaining Natural Vegetation in the Global Biodiversity Hotspots. *Biol. Conserv.* **2014**, *177*, 12–24.

Sommer, U.; Gliwicz, Z. M.; Lampert, W.; Duncan, A. The PEG Model of Seasonal Succession of Planktonic Events in Fresh Waters. *Arch. Hydrobiol.* **1986**, *106*, 433–471.

Tellioğlu, A. *Elazığ Hazar Gölü Zooplanktonu*; Fırat University, 1998. Doctoral Dissertation.

Tellioğlu, A.; Akman, F. A Taxonomical Study on the Rotifera Fauna in Pertek Region of Keban Dam Lake. *Eur. J. Fish. Aquat. Sci.* **2007**, *24*, 135–136.

Ustaoğlu, M. R.; Altındağ, A.; Kaya, M.; Akbulut, N.; Bozkurt, A.; Özdemir Mis, D.; et al. A Checklist of Turkish Rotifers. *Turk. J. Zool.* **2012**, *36*, 607–622.

Vezhnavets, V. V.; Kouraev, A. V.; Gukasyan, E.; Gabrielyan, B. K. Zooplankton Study of Lake Sevan as an Indicator of Ecosystem Stability in the Context of Global Climate Change. *Inland Water Biol.* **2024**, *17*, 48–58.

Wang, C.; Li, E.; Zhang, L.; Wei, H.; Zhang, L.; Wang, Z. Long-Term Succession Characteristics and Driving Factors of Zooplankton Communities in a Typical Subtropical Shallow Lake, Central China. *Environ. Sci. Pollut. Res.* **2023**, *30*, 49435–49449.

Yiğit, S.; Altındağ, A. A Taxonomical Study on the Zooplankton Fauna of Hirfanlı Dam Lake (Kırşehir), Turkey. *Gazi Univ. J. Sci.* **2005**, *18*, 563–567.

Yıldız, Ş. Zernek-Baraj Gölü (Van, Türkiye) Zooplankton Faunası. *Biyol. Bilim. Araştırma Derg.* **2012**, *5*, 57–59.

Yıldız, Ş.; Özgökçe, M. S.; Özgökçe, F.; Karaca, I.; Polat, E. Zooplankton Composition of Van Lake Coastline in Turkey. *Afr. J. Biotechnol.* **2010**, *9*, 8248–8252.

CHAPTER 9

SOIL BIOTA OF THE IRANO-ANATOLIAN HOTSPOT

MERIÇ ÇAKIR*

Department of Soil Science and Ecology, Faculty of Forestry, Çankırı Karatekin University, Çankırı, Türkiye

*Corresponding author

ABSTRACT

The Irano-Anatolian region is a biodiversity hotspot with significant global contributions in soil biota. For example, Mesostigmatid Mites represent 7% of global diversity in the region, while Oribatid Mites account for 5%. These species are critical for ecosystem functions like nutrient cycling, highlighting the area's ecological significance. Collembola (4%) and Isopoda (3%) also hold notable diversity, while Prostigmatic Mites are less represented (1%). The rich biodiversity in the region is threatened by land-use changes, overgrazing, deforestation, climate change, and excessive agricultural chemical use. These factors degrade soil quality, affect the resilience of soil communities, and disrupt critical ecosystem services such as decomposition and nutrient cycling. Climate change, in particular, alters temperature and precipitation patterns, impacting the distribution and abundance of soil organisms. Conservation efforts are critical to safeguarding the region's soil biodiversity. Implementing sustainable land-use practices, improving agricultural management, and developing strategies to mitigate climate change effects are essential. Addressing research gaps, particularly in understudied taxa, is essential to gaining a comprehensive understanding of biodiversity. Collaborative efforts among regional scientists and policymakers could greatly enhance conservation

Biodiversity Hotspot of the Irano-Anatolian Region. T. Pullaiah (Ed.)

DOI: 10.1201/9781779645760-9

strategies and ensure the preservation of the unique ecosystems within the Irano-Anatolian hotspot.

9.1 INTRODUCTION

The Irano-Anatolian Hotspot, a complex and vast system of high mountains and enclosed basins, spans a significant area of 899,773 km^2, encompassing major portions of central and eastern Türkiye, and northern and western Iran (Mittermeier et al., 2004). The hotspot includes diverse ecoregions, such as the Central Anatolian steppe, Central Anatolian deciduous forests, Eastern Anatolian deciduous forests, Eastern Anatolian montane steppe, Elburz Range forest steppe, Kopet Dag woodlands and forest steppe, and Zagros Mountains forest steppe. This topographically complex region supports a rich array of flora and fauna, with 6,000 plant species (2,500 endemic), 364 bird species (116 endemic), and 150 mammals (9 endemic) (Zachos and Habel, 2011).

Terrestrial ecosystems comprise two fundamental subsystems: the producer subsystem (above ground) and the decomposer subsystem (below ground). These components are mutually obligate, in that they are dependent upon one another (Wardle, 2002). In their meta-analysis study, Zhang et al. (2022) demonstrated that an increase in plant diversity leads to a corresponding increase in the abundance and diversity of soil fauna. In this context, the Irano-Anatolian ecoregion is characterized by a high level of plant diversity and a high rate of endemism, which suggests that it will also exhibit a rich soil fauna diversity. The diversity is influenced by a number of factors, including the impact of other organisms, the limitations of population dispersal, the influence of the surrounding environment and the characteristics of the local topography (Yu et al., 2022). The Irano-Anatolian region is of great evolutionary significance due to its diverse topography, encompassing mountainous areas, valleys, and islands (MacArthur and Wilson, 2001) that provide a wide range of habitats for various organisms. Moreover, the region has acted as a refuge for species during glacial periods and serves as a crucial migration route for flora and fauna moving between Europe, Asia, and the Middle East. These distinctive features, in conjunction with the region's climate, are anticipated to have a substantial impact on the diversity of soil organisms. Despite its potential for rich soil fauna diversity, the Irano-Anatolian region lacks sufficient research in this area, hindering definitive conclusions about the diversity or ecological characteristics of its soil fauna.

The Irano-Anatolian region is notable for its rich ecosystem diversity, high endemism, and soil fauna adapted to challenging environmental conditions. The region's diverse habitats support a range of soil organisms crucial for nutrient cycling, organic matter decomposition, and soil fertility. Soil fauna play a vital role in ecosystem functioning, particularly in arid and nutrient-poor areas, making them essential for maintaining soil health and supporting agriculture (Çakır, 2017). These organisms are also sensitive to environmental changes, serving as valuable indicators of the region's ecological health. This chapter aims to provide an overview of the current knowledge on some important soil fauna groups, including Collembola (Springtails), Acari (Mites), Isopoda (Woodlice), Diplopoda (Millipedes), Formicidae (Ants), and Lumbricina (Earthworms) documented in the Irano-Anatolian region, as well as the threats to these organisms.

9.2 COLLEMBOLA (SPRINGTAILS)

The Collembolans represent the most numerous fauna among micro-arthropods, with about 9000 described species (Bellinger et al., 2024). They are distributed globally and occur in all biomes, from tropical to Arctic, and from forest to grassland, desert, and throughout the soil profile. Collembolans are known as "springtails" because of the organs beneath their abdomens that allow them to jump (Figure 9.1). Most species are relatively small, typically measuring only a few millimeters in length. However, there are also brightly colored Collembola species among them. In favorable conditions, the number of Collembola can reach over 100,000 per square meter. Their distribution in the soil varies according to the plant type and soil depth. They are mostly found in the litter and rhizosphere layers. It is typical for surface-dwelling forms to be well equipped with furcula. Residents of deeper soil layers are typically devoid of a furcula, or possess only a rudimentary one. Additionally, they often lack eye pigmentation and eyes (Figure 9.1) (Petersen, 2002). The current system of classification arranges 33 families of Collembolans into four major groups.

Collembola are typically highly specialized feeders on soil micro-biota, including fungi, bacteria, actinomycetes, and algae. They play a crucial role in regulating the population dynamics of these organisms in soil. They also play important roles in plant litter decomposition and contribute significantly to the nutrient cycle (Coleman et al., 2018; Rusek, 1998). Additionally, Collembola have important effects on nitrogen mineralization, soil respiration, leaching of dissolved organic carbon, and plant growth (Filser, 2002). In addition to enhancing soil and plant characteristics, Collembola can also modify the distribution of nutrients in the soil, which subsequently affects the functional attributes of plants related to nutrient absorption, particularly the root architecture (Bellini et al., 2023). Therefore, the density of Collembola may exert a significant influence on the functioning of the soil and the ecosystem services.

According to a literature review, the Irano-Anatolian region of Türkiye is home to 107 species of Collembola, distributed among 16 families (Sevgili and Özata, 2014). However, the research has predominantly focused on the humid regions of Türkiye, with limited investigation conducted in the semi-arid inland areas. Iran, on the other hand, boasts 286 species of Collembola, belonging to 20 families (Mayvan et al., 2023). The density of Collembola exhibits variation across gradients of vegetation and soil profiles. For instance, in semiarid regions of Türkiye, Collembola density was recorded as 2118 ind. m^{-2} under oak shrub (*Quercus pubescens*) and 761 ind. m^{-2} under almond shrub (*Amygdalus orientalis*) (Çakır et al., 2018). Similarly, in a semiarid *Pinus nigra* forest, the density was determined to be 11206 ind. m^{-2} (Çakır, 2019). A study in Iran documented Collembola densities along gradients of soil profiles in an oak (*Q. castaneifolia*) forest as 3556 ind. m^{-2} in litter, 1995 ind. m^{-2} in soil depths of 0–3 cm, and 2047 ind. m^{-2} in soil depths of 3–6 cm (Mayvan et al., 2015).

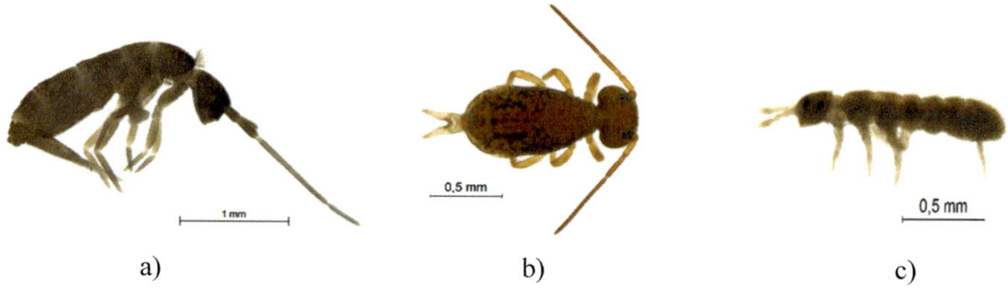

FIGURE 9.1 The most abundant Collembola families in soil: (a) Entomobryidae, (b) Sminthuridae, (c) Isotomidae (Photo: M. Çakır).

9.3 *ACARI* (MITES)

Soil mites (Arachnida: *Acari*), which are related to spiders, represent the most abundant microarthropod group in many ecosystems. It is estimated that up to 250,000 mites may be present in one square meter of forest area. In pastureland, the number of mites may exceed 100,000 per square meter, while in arid areas, the population density may reach 500 mites per square meter (Coleman et al., 2018; Wallwork, 1976). Soil-dwelling mites, in general, are classified into four suborders: Oribatid, Mesostigmata, Prostigmata, and Astigmata mites.

Among the four mite groups, the oribatids are characteristic soil dwelling mites. They are typically fungivorous or detritivores and may also exhibit predatory behavior towards soil nematodes (Schneider et al., 2004). Mesostigmatid mites are known to feed on other microarthropods, with a few species having a dietary preference for fungi (Koehler, 1999). The Prostigmata are a diverse group of mites that exhibit a wide range of feeding habits and strategies. Astigmata mites are found associated with rich, decomposing nitrogen sources and are rarely abundant compared to other mites (Coleman et al., 2018).

9.3.1 ORIBATID MITES

Oribatids are soil-dwelling mites, with approximately 11,000 species currently known (Baran et al., 2018). Oribatids are distinguished from other microarthropods by the presence of a hard, typically calcareous exoskeleton. The majority of oribatids exhibit a brown to tan coloration (Figure 9.2). They obtain the calcium in their exoskeletons by feeding on fungi. Fungi contain calcium oxalate crystals, which may be metabolized by the mites (Norton and Behan-Pelletier, 1991). In forest soils, oribatid densities range from 50,000 to 500,000 individuals; coniferous forests have the highest oribatid densities, followed by deciduous forests, grasslands, and deserts. Oribatid density is significantly impacted by agricultural practices, with an average number of individuals per square meter being as low as 50% (Hülsmann and Wolters, 1998).

Oribatid mites exhibit a certain degree of specialization in their food resources, despite their status as generalists. There is a differentiation in the feeding habits of oribatid mites according to the niche they occupy. Nevertheless, the feeding preferences of oribatid mites are relatively limited (Coleman et al., 2018; Schneider et al., 2004). Synthesizing data from 25 publications reporting on 292 oribatid mite taxa collected from various sites, Maraun et al. (2023) identified six principal trophic groups: moss feeders, lichen feeders, primary decomposers, fungal feeders/secondary decomposers, predators/scavengers, and marine algal feeders. The proportional distribution among these groups revealed that 43.7%

of the taxa were fungal feeders/secondary decomposers, 27.0% were primary decomposers, 15.7% were predators/scavengers, and a small proportion fed on lichens (4.9%), mosses (2.1%), or marine algae (0.7%). Additionally, only 6.1% of the taxa incorporated calcium carbonate ($CaCO_3$) into their skeletons. It has been demonstrated that oribatid mites can affect the decomposition of organic matter and nutrient dynamics in forest floors. However, their impact appears to be indirect, occurring through the grazing of microbial populations or the fragmentation of plant detritus, which in turn influences the decomposition process (Coleman et al., 2018; Potapov et al., 2022).

The oribatid mite fauna of Iran includes 380 species from 86 families (Akrami, 2015). In contrast, no comprehensive studies have been conducted in Türkiye to provide an inclusive record of the total number of species present in the country. However, 24 oribatid species have been recorded in the Çat Forest, Sivas Province, from samples including soil, litter, moss, lichen, and tree bark (Toluk and Akin, 2017). In Mount Erciyes (3917 m a.s.l.), Kayseri Province, soil and litter samples revealed eight oribatid taxa (Toluk et al., 2019). Additionally, five species were identified from Ali Mountain (1870 m a.s.l.) in the same region (Alidağı and Ayyıldız, 2022).

FIGURE 9.2 The most abundant species of oribatid mites in soil (Photo: M. Çakır).

9.3.2 *PROSTIGMATA MITES*

Prostigmata are mites consisting of about 26,000 species and many more undescribed (Zhang et al., 2011). The family of these mites includes species that act as predators, microbial feeders, plant feeders, or parasites (Figure 9.3) (Dindal, 1990). A significant proportion of these species are predators, although some families contain fungal-feeding mites which may become particularly prevalent. The Prostigmata, which are more prevalent in temperate than tropical and subtropical regions, may become more numerous than the oribatids (Coleman et al., 2018). The larger predaceous Prostigmata are known to feed upon a variety of soft-bodied arthropods, with a particular preference for Collembola and their eggs. In contrast, the smaller species are hematophagous. It is hypothesized that certain members of the phytophagous families Linotetranidae and Tuckerellidae may feed on the roots of grasses (Guanilo et al., 2012).

To date, there have been no sufficient studies conducted on Prostigmata mites in the Irano-Anatolian region. Iran is home to 32 species belonging to the Trombidioidea superfamily and 63 species belonging to the Tydeoidea superfamily (Darbemamieh et al., 2021; Saboori and Hakimitabar, 2013). A total of 43 species is currently recognized within the Trombidiidae superfamily in Türkiye (Sevsay, 2017). Furthermore, 15 species belonging to the Erythraeoidea superfamily and 23 species of Microtrombidiidae have been identified in the cities of Diyarbakır and Mardin (Türkiye) (Maral, 2021).

FIGURE 9.3 Some Prostigmata species in soil (Photo: M. Çakır).

9.3.3 MESOSTIGMATA MITES

Almost all soil-dwelling Mesostigmata species are predators and contain fewer soil-dwelling species than Oribatid or Prostigmata (Coleman et al., 2018). The taxonomically diverse group of Mesostigmata mites is known to comprise over 12,000 species (Figure 9.4). These mites are especially abundant and species-rich in forest, grassland, and agricultural soils (Potapov et al., 2022). The presence of a greater diversity of Mesostigmata mites is observed in forest floors with an abundance of surface litter. The results of laboratory experiments indicate that predatory Mesostigmata either feed preferentially on prey with relatively low mobility (e.g., nematodes, enchytraeids, insect larvae, arthropod eggs) or on soft-bodied microarthropods (e.g., Collembola) (Beaulieu and Walter, 2006). It has been demonstrated that Mesostigmata mites have the potential to exert control over the density of soil microarthropods (Schneider and Maraun, 2009).

The most recent data indicates that there are 241 taxa of Mesostigmata present in Türkiye (Erman et al., 2024). In the Iranian fauna, Mesostigmatid mites, excluding the Phytoseiidae family, have been recorded as comprising 616 species (Riahi and Nemati, 2024). The Phytoseiidae family is represented by 103 species (Kazemi et al., 2022).

FIGURE 9.4 Soil-dwelling predatory mesostigmata species (Photo: M. Çakır).

9.4 ISOPODA (WOODLICE)

Terrestrial isopods (woodlice) are typical cryptozoans, occurring in a variety of below-ground habitats, including under rocks and logs (Figure 9.5) (Muchmore, 1990). They are widespread terrestrial crustaceans, with over 4000 species distributed across a range of habitats, including deserts, where they are susceptible to desiccation (Potapov et al., 2022; Sfenthourakis and Taiti, 2015). Isopods are heterotrophic organisms that consume both leaf litter and decaying wood and play an important role in the decomposition of litter. Terrestrial isopods are responsible for the consumption of more than 10% of the annual litter, which results in a four-fold increase in the surface area available to microorganisms (Pokhylenko et al., 2020). Additionally, they are capable of ingesting animal tissue; they also ingest mineral soil (Lam et al., 2024; Yang and Li, 2020). The large amounts of fecal pellets produced by isopods decompose rapidly, returning essential nutrients to the soil (Quadros and Araujo, 2008). The daily food consumption of an isopod depends on the animal's size and developmental stage. Furthermore, the process is influenced by the litter species, litter quality, the degree of decomposition, the moisture content of litter, and the microclimatic conditions present in the habitat (Eisenbeis and Wichard, 2012).

Isopods have a wide distribution area, from humid ecosystems to deserts, from sea level to high mountains and caves (Hornung, 2011). The consumption of phytophagous matter in desert species serves as an adaptation to the moisture deficit that is characteristic of arid climates (Somme, 1995; Wallwork, 1976).

Studies focusing on isopod fauna are scarce in Türkiye, and the available knowledge on the diversity of terrestrial isopods almost entirely relies on studies conducted over 60 years ago (Strouhal, 1962; Verhoeff, 1943, 1948). Verhoeff (1948) stated that the number of terrestrial isopod species in Türkiye is 70. The most recent addition to the Turkish Isopoda fauna was the discovery of a single species, *Schizidium davidi* (Özgen et al., 2018). Further research in this field is necessary to gain a better understanding of the soil biodiversity in the country. On the other hand, Iranian Isopods are represented by 46 species. It was stated that there were a significant number of species that had not yet been identified throughout the country, and that a significant portion of them were waiting to be investigated (Kashani, 2018; Kashani et al., 2021).

FIGURE 9.5 An isopod species that is known to consume litter in significant quantities (Photo: M. Çakır).

9.5 DIPLOPODA (MILLIPEDES)

The Diplopoda (millipedes) constitute a highly diverse group of detritivores, comprising over 11,000 described species (Figure 9.6) (Golovatch and Liu, 2020). Diplopods represent a significant component of the detritivore community in both temperate and tropical hardwood forests, where they are responsible for the consumption of dead vegetable matter. The primary ecological function of Diplopoda is that of "*litter transformers*" (Lavelle and Spain, 2001). Diplopods represent the primary consumers of fallen leaves, facilitating the entry of 15–25% of the calcium present in the litter in broadleaf forests. They accelerate nitrogen mineralization through their role in decomposition (Fujimaki et al., 2010). Millipedes are selective feeders, avoiding leaf litter high in polyphenols and preferring leaves high in calcium. They generally avoid freshly fallen leaves, despite their high assimilation efficiency (David and Gillon, 2002). The Diplopoda are typically saprophagous, feeding on leaf litter that has been colonized by fungi and bacteria. However, some species, genera, and families have more specific diets, including fungal mycelia, algae films, animal remains, other animals, and crops (Koubová et al., 2023).

Despite their dependence on moisture, millipedes are also inhabitants of arid and semi-arid regions. Millipedes are devoid of a waxy layer on their epicuticle, rendering them susceptible to rapid desiccation in environments characterized by low relative humidity (Coleman et al., 2018). In order to protect themselves from the damaging effects of high temperatures and low humidity, they hide under vegetation or debris (Crawford, 1981).

Another soil fauna group that has not been studied sufficiently in the Irano-Anatolian region is the class Diplopoda. In Türkiye, diplopods are represented by 135 species, as documented by Enghoff (2006). A total of 56 species has been recorded in Iran (Golovatch et al., 2022).

FIGURE 9.6 Some *Diplopoda* species in soil (Photo: M. Çakır).

9.6 FORMICIDAE (ANTS)

The Formicidae, or ants, represent a species-rich, geographically widespread, and diverse group of eusocial insects (Kass et al., 2022). Approximately 16,000 ant species have been identified (Potapov et al., 2022). Ants undoubtedly represent the most significant family of soil insects due to their profound influence on soil structure.

Ants are a highly distributed and diverse group of organisms, occurring in a vast range of habitats from the Arctic to the tropics (Schultheiss et al., 2022). Ant communities contain many species, even in arid areas such as deserts

(Crawford, 1981; Whitford, 2000). The influence of ants on their surrounding ecosystems is considerable. Ants are regarded as "*ecosystem engineers*" with the capacity to mobilize considerable quantities of soil, a feat comparable to that of earthworms in some ecosystems (Cammeraat and Risch, 2008; Frouz and Jilková, 2008; Jones et al., 1994). The influence of ants on soil structure is particularly important in arid environments where earthworm abundance is low (Whitford, 2000).

Ants play a significant role in the ecosystem as predators, scavengers, and herbivores, with the capacity to disperse seeds and alter soil structure through their burrowing activities (Tiede et al., 2017). Ants engage in a multitude of interactions with a diverse array of organisms, including plants, fungi, and other arthropods. Ants can shape whole communities of soil microbiota, arthropod communities, and plants by mutualistic interactions, predation, foraging activity, and creating niches in the soil. The activities and effects of ants can influence biodiversity in the ecosystems in which they reside (Parker and Kronauer, 2021).

Türkiye is home to 143 species of ants (Kiran and Karaman, 2020), with 137 of these species found in the Central Anatolian region within the Irano-Anatolian region (Kiran and Karaman, 2021). In contrast, the Iranian ant fauna is believed to encompass a more diverse array of ant species, with an estimation of over 250 species (Rad et al., 2018).

9.7 *LUMBRICINA* (EARTHWORMS)

To date, about 7,000 species of earthworm have been described (Orgiazzi et al., 2016). It has been determined that earthworms are most abundant in the forests and meadows of temperate and tropical regions, while their numbers decrease in arid and cold environments where soil water is unsuitable (Phillips et al., 2021). Across a range of habitats from the tropics to arid areas, earthworms display morphological, physiological, and behavioral adaptations (Edwards et al., 2022). The density of worms in the world can be less than 10 and more than 2000 individuals per m^2. The highest amount was seen in fertile pastures and the lowest in acidic coniferous forest soil and taiga forests where water is unfavorable (Coleman et al., 2018). The classification of earthworms typically follows two criteria: morphology and behavior. These are used to group earthworms into three principal ecological groups: epigenic, anemic, and endogenic (Satchell, 1967). *Epigenic* earthworms are found in the surface layer of soil, where they feed on organic matter such as plant litter and deadwood. These species do not typically construct burrows but instead inhabit and feed on the plant litter. *Endogenic* earthworms are soil-dwelling organisms that feed on the soil itself. To facilitate movement and foraging, endogenic earthworms excavate horizontal burrows in the soil. These burrows are reused to a certain extent. *Anicic* earthworms construct permanent vertical burrows in soil. They consume the foliage that they dislodge from the soil surface and transport into their burrows. Additionally, they deposit casts on the surface, which can frequently be observed in grasslands. Additionally, some anemic earthworm species construct middens, or piles of casts, in the vicinity of their burrow entrances.

The first group of earthworm's feeds on a wide range of decaying plant litter. However, they tend to favor more palatable and nutrient-rich materials with a low C:N ratio (Edwards and Arancon, 2022). The second group of earthworms, which feed on soil, have a considerable digestive capacity, with the ability to digest soil several times their own body mass per day. This is attributed to the low nutritional quality of mineral soil (Curry and Schmidt, 2007). This leads to the definition of earthworms as "*ecosystem engineers*" (Darwin, 1881; Lavelle and Spain, 2001; Le Bayon et al., 2017) based on their impact on mixing of mineral and organic materials, aggregate formation, bioturbation, and macropore construction.

Earthworms are macroinvertebrates and therefore more straightforward to work with and collect than the other groups mentioned above. Furthermore, they have been the subject of more extensive study. It has been stated that there are 87 species of earthworms in Türkiye (Mısırlıoğlu et al., 2019). It has been stated that there are more than 28 species of earthworms in Iran (Mısırlıoğlu et al., 2018). It has been reported that there are 12 species in the Elburz region in the north of Iran (Latif et al., 2009) and 24 species in the Zagros Mountains in the south (Latif et al., 2016).

9.8 OVERVIEW OF THE CURRENT KNOWLEDGE ON SOIL FAUNA GROUPS, DOCUMENTED IN THE IRANO-ANATOLIAN REGION

The data on soil fauna in the Irano-Anatolian region mainly relies on studies conducted in Türkiye and Iran, while contributions from other countries in the region are either lacking or very limited. This can be attributed to the fact that the land areas encompassing the region are larger in these two countries than in others. Still, the available research reviewed in this chapter exposes notable differences in study comprehensiveness even between Türkiye and Iran. This research gap is of considerable importance, as it hinders our comprehensive understanding of soil biodiversity in the region.

Table 9.1 summarizes the current knowledge of several key soil fauna groups in the Irano-Anatolian region, including Collembola (springtails), Acari (mites), Isopoda (woodlice), Diplopoda (millipedes), Formicidae (ants), and Lumbricine (earthworms). This summary emphasizes the necessity for further in-depth studies to address the gaps in understanding soil fauna diversity. Collaborative research efforts among the countries within the Irano-Anatolian region could significantly contribute to providing a more comprehensive understanding of soil biodiversity in the area.

The Collembola (Springtails) constitute 4% of the global diversity observed in the region. With 9,000 species documented globally and 393 in the Irano-Anatolia region, springtails are well-represented, likely due to their ability to thrive in a variety of habitats. They play a significant role in the cycling of nutrients and the maintenance of soil health.

Table 9.1 reveals that Mesostigmata mites constitute the largest group, representing 7% of the total. The identification of 12,000 species of Mesostigmata mites globally, with 857 species present in the Irano-Anatolian region, indicates a significant diversity of these mites in this area. It may be posited that this group has established a strong presence in this region due to the favorable ecological conditions that it has found here. The region also exhibits a high level of oribatid mite diversity, with 5% of global diversity represented by 11,000 species worldwide and 500 in the Irano-Anatolian region. These mites frequently play a pivotal role in soil microfauna, influencing ecosystem processes, which suggests that this region is ecologically diverse for oribatid mites. In contrast, Prostigmata Mites are represented by less than 1% in this region. Despite having a global diversity of 26,000 species, only 138 species are identified in Irano-Anatolia. This could imply that this mite group is less diverse in this specific region or that their presence has not been as thoroughly studied.

The Isopoda (Woodlice) are represented at a lower percentage, at 3% in the Irano-Anatolian region. Despite the global diversity of isopods, with 4,000 species, the 117 species found in this region indicate a limited distribution in Irano-Anatolia. However, these Isopoda play an important role in the region's ecological niche. Diplopoda (Millipedes), Formicidae (Ants), and Lumbricina (Earthworms) are each represented by 2% of

TABLE 9.1 Summary of Key Soil Fauna Groups in the Irano-Anatolian Region: Species Richness and Knowledge Gaps

Soil Fauna Group	Number of Identified Species		Reference(s)	Additional Information/ Knowledge Gaps
	Global	Irano-Anatolian (Approximately)		
Collembola (Springtails)	9000	393+	Bellinger et al. (2024); Çakır (2019); Mayvan et al. (2015); Sevgili and Özata (2014)	More research needed in semi-arid areas of Türkiye and other Irano-Anatolian countries
Oribatid Mites	11000	500+	Akrami (2015); Alidağı and Ayyıldız (2022); Baran et al. (2018); Toluk and Akin (2017)	Comprehensive study lacking in Türkiye and other Irano-Anatolian countries
Prostigmata Mites	26000	138+	Saboori and Hakimitabar (2013); Sevsay (2017); Zhang et al. (2011)	Countries need more extensive research
Mesostigmata Mites	12000	857	Coleman et al. (2018); Erman et al. (2024); Riahi and Nemati (2024)	Iran better studied; more data needed in Türkiye and other Irano-Anatolian countries
Isopoda (Woodlice)	4000	117+	Kashani et al. (2021); Potapov et al. (2022); Verhoeff (1948); Özgen et al. (2018)	Türkiye's data outdated, major gaps in other Irano-Anatolian countries
Diplopoda (Millipedes)	11000	191+	Enghoff (2006); Golovatch and Liu (2020); Golovatch et al. (2022)	Irano-Anatolian countries understudied
Formicidae (Ants)	16000	393+	Kass et al. (2022); Kiran and Karaman (2020); Pashaei Rad et al. (2018)	Türkiye well studied, but more work needed in central region and other Irano-Anatolian countries
Lumbricina (Earthworms)	7000	115+	Latif et al. (2016); Mısırlıoğlu et al. (2019); Orgiazzi et al. (2016)	More research needed in Iran and other Irano-Anatolian countries

their global diversity. While these groups are widespread globally, their representation in this region is relatively low. Nonetheless, ants and earthworms, for instance, play crucial roles in soil structure and ecosystem functioning, making their presence ecologically significant despite the lower percentages.

The Irano-Anatolian region exhibits a higher-than-average diversity of Mesostigmata mites and Oribatid mites, which suggests that the region plays a significant ecological role for these taxa. Conversely, groups such as Prostigmata mites and others with lower percentages (e.g., Lumbricina, Formicidae) may be less well-adapted to the ecological conditions in this region, or there may be research gaps leading to the lower representation. These percentages demonstrate how the Irano-Anatolian region contributes to global biodiversity for specific taxa and provide insights for biogeographical analysis.

9.9 MAJOR THREATS TO SOIL BIOTA IN THE IRANO-ANATOLIAN

A significant reduction in soil biodiversity can be attributed to various factors, including alterations in land use, shifts in vegetation, and transformations in the soil water budget (Geisen et al., 2019). Several factors threaten the soil biota in the Irano-Anatolian hotspot. The most significant threat to the Turkish portion of this region is the expansion of irrigation infrastructure (Buyukcangaz and Korukcu, 2007). Excessive exploitation of water resources for agricultural purposes has led to a substantial reduction of extensive steppe regions and the drying up of numerous closed-basin lakes (Dursun et al., 2012). In the Ararat Valley (Türkiye), 1,500 km^2 of marshland has been drained and converted into agricultural land. Similarly, the Javakheti mountain wetlands (Georgia) and Lake Sevan (Armenia) have been largely destroyed. In Iran, groundwater resources are depleting as agricultural areas expand (Zaki, 2020). Additionally, overgrazing and the loss of shrub and forest areas are major concerns. In Türkiye, the combined effects of a growing population and overgrazing have led to soil erosion, which has reduced forest cover in Central Anatolia and degraded soil quality (Gökbulak et al., 2018). In Iran, illegal thinning and increased road access are causing forests to shrink and deteriorate (Shahnaseri et al., 2023).

Climate change also has a substantial impact on soil arthropods in the Irano-Anatolian region, affecting their distribution, abundance, and ecosystem interactions. As temperatures rise and precipitation patterns shift, the structure of soil communities, including arthropods and their interactions with microbes, can be disrupted (Eisenhauer et al., 2012). In regions like the Irano-Anatolian zone, where ecosystems are already vulnerable due to arid conditions, climate change exacerbates these stresses (Ergüner et al., 2019; Yousefi et al., 2019). Soil arthropods play key roles in processes such as decomposition and nutrient cycling, so changes in their community structure can have broader ecological consequences. Warming and altered moisture conditions may reduce soil biodiversity, negatively affecting the resilience of these communities (Aupic-Samain et al., 2021). Global environmental changes may also reduce the functional diversity of soil microarthropods, impacting their interactions with soil microorganisms (Liu and Yang, 2023). This can undermine ecosystem stability and functionality, including nutrient cycling and soil structure maintenance. Some studies indicate that increased temperatures and moisture variability can trigger declines in key species and alter trophic interactions within soil food webs. Understanding how these changes will unfold under future climate scenarios is crucial for conserving the biodiversity and ecosystem services provided by soil arthropods.

Pesticides, including insecticides and herbicides, are designed to eliminate unwanted organisms but often have broad-spectrum effects, harming non-target species such as soil arthropods, nematodes, and beneficial microbes. In the Irano-Anatolian region, where agriculture is a dominant activity, excessive pesticide use can significantly reduce the diversity and abundance of essential soil fauna (Mayvan et al., 2022; Ustuner et al., 2020). The use of chemical fertilizers, particularly those rich in nitrogen, phosphorus, and potassium (NPK), can also alter the nutrient balance of the soil and negatively affect soil fauna. While fertilizers are intended to boost plant growth, their overuse can lead to nutrient leaching, soil acidification, and the disruption of microbial communities. In the Irano-Anatolian region, which already faces challenges related to aridity and poor soil quality, chemical fertilizers can exacerbate these issues by killing off sensitive soil organisms (Karimifard et al., 2024). Fertilizers can cause shifts in microbial populations, favoring some species over others, thereby reducing overall biodiversity. This disrupts symbiotic relationships between plants and soil organisms, such as mycorrhizal fungi, which help plants access nutrients (Sabir et al., 2021). As soil fauna become less diverse, the ecosystem's ability to retain moisture, recycle nutrients, and support plant life diminishes. In summary, land-use changes, global warming, excessive pesticide use, and reliance on chemical fertilizers are likely to have detrimental effects on soil fauna in the Irano-Anatolian region. These practices have the potential to reduce biodiversity, disrupt ecological functions, and degrade soil quality.

The Irano-Anatolian region is known for providing refuge for various species during glacial periods and serving as a crucial intersection for the migration of flora and fauna between Europe, Asia, and the Middle East. Its unique climate and topography have led to diverse biodiversity, essential for understanding biogeographical patterns and species diversification. However, this biodiversity is threatened by factors such as the expansion of irrigation infrastructure, overgrazing, deforestation, climate change, and excessive use of pesticides and fertilizers. The challenges noted contribute to the decline of soil biodiversity, disrupt ecological functions, and degrade soil quality. Taxonomic surveys and ecological assessments are necessary to improve our understanding of soil biodiversity and develop effective conservation strategies to tackle these challenges. Furthermore, promoting collaboration among researchers from both countries could lead to a more cohesive approach to researching and conserving the unique soil fauna of the Irano-Anatolian region, thus contributing to the conservation of biodiversity in this ecologically important area.

KEYWORDS

- **biodiversity**
- **conservation strategies**
- **ecosystem services**
- **Irano-Anatolian hotspot**
- **microarthropods**
- **soil fauna**

REFERENCES

Akrami, M. A. An Annotated Checklist of Oribatid Mites (Acari: Oribatida) of Iran. *Zootaxa* **2015**, *3963*, 451–501.

Alidağı, H.; Ayyıldız, N. Taxonomic Investigations on the Oppiid Mites (Acari: Oppiidae) of Ali Mountain (Kayseri). *Acarol. Stud.* **2022**, *4*, 104–108.

Aupic-Samain, A.; Baldy, V.; Delcourt, N.; Krogh, P. H.; Gauquelin, T.; Fernandez, C.; Santonja, M. Water Availability Rather than Temperature Control Soil Fauna Community Structure and Prey–Predator Interactions. *Funct. Ecol.* **2021**, *35* (7), 1550–1559.
Baran, Ş.; Bezci, T.; Ayyıldız, N. Supplementary Checklist of Oribatid Mites (Acari) from Turkey. *Munis Entomol. Zool.* **2018**, *13*, 91–97.
Beaulieu, F.; Walter, D. E. Predation in Suspended and Forest Floor Soils: Observations on Australian Mesostigmatic Mites. *Acarologia* **2006**, *47*, 43–54.
Bellinger, P. F.; Christiansen, K. A.; Janssens, F. Checklist of the Collembola of the World. **2024**. http://www.collembola.org (accessed April 21, 2026).
Bellini, B. C.; Weiner, W. M.; Winck, B. R. *Systematics, Ecology and Taxonomy of Collembola*: *Introduction to the Special Issue*; MDPI, 2023; p 221.
Buyukcangaz, H.; Korukcu, A. Integrated Approach for Water Resources and Irrigation Management in Turkey. *Water Int.* **2007**, *32* (S1), 710–719.
Çakır, M. The Importance of Soil Fauna in Arid Ecosystems. *Anatol. J. For. Res.* **2017**, *3* (1), 67–78.
Çakır, M. The Negative Effect of Wood Ants (*Formica rufa*) on Microarthropod Density and Soil Biological Quality in a Semi-Arid Pine Forest. *Pedobiologia* **2019**, *77*, 150593.
Çakır, M.; Tanı, M.; Tunç, T. Bozkır-Orman Geçiş Kuşağındaki Çalı Türlerinin Toprak Biyoçeşitliliğine Etkisi. *J. Bartın Fac. For.* **2018**, *20*, 583–589.
Cammeraat, E.; Risch, A. The Impact of Ants on Mineral Soil Properties and Processes at Different Spatial Scales. *J. Appl. Entomol.* **2008**, *132*, 285–294.
Coleman, D. C.; Callaham, M. A.; Crossley, D. *Fundamentals of Soil Ecology*, 3rd ed.; Academic Press: USA, 2018.
Crawford, C. S. *Biology of Desert Invertebrates*; Springer: Berlin, 1981.
Curry, J. P.; Schmidt, O. The Feeding Ecology of Earthworms—A Review. *Pedobiologia* **2007**, *50*, 463–477.
Darbemamieh, M.; Ahadiyat, A.; Farmahiny Farahani, V. R.; Sharifian, A. Iranian Checklist of Tydeoidea (Trombidiformes: Prostigmata) until the End of 2020. *Int. J. Acarol.* **2021**, *47*, 592–602.
Darwin, C. *The Formation of Vegetable Mold through the Action of Worms, with Observations on Their Habits*; J. Murray: London, 1881.
David, J. F.; Gillon, D. Annual Feeding Rate of the Millipede *Glomeris marginata* on Holm Oak (*Quercus ilex*) Leaf Litter under Mediterranean Conditions. *Pedobiologia* **2002**, *46*, 42–52.
Dindal, D. L. *Soil Biology Guide*; Wiley: New York, 1990.
Dursun, S.; Onder, S.; Acar, R.; Direk, M.; Mucehver, O. Effect of Environmental and Socioeconomic Change on Agricultural Production in Konya Region. In *International Conference on Applied Life Sciences*; IntechOpen, 2012; pp 19–36.
Edwards, C. A.; Arancon, N. Q. The Role of Earthworms in Organic Matter and Nutrient Cycles. In *Biology and Ecology of Earthworms*; Springer, 2022; pp 233–274.
Edwards, C. A.; Arancon, N. Q.; Bohlen, P. J.; Hendrix, P. *Biology and Ecology of Earthworms*; Springer: New York, 2022.
Eisenbeis, G.; Wichard, W. *Atlas on the Biology of Soil Arthropods*; Springer Science & Business Media, 2012.
Eisenhauer, N.; Cesarz, S.; Koller, R.; Worm, K.; Reich, P. B. Global Change Belowground: Impacts of Elevated CO_2, Nitrogen, and Summer Drought on Soil Food Webs and Biodiversity. *Glob. Change Biol.* **2012**, *18* (2), 435–447.

Enghoff, H. The Millipedes of Turkey (Diplopoda). *Steenstrupia* **2006**, *29* (2), 175–198.

Ergüner, Y.; Kumar, J.; Hoffman, F. M.; Dalfes, H. N.; Hargrove, W. W. Mapping Ecoregions under Climate Change: A Case Study from the Biological "Crossroads" of Three Continents, Turkey. *Landscape Ecol.* **2019**, *34*, 35–50.

Erman, O.; Doğan, S.; Ayyıldız, N.; Özkan, M. Checklist of the Mites (Acari) of Türkiye. Third Supplement. *Acarol. Stud.* **2024**, *6*, 81–111.

Filser, J. The Role of Collembola in Carbon and Nitrogen Cycling in Soil. In *Apterygota at the Beginning of the Third Millennium: Proceedings of the Xth International Colloquium on Apterygota, České Budějovice 2000*; *Pedobiologia* **2002**, *46*, 234–245.

Frouz, J.; Jilková, V. The Effect of Ants on Soil Properties and Processes (Hymenoptera: Formicidae). *Myrmecol. News* **2008**, *11*, 191–199.

Fujimaki, R.; Sato, Y.; Okai, N.; Kaneko, N. The Train Millipede (*Parafontaria laminata*) Mediates Soil Aggregation and N Dynamics in a Japanese Larch Forest. *Geoderma* **2010**, *159*, 216–220.

Geisen, S.; Wall, D. H.; van der Putten, W. H. Challenges and Opportunities for Soil Biodiversity in the Anthropocene. *Curr. Biol.* **2019**, *29* (19), R1036–R1044.

Gökbulak, F.; Erdoğan, B. U.; Yıldırım, H. T.; Özçelik, M. S. Causes of Land Degradation and Rehabilitation Efforts of Rangelands in Turkey. *Forestist* **2018**, *68* (2), 106–113.

Golovatch, S. I.; Izadi, M.; Habashi, H.; Shayanmehr, M.; Rahmani, R.; Rafiee, F. An Updated Checklist of the Millipedes of Iran (Diplopoda), with Special Emphasis on the Fauna of Hyrcania, Including Noteworthy Records of Three Species from Its Easternmost Part. *Arthropoda Selecta* **2022**, *31*, 265–270.

Golovatch, S. I.; Liu, W. Diversity, Distribution Patterns, and Faunogenesis of the Millipedes (Diplopoda) of Mainland China. *ZooKeys* **2020**, *930*, 153–198.

Guanilo, A. D.; de Moraes, G. J.; Flechtmann, C. H.; Knapp, M. Phytophagous and Fungivorous Mites (Acari: Prostigmata, Astigmata) from Peru. *Int. J. Acarol.* **2012**, *38*, 120–134.

Hornung, E. Evolutionary Adaptation of Oniscidean Isopods to Terrestrial Life: Structure, Physiology and Behavior. *Terrestrial Arthropod Rev.* **2011**, *4*, 95–130.

Hülsmann, A.; Wolters, V. The Effects of Different Tillage Practices on Soil Mites, with Particular Reference to Oribatida. *Appl. Soil Ecol.* **1998**, *9* (1–3), 327–332.

Jones, C. G.; Lawton, J. H.; Shachak, M. Organisms as Ecosystem Engineers. *Oikos* **1994**, 373–386.

Karimifard, S.; Saberi-Pirooz, R.; Ahmadzadeh, F.; Aghamir, F. Investigating the Impacts of Agricultural Land Use on Soil Earthworm Communities: A Case Study of Northern Zagros Mountains of Iran. *Helion* **2024**, *10* (1).

Kashani, G.; Eshaghi, B.; Abedini, A. New Data on the Terrestrial Isopod Fauna of Iran (Isopoda: Oniscidea) with the First Description of the Male of *Schizidium golovatchi* Schmalfuss. *Iran. J. Anim. Biosyst.* **2021**, *17*, 71–78.

Kashani, M. G. A Preliminary Checklist of Terrestrial Isopods (Isopoda: Oniscidea) from Iran. *Iran. J. Anim. Biosyst.* **2018**, *14*, 123–130.

Kass, J. M.; Guénard, B.; Dudley, K. L.; Jenkins, C. N.; Azuma, F.; Fisher, B. L.; Parr, C. L.; Gibb, H.; Longino, J. T.; Ward, P. S.; Chao, A.; Lubertazzi, D.; Weiser, M.; Jetz, W.; Guralnick, R.; Blatrix, R.; Lauriers, J. D.; Donoso, D. A.; Georgiadis, C.; Gomez, K.; Hawkes, P. G.; Johnson, R. A.; Lattke, J. E.; MacGown, J. A.; Mackay, W.; Robson,

S.; Sanders, N. J.; Dunn, R. R.; Economo, E. P. The Global Distribution of Known and Undiscovered ant Biodiversity. *Science Advances* **2022**, *8* (31).

Kazemi, S.; Mohammad-Doustaresharaf, M.; Döker, I. An Annotated Checklist of the Iranian Phytoseiidae (Acari: Mesostigmata), with an Updated Key to the Species. *Syst. Appl. Acarol.* **2022**, *27* (4), 697–748.

Kiran, K.; Karaman, C. Additions to the Ant Fauna of Turkey (Hymenoptera, Formicidae). *Zoosystema* **2020**, *42*, 285–329.

Kiran, K.; Karaman, C. Ant Fauna (Hymenoptera: Formicidae) of Central Anatolian Region of Turkey. *Turk. J. Zool.* **2021**, *45*, 161–196.

Koehler, H. H. Predatory Mites (Gamasina, Mesostigmata). In *Invertebrate Biodiversity as Bioindicators of Sustainable Landscapes: Practical Use of Invertebrates to Assess Sustainable Land Use*; Paoletti, M. G., Ed.; Elsevier: Netherlands, 1999; pp 395–410.

Koubová, A.; Lorenc, F.; Horváthová, T.; Chroňáková, A.; Šustr, V. Millipede Gut-Derived Microbes as a Potential Source of Cellulolytic Enzymes. *World J. Microbiol. Biotechnol.* **2023**, *39*, 169.

Lam, W. N.; Slade, E. M.; Wardle, D. A. Effects of Leaf Litter Traits on Terrestrial Isopod and Millipede Consumption, Assimilation and Growth. *Funct. Ecol.* **2024**, *38*, 1018–1031.

Latif, R.; Ezzatpanah, S.; Malek, M.; Parsa, H. Earthworms of the Central Elburz Mountains, Iran. *Iran. J. Anim. Biosyst.* **2009**, *5*, 1–15.

Latif, R.; Malek, M.; Csuzdi, C. New Earthworm Records from the Central Zagros Mountain, Iran. *North-West. J. Zool.* **2016**, *13*, 326–336.

Lavelle, P.; Spain, A. V. *Soil Ecology*; Kluwer Academic, 2001.

Le Bayon, R. C.; Bullinger-Weber, G.; Schomburg, A.; Turberg, P.; Schlaepfer, R.; Guenat, C. Earthworms as Ecosystem Engineers: A Review. In *Earthworms: Types, Roles and Research*; 2017; pp 129–178.

Liu, R.; Yang, B. Soil Microbe–Arthropod Interactions under Global Change. *Frontiers Media SA* **2023**, *14*, 1280103.

MacArthur, R. H.; Wilson, E. O. *The Theory of Island Biogeography*; Princeton University Press: New Jersey, 2001.

Maral, H. Contributions to the Terrestrial Parasitengona Fauna (Acari: Trombidiformes: Prostigmata) of Diyarbakır and Mardin (Turkey). *Int. J. Agric. Environ. Food Sci.* **2021**, *5*, 393–397.

Maraun, M.; Thomas, T.; Fast, E.; Treibert, N.; Caruso, T.; Schaefer, I.; Lu, J. Z.; Scheu, S. New Perspectives on Soil Animal Trophic Ecology through the Lens of C and N Stable Isotope Ratios of Oribatid Mites. *Soil Biol. Biochem.* **2023**, *177*, 108890.

Mayvan, M. M.; Greenslade, P.; Sadeghi-Namaghi, H. An Annotated Checklist of the Collembola (Hexapoda) from Iran. *Zootaxa* **2023**, *5275* (1), 1–101.

Mayvan, M. M.; Namaghi, H. S.; Shayanmehr, M.; Greenslade, P.; Palmer, G. Density, Diversity, and Seasonal Fluctuations in Soil Collembola in Three Differently Managed Ecosystems in North Khorasan, Iran. *Turk. J. Zool.* **2022**, *46* (1), 115–128.

Mayvan, M. M.; Shayanmehr, M.; Scheu, S. Depth Distribution and Inter-Annual Fluctuations in Density and Diversity of Collembola in an Iranian Hyrcanian Forest: Additional Contribution. *Soil Org.* **2015**, *87*, 239–247.

Mısırlıoğlu, M.; Valchovski, H.; Reynolds, J. W. Updated List of Earthworms (Clitellata, Megadrili) from Turkey. *Megadrilogica* **2019**, *24*, 99–106.

Mısırlıoğlu, M.; Valchovski, H.; Stojanović, M. Review of the Earthworm Biodiversity of Turkey and Its Neighboring Countries (Clitellata, Megadrili). *Opusc. Zool.* **2018**, *49*, 141–149.

Mittermeier, R. A.; Gil, P. R.; Hoffmann, M.; Pilgrim, J.; Brooks, T.; Mittermeier, C. G.; Lamoreux, J.; Fonseca, G. A. B. D. *Hotspots Revisited: Earth's Biologically Richest and Most Endangered Terrestrial Ecoregions*; Cemex, 2004.

Muchmore, W. B. Terrestrial Isopoda. In *Soil Biology Guide*; Dindal, D. L., Ed.; Wiley: New York, 1990; pp 805–818.

Norton, R. A.; Behan-Pelletier, V. M. Calcium Carbonate and Calcium Oxalate as Cuticular Hardening Agents in Oribatid Mites (Acari: Oribatida). *Can. J. Zool.* **1991**, *69*, 1504–1511.

Orgiazzi, A.; Bardgett, R. D.; Barrios, E.; Behan-Pelletier, V.; Briones, M. J. I.; Chotte, J. L.; De Deyn, G. B.; Eggleton, P.; Fierer, N.; Fraser, T.; Hedlund, K.; Jeffery, S.; Johnson, N. C.; Jones, A.; Kandeler, E.; Kaneko, N.; Lavelle, P.; Lemanceau, P.; Miko, L.; Montanarella, L.; Moreira, F. M. S.; Ramirez, K. S.; Scheu, S.; Singh, B. K.; Six, J.; van der Putten, W. H.; Wall, D. H. *Global Soil Biodiversity Atlas*; Publications Office of the European Union: Luxembourg, 2016.

Özgen, İ.; Kebapçı, Ü.; Khisametdinova, D. D. Contribution to the Knowledge of the Oniscidea (Crustacea: Isopoda) Fauna of Turkey with a New Record: *Schizidium davidi* (Dolfuss, 1887). *Int. J. Fauna Biol. Stud.* **2018**, *5*, 8–11.

Parker, J.; Kronauer, D. J. How Ants Shape Biodiversity. *Curr. Biol.* **2021**, *31*, R1208–R1214.

Pashaei Rad, S.; Taylor, B.; Torabi, R.; Aram, E.; Abolfathi, G.; Afshari, R.; Borjali, F.; Ghatei, M.; Hediary, F.; Jazini, F. Further Records of Ants (Hymenoptera: Formicidae) from Iran. *Zool. Middle East* **2018**, *64*, 145–159.

Petersen, H. General Aspects of Collembolan Ecology at the Turn of the Millennium. *Pedobiologia* **2002**, *46*, 246–260.

Phillips, H. R.; Bach, E. M.; Bartz, M. L.; Bennett, J. M.; Beugnon, R.; Briones, M. J.; Brown, G. G.; Ferlian, O.; Gongalsky, K. B.; Guerra, C. A. Global Data on Earthworm Abundance, Biomass, Diversity and Corresponding Environmental Properties. *Sci. Data* **2021**, *8*, 136.

Pokhylenko, A.; Didur, O.; Kulbachko, Y.; Bandura, L.; Chernykh, S. Influence of Saprophages (Isopoda, Diplopoda) on Leaf Litter Decomposition under Different Levels of Humidification and Chemical Loading. *Biosyst. Divers.* **2020**, *28*, 384–389.

Potapov, A. M.; Beaulieu, F.; Birkhofer, K.; Bluhm, S. L.; Degtyarev, M. I.; Devetter, M.; Goncharov, A. A.; Gongalsky, K. B.; Klarner, B.; Korobushkin, D. I. Feeding Habits and Multifunctional Classification of Soil-Associated Consumers from Protists to Vertebrates. *Biol. Rev.* **2022**.

Quadros, A. F.; Araujo, P. B. An Assemblage of Terrestrial Isopods (Crustacea) in Southern Brazil and Its Contribution to Leaf Litter Processing. *Revista Brasileira de Zoologia* **2008**, *25*, 58–66.

Riahi, E.; Nemati, A. A Catalogue of the Iranian Mesostigmata (Acari) Excluding Phytoseiidae. *Persian Journal of Acarology* **2024**, *13*.

Rusek, J. Biodiversity of Collembola and Their Functional Role in the Ecosystem. *Biodivers. Conserv.* **1998**, *7* (9), 1207–1219.

Sabir, M. S.; Shahzadi, F.; Ali, F.; Shakeela, Q.; Niaz, Z.; Ahmed, S. Comparative Effect of Fertilization Practices on Soil Microbial Diversity and Activity: An Overview. *Curr. Microbiol.* **2021**, *78*, 3644–3655.

Saboori, A.; Hakimitabar, M. A Checklist of the Trombidioidea (Acari: Prostigmata) of Iran. *J. Crop Prot.* **2013**, *2*, 33–42.
Satchell, J. Lumbricidae. In *Soil Biology*; Burges, A.; Raw, F., Eds.; Academic Press: London, 1967.
Schneider, K.; Maraun, M. Top-Down Control of Soil Microarthropods—Evidence from a Laboratory Experiment. *Soil Biol. Biochem.* **2009**, *41* (1), 170–175.
Schneider, K.; Renker, C.; Scheu, S.; Maraun, M. Feeding Biology of Oribatid Mites: A Minireview. *Phytophaga* **2004**, *14*, 247–256.
Schultheiss, P.; Nooten, S. S.; Wang, R.; Wong, M. K.; Brassard, F.; Guénard, B. The Abundance, Biomass, and Distribution of Ants on Earth. *Proc. Natl. Acad. Sci. U.S.A.* **2022**, *119*, e2201550119.
Sevgili, H.; Özata, M. A. Checklist of the Springtails (Hexapoda: Collembola) of Turkey. *Zool. Middle East* **2014**, *60*, 162–168.
Sevsay, S. A Checklist of the Erythraeoidea and Trombidioidea (Actinotrichida: Prostigmata) of Turkey. *Turk. Entomol. Bull.* **2017**, *7*, 91–112.
Sfenthourakis, S.; Taiti, S. Patterns of Taxonomic Diversity among Terrestrial Isopods. *ZooKeys* **2015**, *515*, 13–25.
Shahnaseri, G.; Malekian, M.; Pourmoghadam, K. Habitat Loss of the Chestnut-Leaved Oak (*Quercus castaneifolia*) in the Hyrcanian Forests of Iran: Impacts of Anthropogenic Factors on Forest Thinning and Degradation. *Glob. Ecol. Conserv.* **2023**, *46*, e02600.
Somme, L. *Invertebrates in Hot and Cold Arid Environment*; Springer: Berlin, 1995.
Strouhal, H. Die *Haplophthalmus*-Arten der Türkei (Isop. Terr.) (4. Beitrag zur Kenntnis der türkischen Isopoden). *Ann. Naturhist. Mus. Wien* **1962**, 385–406.
Tiede, Y.; Schlautmann, J.; Donoso, D. A.; Wallis, C. I.; Bendix, J.; Brandl, R.; Farwig, N. Ants as Indicators of Environmental Change and Ecosystem Processes. *Ecol. Indic.* **2017**, *83*, 527–537.
Toluk, A.; Akin, A. T. Oribatid Mite Fauna (Acari) of Çat Forest, Sivas Province, Turkey. *Turk. J. Entomol.* **2017**, *41*, 293–307.
Toluk, A.; Yanmaz, S. B.; Efe, Z. Ş. Oppioid Oribatid Mites (Acari: Oribatida) of Erciyes Mountain (Kayseri, Turkey). *Plant Prot. Bull.* **2019**, *59*, 9–14.
Ustuner, T.; Sakran, A.; Almhemed, K. Effect of Herbicides on Living Organisms in the Ecosystem and Available Alternative Control Methods. *Int. J. Sci. Res. Publ.* **2020**, *10* (8), 633–641.
Verhoeff, K. W. Über Land-Isopoden aus der Türkei. II. *İstanbul Univ. Fen Fak. Mecm. Ser. B* **1943**, *8*, 1–29.
Verhoeff, K. W. Über Land-Isopoden aus der Türkei. III. *İstanbul Univ. Fen Fak. Mecm. Ser. B* **1949**, *14*, 21–48.
Wallwork, J. A. *The Distribution and Diversity of Soil Fauna*; Academic Press: London, 1976.
Wardle, D. A. *Communities and Ecosystems: Linking the Aboveground and Belowground Components*; Princeton University Press: Princeton, 2002.
Whitford, W. G. Keystone Arthropods as Webmasters in Desert Ecosystems. In *Invertebrates as Webmasters in Ecosystems*; Coleman, D. C.; Hendrix, P. F., Eds.; CABI Publishing, 2000; pp 25–41.
Yang, X.; Li, T. Effects of Terrestrial Isopods on Soil Nutrients during Litter Decomposition. *Geoderma* **2020**, *376*, 114546.

Yousefi, M.; Kafash, A.; Valizadegan, N.; Ilanloo, S. S.; Rajabizadeh, M.; Malekoutikhah, S.; Yousefkhani, S. S. H.; Ashrafi, S. Climate Change Is a Major Problem for Biodiversity Conservation: A Systematic Review of Recent Studies in Iran. *Contemp. Probl. Ecol.* **2019**, *12*, 394–403.

Yu, S.; Wu, Z.; Xu, G.; Li, C.; Wu, Z.; Li, Z.; Chen, X.; Lin, M.; Fang, X.; Lin, Y. Inconsistent Patterns of Soil Fauna Biodiversity and Soil Physicochemical Characteristics along an Urbanization Gradient. *Front. Ecol. Evol.* **2022**, *9*, 824004.

Zachos, F. E.; Habel, J. C. *Biodiversity Hotspots: Distribution and Protection of Conservation Priority Areas*; Springer-Verlag, 2011.

Zaki, N. A. The Role of Agriculture Expansion in Water Resources Depletion in Central Iran; University of Oulu: Tampere, 2020.

Zhang, Y.; Peng, S.; Chen, X.; Chen, H. Y. Plant Diversity Increases the Abundance and Diversity of Soil Fauna: A Meta-Analysis. *Geoderma* **2022**, *411*, 115694.

Zhang, Z. Q.; Fan, Q. H.; Pešić, V.; Smit, H.; Bochkov, A. V.; Khaustov, A. A.; Baker, A.; Wohltmann, A.; Wen, T. H.; Amrine, J. W.; Beron, P.; Lin, J. Z.; Gąbryś, G.; Husband, R. Order Trombidiformes Reuter, 1909. In *Animal Biodiversity: An Outline of Higher-Level Classification and Survey of Taxonomic Richness*; Zhang, Z. Q., Ed.; Magnolia Press, 2011; pp 129–138.

CHAPTER 10

CLIMATE CHANGE AND BIODIVERSITY OF THE IRANO-ANATOLIAN HOTSPOT

MILAD NOURI*

Soil and Water Research Institute, Agricultural Research, Education and Extension Organization (AREEO), Karaj, Iran

*Corresponding authors

ABSTRACT

The Irano-Anatolian biodiversity hotspot, spanning parts of Armenia, Azerbaijan, Georgia, Iran, Iraq, Turkey, and Turkmenistan, is ecologically vital due to its diverse topography and climate, which support a rich variety of habitats and species. It has experienced warming and drying trends that are expected to continue. Most areas with dry-summer continental and humid continental climate regimes, currently prevalent across the eastern Anatolia and Zagros mountainous regions, will shift to cold semi-arid and Mediterranean climates by 2100 under RCP8.5. Moreover, the areas with hot arid desert climate are likely to expand in the central and southern Zagros, the Kopet-Dog ranges, and eastern Alborz. Climate change will exert significant impacts on the biodiversity within the Irano-Anatolian ecoregions via changing phenology, destructing and fragmenting habitats, and altering the distribution and abundance of many species. Strategies such as adjusting protection networks, tackling additional stress, and restoring habitats are suggested to improve the resilience, resistance, and adaptive capacity of Irano-Anatolian ecosystems. Such adaptation and mitigation plans should be implemented through a participatory approach, involving all relevant stakeholders. Overall, it seems that climate change would endanger the biodiversity within the Irano-Anatolian ecoregions,

Biodiversity Hotspot of the Irano-Anatolian Region. T. Pullaiah (Ed.)

DOI: 10.1201/9781779645760-10

underlining the importance of developing robust strategies to mitigate the adverse impacts.

10.1 CURRENT CLIMATIC CONDITIONS

The Irano-Anatolian biodiversity hotspot includes seven ecoregions of the eastern Anatolian montane steppe (1), the eastern Anatolian deciduous steppe (2), the central Anatolian steppe (3), the central Anatolian steppe and woodlands (4), the Alborz range forest steppe (5), the Kopet-Dag woodlands and forest steppe (6), and the Zagros Mountains Forest steppe (7) (Figure 10.1). Here, a brief description of the present climatic conditions (1980–2016) in these ecoregions is given based on the updated Köppen-Geiger climate classification explained in Beck et al. (2018) and Peel et al. (2007). For further details on climate classifications, please refer to https://www.gloh2o.org/koppen/.

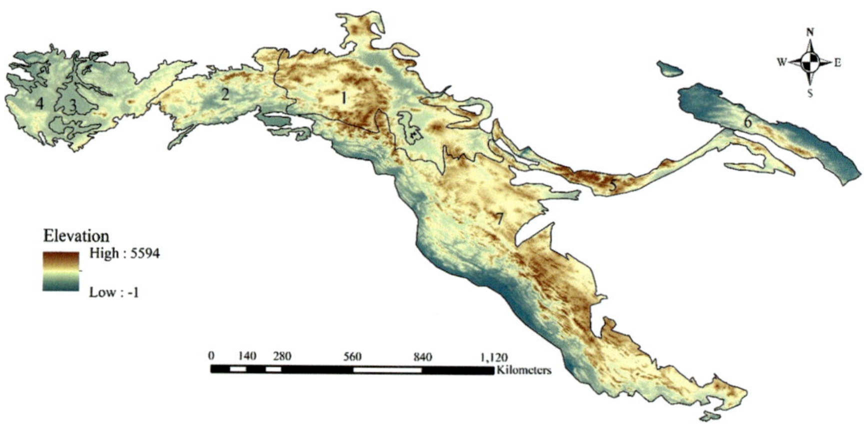

FIGURE 10.1 Map of the Irano-Anatolian ecoregions. Elevation data were derived from NASA Shuttle Radar Topography Mission (SRTM), using the SRTMGL1 Version 3.0 dataset, which provides digital elevation models at ~30-meter spatial resolution.

Source: Farr et al. (2007). Data available from NASA LP DAAC: https://lpdaac.usgs.gov/products/srtmgl1v003/. Image courtesy of NASA/JPL-Caltech.

10.1.1 THE EASTERN ANATOLIAN MONTANE STEPPE

The ecoregion exhibits diverse climatic conditions due to the existence of varied elevations. The northern highlands of the ecoregion contain several protected areas, including the Sevani National Park, Lake Arpi

National Park, and Javarkheti National Park, featuring humid continental and sub-arctic (Dfb and Dfc) climates. These areas experience very cold winters with annual precipitation frequently exceeding 1000 mm. It is noteworthy that snowfall is the predominant precipitation type during winter in this ecoregion. The western highlands contain the mountainous terrain of Turkey, including notable features such as Mount Ararat and the Bingöl Mountains, with dry-summer continental and subarctic (Dsa, Dsb, and Dsc) and humid continental (Dfb) climatic regimes. Conversely, the southern and eastern areas located in the intermediate elevations are primarily characterized by cold arid desert and semi-arid (BWk and BSk) climates, and generally receive scant precipitation, with an average amount rarely exceeding 400 mm.

10.1.2 THE EASTERN ANATOLIAN DECIDUOUS STEPPE

Located in eastern Turkey, this ecoregion is predominantly characterized by a varied topography of rolling hills, plateaus, and valleys. This ecoregion is bounded by the Pontic Mountains in the north and the Anti-Taurus Mountains in the west. The prevailing climatic conditions are dry-summer continental climates (Dsa and Dsb), with snowfall being the predominant type of precipitation during wintertime. However, the southern strip of this ecoregion features a Mediterranean climate regime (Csa).

10.1.3 THE CENTRAL ANATOLIAN STEPPE

It is the smallest ecoregion featured by Lake Tuz, located in the south of Ankara. A cold, semi-arid climate (BSk), with annual precipitation mostly less than 300 mm, is prevalent in this ecoregion (Beck et al., 2018).

10.1.4 THE CENTRAL ANATOLIAN STEPPE AND WOODLANDS

This ecoregion is situated in the heart of Anatolia and is surrounded by several mountain ranges, including the Taurus Mountains to the south, the Pontic Mountains to the north, and the Anti-Taurus Mountains to the east. The prevailing climate in this region is cold semi-arid, classified as BSk, and annual precipitation is typically below 300 mm. Due to the aridity of this plateau, it is susceptible to intense winds, which often result in sandstorms and dust storms. However, in some areas, particularly on the northern flanks of the Taurus and Anti-Taurus, a dry-summer continental climate (Dsb) as well as Mediterranean climate classes (Csa and Csb) dominate, with annual precipitation sometimes surpassing 1000 mm.

10.1.5 THE ALBORZ RANGE FOREST STEPPE

This ecoregion extends from northwestern to northeastern Iran, covering the southern and eastern slopes of the Alborz Mountains, which include the highest peak in the Irano-Anatolian region, the Damavand peak. In the eastern parts of this ecoregion, a cold semi-arid climate (BSk) prevails, with annual precipitation typically less than 400 mm. However, the central and eastern portions are primarily characterized by a cold arid desert climate (BWk) in the lowlands, while a dry-summer continental climate (Dsb) predominates in the highlands.

10.1.6 THE KOPET-DAG WOODLANDS AND FOREST STEPPE

This ecoregion is situated in Central Asia, specifically in southern Turkmenistan and extending into northeastern Iran. It includes the Kopet-Dag ranges, which run parallel to the Iran-Turkmenistan border, and also features a separate section further north in the Uly Balkan Mountain range. It is characterized by cold arid desert (BWk) and semi-arid (BSk) climates, with annual precipitation rarely exceeding 400 mm.

10.1.7 THE ZAGROS MOUNTAINS FOREST STEPPE

This is the largest ecoregion that encompasses the Zagros Mountain range, spanning southeast Turkey, northern Iraq, and western Iran. The snow-capped sub-ranges, for example, Dena and Zardkouh, are crucial water towers, feeding and recharging some of the most significant rivers in Iran, such as the Zayandehroud, Karoun, and Dez. The highlands of the Zagros ranges have continental climates with dry summers (Dsa and Dsb) where annual precipitation higher than 600 mm is expected. The precipitation type during winter is mostly snowfall. Moreover, a Mediterranean climate (Csa) is dominant in western flanks of the northern and central Zagros, while a hot semi-arid climate (BSh) is predominant in western flanks of the southern Zagros. A cold semi-arid climate (BSk) dominates over the eastern flanks of Zagros ranges. There are some areas located in Kerman province of Iran with a hot desert climate (BWh) where annual precipitation hardly surpasses 300 mm.

10.2 CLIMATE CHANGE IMPACTS

Since the pre-industrial period, the increase in atmospheric concentrations of greenhouse gases and aerosols owing to anthropogenic activities such as overusing fossil fuels, land use/cover change, and agricultural practices, has resulted in global warming, and consequently other changes in the climate system (IPCC, 2021). A warming trend has been reported for different regions within the Irano-Anatolian hotspot, such as western Iran (Ahmadi et al., 2018; Tabari et al., 2011) and eastern Anatolia (Hadi and Tombul, 2018; Toros, 2012) during the last half century. Climate change has not only affected mean temperature, but also temperature extremes (Balling et al., 2016; Toros, 2012) and atmospheric water demand (Dadaser-Celik et al., 2016; Nouri et al., 2022). Additionally, climate change has caused significant changes in the distribution, amount, and type of precipitation. These changes are primarily driven by alterations in large-scale storm tracks and atmospheric-ocean teleconnections (Evans, 2009). Some regions within the Irano-Anatolian hotspot, such as western Iran (Nouri, 2023, 2025) and eastern Anatolia (Hadi and Tombul, 2018), have experienced a drying trend with serious implications for water resources.

The Irano-Anatolian ecoregions are typically composed of mountain-plain systems where the runoff that feeds rivers and downstream freshwater resources is mainly generated by snowmelt (Qin et al., 2020). It is, therefore, crucial to conduct an evaluation of the climate change impacts on snow metrics in this region. Nouri and Homaee (2021b) concluded that snow depth, snow duration, and the ratio of snowy day to wet day during winter and spring were reduced over the majority of areas lying in the Zagros and Alborz ranges, indicating the impacts of global warming.

An increasing trend in temperature and temperature-related extremes (Evans, 2009; Fallah-Ghalhari et al., 2019; Lelieveld et al., 2012; Melkonyan, 2015; Vaghefi et al., 2019) and atmospheric water demand (Nouri and Veysi, 2024) was projected across the Irano-Anatolian region for the future. Although the future precipitation projections were highly uncertain (Turkes et al., 2020; Waha et al., 2017), it is expected that the Irano-Anatolian ecoregions will face increasingly more arid conditions (Rahimi et al., 2013; Turkes et al., 2020; Waha et al., 2017).

Here, the projections based on https://www.gloh2o.org/koppen/ were applied for the period 2071–2100 under RCP8.5 (the worst-case scenario) in the Irano-Anatolian hotspot. The continental and humid continental climates (Dfb, Dfc, Dsa, and Dsb) dominant in the mountainous regions of eastern Turkey and Armenia are expected to shrink and predominantly change to the cold semi-arid regime (BSk) within the eastern Anatolian montane steppe ecoregion. Additionally, it is projected that the desert arid regimes (BWh and BWk) surrounding the Urmia Lake and the Aras River along the Iran-Azerbaijan borders would expand in the late 21st century. In the eastern Anatolian deciduous steppe ecoregion, the continental climates (Dsa and Dsb) prevalent across the Anatolian plateau and the Anti-Taurus Mountains areas will change to a Mediterranean climate (Csa), particularly in the southern regions. The continental climates across the highlands of the Taurus Mountains, located in the southern regions of the central Anatolian steppe and woodlands ecoregion, are expected to vanish during the late 21st century and would change to the Mediterranean and semi-arid climatic regimes (Csa and BSk). The semi-arid climate (BSk) in the southern regions of the central Anatolian steppe and woodlands, as well as the central Anatolian steppe, is projected to shift to a cold desert regime (BWk) under the RCP8.5 scenario in 2071–2100. The continental climate classes (Dsa and Dsb) in the northern Zagros highlands will also undergo a shift to a Mediterranean climate (Csa) in the late 21st century. Moreover, the southern and central Zagros highlands will be notably influenced by climate change as the continental climates (Dsa and Dsb) will change to the Mediterranean (Csa) and hot semi-arid (BSh) regimes. There will also be an expansion of the hot arid desert regime (BWh) in the southern regions of the Zagros Mountains Forest steppe ecoregion. Global warming is expected to cause a change from cold (BWk) to hot arid desert (BWh) regime across the Kopet-Dag woodlands and forest steppe ecoregion. Climate change seems to impact the Alborz range forest steppe quite differently. A Mediterranean climatic regime (Csa) in western parts and cold continental and semi-arid climate classes (Dsa, Dsb, and BSk) in the central portions of this ecoregion will expand. However, a shift from cold (BWk) to hot desert (BWh) climate would be experienced in the eastern parts of the Alborz range forest steppe.

Overall, it can be concluded that the Irano-Anatolian hotspot will experience a notable warming and drying trend during the current century. Climate change is anticipated to shift continental climates towards Mediterranean and semi-arid climates in highlands, while the hot arid desert regime is likely to become widespread in more arid conditions.

10.3 WHY WE SHOULD BE MORE WORRIED ABOUT THE CLIMATE CHANGE IMPACTS IN THE IRANO-ANATOLIAN HOTSPOT?

As discussed earlier, the Irano-Anatolian hotspot is characterized by mountainous ecosystems, classified as the Caucasus/Middle East mountainous region (Hock et al., 2019). Mountainous areas are more sensitive to climate change and global warming as explained by the elevation-dependent warming (EDW). It refers to the phenomenon in which air temperature rises more rapidly at higher elevations than lower elevations (Mountain Research Initiative E. D. W. Working Group, 2015; Pepin and Lundquist, 2008; Pepin et al., 2022; You et al., 2020). There are some mechanisms causing EDW, which have been discussed below in sub sections.

10.3.1 ALBEDO EFFECT

The albedo indicates how much radiation is reflected; therefore, surfaces with a high albedo reflect more solar radiation, resulting in a cooler surface. On the other hand, surfaces with low albedo absorb more radiation and have a higher surface temperature. In mountainous regions, the albedo effect is particularly important because of the presence of snow and ice cover. Snow cover is of high albedo, ranging from 0.5–0.9, which is much higher than other surfaces such as water bodies, soil, and vegetation. This implies that snow cover reflects most of the incoming sunlight back into space. This reflective property of snow cover plays a crucial role in the energy balance of the earth by reflecting a significant amount of the incoming solar radiation back into space and thereby reducing the amount of absorbed energy. However, snow depth and cover have experienced a decreasing trend under global warming in most mountainous areas in the Irano-Anatolian hotspot, such as the Zagros (Nouri and Homaee, 2021b), leading to a reduction in the albedo effect. This decrease in albedo leads to increased absorption of solar radiation and more energy being absorbed near the surface, causing the temperature to rise even further. This process creates a positive feedback loop, whereby warming temperature causes a reduction in snow cover and depth, which in turn leads to more warming (Mountain Research Initiative E. D. W. Working Group, 2015; Pepin et al., 2022).

10.3.2 CLOUD COVER

The water vapor in the atmosphere, as a potent greenhouse gas, increases under global warming in mountainous environments. It is worth noticing that at high elevations, the air is cooler and has a lower moisture-holding capacity of saturated air, resulting in the formation of clouds. Warmer air can hold more moisture; therefore, specific humidity tends to increase in case of water availability under global warming. Consequently, the atmosphere is expected to be more humid in warming mountainous areas, which can enhance the greenhouse effect and contribute to EDW. Enhanced cloud cover causes an increase in nighttime minimum temperature (Mountain Research Initiative E. D. W. Working Group, 2015; Rangwala and Miller, 2012). This creates a positive feedback loop, wherein enhanced temperature results in more atmospheric water vapor, leading to even warmer temperatures.

10.3.3 WATER VAPOR MODULATION OF LONGWAVE HEATING

This is an important mechanism deriving EDW in highlands. The longwave downwelling radiation (LDR) reaching the earth's surface depends on the absorption and scattering of longwave radiation by atmospheric constituents such as water vapor, carbon dioxide, and other greenhouse gases. As discussed previously, the augmented air moisture holding capacity increases the atmospheric water vapor concentration under global warming. Consequently, the longwave radiation absorbed or trapped by water vapor is notably enhanced, causing an increased minimum temperature during the night (Rangwala and Miller, 2012; You et al., 2020). Temperature warming as a consequence of water vapor absorption is particularly pronounced in mountainous regions.

10.3.4 AEROSOLS

The presence of air pollutants, such as absorbing (black carbon and dust) and reflective (sulfate oxides) aerosols, can have significant impacts on the near-surface radiation balance (Rangwala and Miller, 2012). These contaminants can absorb and scatter incoming solar radiation, resulting in reduced shortwave solar radiation reaching the surface. This phenomenon is known as the surface dimming effect (Mountain Research Initiative E. D. W. Working Group, 2015). Studies have shown that the majority of atmospheric aerosol pollutants is concentrated at relatively low elevations, typically below 3 km (Ramanathan and Carmichael, 2008). This indicates that lower elevations are likely to experience a greater reduction in shortwave radiation due to the presence of air pollutants compared to higher elevations.

The existence of drylands is another reason explaining why the Irano-Anatolian region is more vulnerable to the impacts of climatic changes. Drylands are the water-limited areas, characterized by low precipitation levels and high evaporation rates. Arid deserts and semi-arid steppe areas are some common types of drylands (Middleton and Sternberg, 2013). Drylands are more susceptible to temperature warming relative to humid areas. Soil moisture plays a crucial role in regulating near-surface temperature by affecting land-atmosphere interactions (Hirsch et al., 2014; Holmes et al., 2017; Nouri and Homaee, 2021a; Seneviratne et al., 2010). Evapotranspiration of soil moisture helps to regulate the temperature by dissipating latent heat, known as the cooling effect. The soil moisture deficit reduces the cooling effect of evapotranspiration (Seneviratne et al., 2006, 2010). In drylands, the lack of soil moisture limits the amount of water available for evapotranspiration, reducing the cooling effect and making these regions more vulnerable to temperature warming (Huang et al., 2017a, b). To put it in more detail, any additional heat as a result of global warming cannot be sufficiently compensated by evapotranspiration considering limited soil moisture resources in drylands. On the other hand, the extra heat can be used for evapotranspiration in humid lands, leading to a decrease in surface warming. Therefore, the same amount of heat will result in a larger warming in drylands compared to humid lands. These physical processes will likely influence the effects of greenhouse gas-induced warming in the future across the drylands (Huang et al., 2017a, b). The Paris Agreement aims to limit the increase in global temperature to well below 2.0°C above the pre-industrial level (Schleussner et al., 2016). However, due to soil drought weakening the cooling effect, drylands may experience a warming of 3.2–4.0°C at a global warming threshold of 2.0°C, which is 44% greater than the warming expected in humid environments. If global warming is limited to 1.5°C, the mean warming of drylands may, however, remain within 3.0°C. Thus, to prevent catastrophic consequences in drylands, it is crucial to limit global warming to 1.5°C (Huang et al., 2017a b).

Overall, due to the existence of mountainous terrains and drylands, the Irano-Anatolian region is considered to be more prone to the stresses emanating from climate change.

10.4 IMPLICATIONS OF CLIMATE CHANGE FOR BIODIVERSITY

The Irano-Anatolian region displayed a drying and warming trend, which is projected to persist in the upcoming years, thereby leading to an even warmer and more arid climate with more frequent and prolonged dry spells. These changes are likely to have significant consequences for biodiversity. I, here, present a succinct overview of the effects of climate change and global warming on the Irano-Anatolian biodiversity.

10.4.1 PHENOLOGY

The phenology of species, that is, the timing of key life cycle events such as reproduction, nesting, and hibernation, can be notably affected by climate change (Oliver and Morecroft, 2014; Walther et al., 2002). The timing of life phases may shift in changing climates, leading to mismatches between species and their environment. For example, some plant species may start to flower earlier under global warming. Consequently, the species depending on these plants for food may not adjust their migration patterns to match the new timing, leading to a decline in their populations (Oliver and Morecroft, 2014). Changes in precipitation patterns and snow phenology can affect the timing of water availability, which can impact the phenology of species (Marsh et al., 2019). For example, some mammals may delay hibernation to take advantage of additional foraging opportunities where winter snow cover decreased. This can also have unintended consequences, such as a mismatch between predator and prey timing, resulting in a decline in biodiversity. Despite some studies implying the influence of climate change on the phenology during the recent decades (Fitchett et al., 2014; Kiapasha et al., 2017; Marsh et al., 2019), no explicit investigation has yet been conducted to evaluate potential phenological changes in the Irano-Anatolian ecoregions under future climate change.

10.4.2 HABITAT LOSS AND FRAGMENTATION

The habitat conditions for certain species may no longer exist within their geographic ranges under climate change (Oliver and Morecroft, 2014). Furthermore, climate change can potentially fragment habitats, leaving populations isolated in patches of suitable habitat (Holyoak and Heath, 2016; Oliver et al., 2015). This poses a significant threat to biodiversity and ecosystem stability in the Irano-Anatolian region, potentially resulting in the loss of important ecosystem services and cascading effects on other species. A number of studies have projected the habitat loss and fragmentation and the negative consequences for several species, including the Anatolian ground squirrel, the Azarbaijan Mountain Newt, wild goats, Urial wild sheep, Armenian wild sheep, Goitered gazelle, and Jebeer, in the Irano-Anatolian ecoregions during the 21st century under various climate change scenarios (Table 10.1).

As previously stated, climate change is expected to result in elevated temperature and decreased precipitation levels, leading to an increase in arid regions in the Irano-Anatolian hotspot. In addition, the snow cover duration and precipitation type are anticipated to be significantly impacted by global warming. Consequently, the replenishment of water resources through snowmelt is likely to be heavily impacted as a result of climate change. In addition to temperature warming, the reduction in water availability and changes in the timing of water availability can be considered major factors contributing to the habitat loss and fragmentation within the Irano-Anatolian hotspot.

10.4.3 DISTRIBUTION SHIFTS

As a consequence of climate change, conditions previously favorable for some species may become uninhabitable; thus, some species would migrate in search of suitable habitats (Oliver and Morecroft, 2014). The affected species may move to higher elevations, known as upward altitudinal expansion, where they can still find suitable habitats offering the necessary resources for their survival. As a result, habitats may become fragmented as populations become confined to smaller and patchy areas. Table 10.1 provides a list of several species, including Persian Leopards, Iranian lizards, and endemic salamanders, projected to undergo upward altitudinal expansion in the Irano-Anatolian ecoregions. This shift in the species distribution has significant ecological implications, such as inter-species competition for resources or adaptation to new environmental circumstances.

It is worth noticing that modeling climate change impacts on biodiversity is inherently uncertain. This uncertainty stems from a number of sources, including model structure, parameters, data limitation, resolution mismatch between the ecological processes and data, lack of understanding of species interactions, and future climate change scenarios (Kujala et al., 2012; Tabor and Williams, 2010). Therefore, care should be taken to deal with the results given in Table 10.1.

TABLE 10.1 Impact Assessments of Climate Change on Biodiversity in the Irano-Anatolian Region

Species	Results	Time Slice	Scenario1	Ecoregions2	References
Persian Leopards	Upward altitudinal expansion	2050	RCP2.6 and RCP8.5	1, 5, 6, and 7	Shahsavarzadeh et al. (2023)
Iranian lizards	Upward altitudinal expansion	2070	RCP2.6 and RCP8.5	1, 5, 6, and 7	Vaissi (2022)
endemic salamander	Upward altitudinal shift and habitat loss	2050 and 2070	RCP2.6, RCP4.5, RCP6.0 and RCP8.5	7	Ashrafzadeh et al. (2019)
Rhinopoma muscatellum	Increased distribution range	2070	RCP2.6, RCP4.5, RCP6.0 and RCP8.5	1, 5, 6, and 7	Kafaei et al. (2021)
Brown bear	Increased distribution range	2050 and 2070	RCP2.6 and RCP8.5	7	Ashrafzadeh et al. (2022)
Anatolian ground squirrels	Range contraction	2030 and 2050	SSP1/RCP2.6, SSP2/RCP4.5, SSP3/RCP7.0, and SSP5/RCP8.5	1, 2, 3, and 4	Gür (2022)
Eurasian Lynx	Habitat loss and reduced distribution range	2065 and 2100	RCP2.6 and RCP8.5	1, 5, 6, and 7	Mahdavi et al. (2020)
Daphne mucronata	Range contraction in lower-elevations	2030 and 2080	A2a/HadCM3	7	Abolmaali et al. (2018)
Azarbaijan Mountain Newt	Habitat loss and fragmentation, and reduced range	2050 and 2070	RCP2.6 and RCP8.5	1, 2, and 7	Ebrahimi et al. (2022)
wild goat Urial wild sheep Armenian wild sheep Goitered gazelle Jebeer	Habitat loss and fragmentation, and reduced range	2080	RCP2.6, RCP4.5, RCP6.0 and RCP8.5	1, 5, 6, and 7	Ebrahimi et al. (2019)

TABLE 10.1 *(Continued)*

Species	Results	Time Slice	Scenario1	Ecoregions2	References
Leopard geckos	Habitat expansion, and increased distribution range	2070	RCP2.6, RCP4.5, RCP6.0 and RCP8.5	1, 5, 6, and 7	Hosseinian Yousefkhani and Nabizadeh (2022)
Yellow-winged darter	Disappearance and range contraction	2021–2040, 2041–2060, 2061–2080, and 2081–2100	SSP245, SSP370 and SSP585	1	Rincón et al. (2023)
Acanthophyllum squarrosum	Range shift and expansion	2030–2080	RCP2.6 and RCP8.5	1, 5, 6, and 7	Mahmoudi Shamsabad et al. (2018)
Scots pine	Reduced distribution range	2065–2100	RCP2.6, RCP4.5, and RCP8.5	1	Dering et al. (2021)
Nepeta binaloudensis	Range expansion	2050–2070	RCP 2.6 and RCP 8.5	6	Erfanian et al. (2021)
Phlomoides binaludensis	Range contraction	2050–2070	RCP 2.6 and RCP 8.5	6	Erfanian et al. (2021)

[1]*RCP*: Representative Concentration Pathway and SSP: Shared Socioeconomic Pathways.

[2]The number of ecoregions is given in Figure 10.1.

10.5 GENERAL STRATEGIES TO COMBAT THE ADVERSE IMPACTS OF CLIMATE CHANGE

As discussed earlier, the impact of climate change on biodiversity necessitates the development of mitigation and adaptation strategies. These strategies typically emphasize the promotion of the resistance, resilience, and change capacities of ecosystems (Lawler, 2009). Resistance refers to the ability of a system to remain unaffected by external forces, while resilience describes the capacity to recover from disturbances. Resilient systems can adapt to changing climates while maintaining their essential functions, whereas less resilient systems may experience a substantial reduction in ecosystem functioning, population declines, and species extinction. Strategies focused on change facilitate the transition of a system from one state to another. The most commonly recommended strategies for promoting resistance, resilience, and change within the Irano-Anatolian region are briefly discussed below.

10.5.1 ADJUSTING PROTECTION NETWORKS

The current protected areas originally targeting endemic species may become unsuitable in the future as species range shifts, and establishing the same areas for protection will no longer assure biodiversity conservation. This is more highlighted in the Irano-Anatolian hotspot, where the current conservation is insufficient to protect biodiversity, particularly in mountainous areas (Noroozi et al., 2019, 2022). Ludovicy et al. (2022) revealed that the protected area network in the Irano-Anatolian region inadequately represents the climatic niches of endemic plant species, especially the species located at the lowest and highest elevations and those with narrow ranges. These species were projected to endure significant losses if the current protected area network will not be adjusted in the future. To effectively align the delineation of protected areas with the goals of biodiversity conservation under climate changes, the following should be given the highest attention by the decision-makers:

- Identifying and prioritizing critical and vulnerable species are vital steps in adjusting protected areas under future climate changes. Species with high sensitivity to climatic changes are at risk of extinction if appropriate management actions are not taken. Therefore, it is crucial to identify, assess, and prioritize species at risk to ensure their conservation needs are adequately addressed.

- Despite the ecological significance of species inhabiting mountainous areas for biodiversity, the current protected area network in the Irano-Anatolian hotspot is characterized by an overrepresentation of climatically homogenous desert ecosystems, while it fails to prioritize protection for endemism hotspots located in high mountain chains (Ludovicy et al., 2022; Noroozi et al., 2019, 2022). This highlights the need for new nature reserves in mountainous ranges, particularly across eastern Anatolia, Alborz, Zagros, and Kopet-Dag Mountains.
- Ensuring connectivity and corridors in new protected areas in the face of changing climates is crucial to facilitate species adaptation and mobility, support ecosystem functions, and improve resilience (Lawler, 2009). This has been suggested to mitigate the adverse effects on the Persian leopard, Eurasian lynx, Pallas's cat, and brown bear (Ashrafzadeh et al., 2022; Mohammadi et al., 2021, 2022) in the Irano-Anatolian ecoregions.
- The expansion of reserve networks and no-hunting zones through enlarging existing reserves and adding buffer zones surrounding these reserves is a prime strategy for biodiversity conservation in changing climates. The augmentation of reserve networks has the potential for sustaining viable populations of brown bears and promoting gene flow between these populations in the Zagros and Alborz ranges and northwestern Iran, while also facilitating the movement through the corridors connecting distinct habitat patches (Ashrafzadeh et al., 2022; Mohammadi et al., 2021).
- In order to achieve successful conservation, the planning process should involve evaluating the efficiency of previously established protected areas and placing emphasis on the creation of new protected areas in strategic locations (Mohammadi et al., 2022; Noroozi et al., 2019). This is more highlighted for the Irano-Anatolian hotspot, as its conservation efficiency is lower than the global average (Noroozi et al., 2022).

10.5.2 REMOVING ADDITIONAL STRESSES

Climate change may exacerbate the effects of other threats, including overgrazing, cropland expansion, excessive water use, and soil contamination. One of the main non-climate-related threats is crop expansion to maintain food security in the future, causing habitat loss and fragmentation.

Molotoks et al. (2018) showed that the Irano-Anatolian region is anticipated to experience the third largest cropland extension after the Indo-Burma and Philippines hotspots, with 6.66% of the entire area converted to cropland by 2050. Moreover, crop expansion may not only reduce the areas suitable for habitat, but also may compete with the environmental sector on soil and water resources, further endangering biodiversity in the Irano-Anatolian hotspot. Damming for irrigation expansion and hydropower, as well as drainage projects, are widely recognized as highly detrimental activities resulting in significant habitat loss, particularly wetlands and river systems, across various regions of the Irano-Anatolian hotspot (Gross, 2012; Naboureh et al., 2021; Şekercioğlu et al., 2011; Zeinoddini et al., 2015). The strategies, except agricultural land expansion, to both conserve biodiversity and maintain food security of the growing population can enhance the resilience of species to climate change in the Irano-Anatolian region (Fader et al., 2013; Willett et al., 2019).

10.5.3 RESTORING HABITATS

Habitat restoration involves efforts to repair, rehabilitate, and restore degraded natural systems (Harris et al., 2006; Renton et al., 2012). The principal objective of habitat restoration is to reinstate the fundamental natural framework, functionality, and biodiversity of ecosystems, thereby fostering increased resilience. When habitats are restored, more resources are made available for the species, and the populations can become more stable and resilient under climate change (Lawler, 2009). Nevertheless, the successful restoration of degraded ecosystems is a demanding and financially burdensome endeavor that requires a good governance structure to ensure its effectiveness. It is noteworthy that restoration projects can become more challenging and uncertain under climate change. An illustration of this challenge can be seen in the case of the Urmia Lake, where the persistence of a negative long-term trend in the lake water level, despite sporadic stabilization at some intervals, underscores the difficulties encountered in efforts to restore this hypersaline ecosystem in the Irano-Anatolian hotspot (Saemian et al., 2020).

The Irano-Anatolian hotspot spans seven different countries, highlighting the importance of collaboration and participation between governmental and non-governmental agencies in order to effectively conserve biodiversity in the face of climate change in this region. Hence, it is crucial for these countries to work together, share data and information, and design shared strategies in order to effectively adjust protection networks, remove additional stresses, and restore habitats. Additionally, engaging local communities in the development of strategies is critical to improving the sense of responsibility toward biodiversity conservation.

10.6 SUMMARY

The Irano-Anatolian region exhibited a warming and drying trend, causing an increase in more arid areas in the majority of the region. The current climate characterized by dry-summer continental and humid continental conditions in highlands was projected to shift towards cold semi-arid and Mediterranean regimes in the late 21st century, underlining the high vulnerability of mountainous regions to climate change. Moreover, there will be an expansion of the hot arid desert climate in more arid areas. Climate change is expected to adversely affect biodiversity across the Irano-Anatolian hotspot via changing phenology, range, and habitat suitability. Mitigating these impacts will require developing and implementing effective strategies, such as adjusting protection networks, addressing additional stressors, and restoring habitats through collaborative efforts.

KEYWORDS

- **distribution shift**
- **ecosystem resilience**
- **habitat loss**
- **mountainous regions**
- **protection network adjustment**
- **resiliency**

REFERENCES

Abolmaali, S. M. R.; Tarkesh, M.; Bashari, H. MaxEnt Modeling for Predicting Suitable Habitats and Identifying the Effects of Climate Change on a Threatened Species, *Daphne mucronata*, in Central Iran. *Ecol. Inform.* **2018**, *43*, 116–123.

Ahmadi, F.; Nazeri Tahroudi, M.; Mirabbasi, R.; Khalili, K.; Jhajharia, D. Spatiotemporal Trend and Abrupt Change Analysis of Temperature in Iran. *Meteorol. Appl.* **2018**, *25*, 314–321.

Ashrafzadeh, M. R.; Khosravi, R.; Mohammadi, A.; Naghipour, A. A.; Khoshnamvand, H.; Haidarian, M.; Penteriani, V. Modeling Climate Change Impacts on the Distribution of an Endangered Brown Bear Population in Its Critical Habitat in Iran. *Sci. Total Environ.* **2022**, *837*, 155753.

Ashrafzadeh, M. R.; Naghipour, A. A.; Haidarian, M.; Kusza, S.; Pilliod, D. S. Effects of Climate Change on Habitat and Connectivity for Populations of a Vulnerable, Endemic Salamander in Iran. *Glob. Ecol. Conserv.* **2019**, *19*.

Balling, R. C.; Kiany, M. S. K.; Roy, S. S. Anthropogenic Signals in Iranian Extreme Temperature Indices. *Atmos. Res.* **2016**, *169*, 96–101.

Beck, H. E.; Zimmermann, N. E.; McVicar, T. R.; Vergopolan, N.; Berg, A.; Wood, E. F. Present and Future Köppen–Geiger Climate Classification Maps at 1-km Resolution. *Sci. Data* **2018**, *5*, 180214.

Dadaser-Celik, F.; Cengiz, E.; Guzel, O. Trends in Reference Evapotranspiration in Turkey: 1975–2006. *Int. J. Climatol.* **2016**, *36*, 1733–1743.

Dering, M.; Baranowska, M.; Beridze, B.; Chybicki, I. J.; Danelia, I.; Iszkulo, G.; Kvartskhava, G.; Kosinski, P.; Raczka, G.; Thomas, P. A.; Tomaszewski, D.; Walas, L.; Sekiewicz, K. The Evolutionary Heritage and Ecological Uniqueness of Scots Pine in the Caucasus Ecoregion Is at Risk of Climate Changes. *Sci. Rep.* **2021**, *11*, 22845.

Ebrahimi, A.; Sardari, P.; Safavian, S.; Jafarzade, Z.; Bashghareh, S.; Khavari, Z. Climate Change Effects on Species of Bovidae Family in Iran. *Environ. Earth Sci.* **2019**, *78*.

Ebrahimi, E.; Ranjbaran, Y.; Sayahnia, R.; Ahmadzadeh, F. Assessing the Climate Change Effects on the Distribution Pattern of the Azerbaijan Mountain Newt (*Neurergus crocatus*). *Ecol. Complex.* **2022**, *50*.

Erfanian, M. B.; Sagharyan, M.; Memariani, F.; Ejtehadi, H. Predicting Range Shifts of Three Endangered Endemic Plants of the Khorassan–Kopet Dagh Floristic Province under Global Change. *Sci. Rep.* **2021**, *11*, 9159.

Evans, J. P. 21st Century Climate Change in the Middle East. *Clim. Change* **2009**, *92*, 417–432.

Fader, M.; Gerten, D.; Krause, M.; Lucht, W.; Cramer, W. Spatial Decoupling of Agricultural Production and Consumption: Quantifying Dependences of Countries on Food Imports Due to Domestic Land and Water Constraints. *Environ. Res. Lett.* **2013**, *8*, 014046.

Fallah-Ghalhari, G.; Shakeri, F.; Dadashi-Roudbari, A. Impacts of Climate Changes on the Maximum and Minimum Temperature in Iran. *Theor. Appl. Climatol.* **2019**, *138*, 1539–1562.

Farr, T. G.; Rosen, P. A.; Caro, E.; Crippen, R.; Duren, R.; Hensley, S.; Kobrick, M.; Paller, M.; Rodriguez, E.; Roth, L.; Seal, D.; Shaffer, S.; Shimada, J.; Umland, J.; Werner, M.; Oskin, M.; Burbank, D.; Alsdorf, D. The Shuttle Radar Topography Mission. *Rev. Geophys.* **2007**, *45*.

Fitchett, J. M.; Grab, S. W.; Thompson, D. I.; Roshan, G. Spatiotemporal Variation in Phenological Response of Citrus to Climate Change in Iran: 1960–2010. *Agric. For. Meteorol.* **2014**, *198, 199*, 285–293.

Gross, M. Turkey's Biodiversity at the Crossroads. *Curr. Biol.* **2012**, *22*, R503–R505.

Gür, H. The Future Impact of Climate and Land-Use Changes on Anatolian Ground Squirrels under Different Scenarios. *Ecol. Inform.* **2022**, *70*.

Hadi, S. J.; Tombul, M. Long-Term Spatiotemporal Trend Analysis of Precipitation and Temperature over Turkey. *Meteorol. Appl.* **2018**, *25*, 445–455.

Harris, J. A.; Hobbs, R. J.; Higgs, E.; Aronson, J. Ecological Restoration and Global Climate Change. *Restor. Ecol.* **2006**, *14*, 170–176.

Hirsch, A. L.; Pitman, A. J.; Kala, J. The Role of Land Cover Change in Modulating the Soil Moisture–Temperature Land–Atmosphere Coupling Strength over Australia. *Geophys. Res. Lett.* **2014**, *41*, 5883–5890.

Hock, R.; Rasul, G.; Adler, C.; Cáceres, B.; Gruber, S.; Hirabayashi, Y.; Jackson, M.; Kääb, A.; Kang, S.; Kutuzov, S.; Milner, A.; Molau, U.; Morin, S.; Orlove, B.; Steltzer, H. High Mountain Areas. In *IPCC Special Report on the Ocean and Cryosphere in a Changing Climate*; Pörtner, H. O.; Roberts, D. C.; Masson-Delmotte, V.; Zhai, P.; Tignor, M.; Poloczanska, E.; Mintenbeck, K.; Alegría, A.; Nicolai, M.; Okem, A.; Petzold, J.; Rama, B.; Weyer, N. M., Eds.; Working Group II Technical Support Unit, Intergovernmental Panel on Climate Change, 2019.

Holmes, A.; Rüdiger, C.; Mueller, B.; Hirschi, M.; Tapper, N. Variability of Soil Moisture Proxies and Hot Days across the Climate Regimes of Australia. *Geophys. Res. Lett.* **2017**, *44*, 7265–7275.

Holyoak, M.; Heath, S. K. The Integration of Climate Change, Spatial Dynamics, and Habitat Fragmentation: A Conceptual Overview. *Integr. Zool.* **2016**, *11*, 40–59.

Hosseinian Yousefkhani, S.; Nabizadeh, H. Response to Future Climate Change Effect of Two Leopard Geckos' Distribution (Sauria: Eublepharidae) in Iran. *J. Asia-Pac. Biodivers.* **2022**, *15*, 354–358.

Huang, J.; Li, Y.; Fu, C.; Chen, F.; Fu, Q.; Dai, A.; Shinoda, M.; Ma, Z.; Guo, W.; Li, Z.; Zhang, L.; Liu, Y.; Yu, H.; He, Y.; Xie, Y.; Guan, X.; Ji, M.; Lin, L.; Wang, S.; Yan, H.; Wang, G. Dryland Climate Change: Recent Progress and Challenges. *Rev. Geophys.* **2017a**, *55*, 719–778.

Huang, J.; Yu, H.; Dai, A.; Wei, Y.; Kang, L. Drylands Face Potential Threat under 2°C Global Warming Target. *Nat. Clim. Change* **2017**, *7*, 417.

IPCC. *Climate Change 2021: The Physical Science Basis. Contribution of Working Group I to the Sixth Assessment Report of the Intergovernmental Panel on Climate Change*; Cambridge University Press: Cambridge, U.K.; New York, NY, 2021.

Kafaei, S.; Karami, P.; Mehdizadeh, R.; Akmali, V. Relationship between Niche Breadth and Range Shifts of *Rhinopoma muscatellum* (Chiroptera: Rhinopomatidae) in Climate Change Scenarios in Arid and Semiarid Mountainous Region of Iran. *J. Mt. Sci.* **2021**, *18*, 2357–2376.

Kiapasha, K.; Darvishsefat, A. A.; Julien, Y.; Sobrino, J. A.; Zargham, N.; Attarod, P.; Schaepman, M. E. Trends in Phenological Parameters and Relationship between Land Surface Phenology and Climate Data in the Hyrcanian Forests of Iran. *IEEE J. Sel. Top. Appl. Earth Obs. Remote Sens.* **2017**, *10*, 4961–4970.

Kujala, H.; Burgman, M. A.; Moilanen, A. Treatment of Uncertainty in Conservation under Climate Change. *Conserv. Lett.* **2012**, *6*, 73–85.

Lawler, J. J. Climate Change Adaptation Strategies for Resource Management and Conservation Planning. *Ann. N. Y. Acad. Sci.* **2009**, *1162*, 79–98.

Lelieveld, J.; Hadjinicolaou, P.; Kostopoulou, E.; Chenoweth, J.; El Maayar, M.; Giannakopoulos, C.; Hannides, C.; Lange, M. A.; Tanarhte, M.; Tyrlis, E.; Xoplaki, E. Climate Change and Impacts in the Eastern Mediterranean and the Middle East. *Clim. Change* **2012**, *114*, 667–687.

Ludovicy, S.; Noroozi, J.; Semenchuk, P.; Moser, D.; Wessely, J.; Talebi, A.; Dullinger, S. Protected Area Network Insufficiently Represents Climatic Niches of Endemic Plants in a Global Biodiversity Hotspot. *Biol. Conserv.* **2022**, *275*.

Mahdavi, T.; Shams-Esfandabad, B.; Toranjzar, H.; Abdi, N.; Ahmadi, A. Potential Impact of Climate Change on the Distribution of the Eurasian Lynx (*Lynx lynx*) in Iran (Mammalia: Felidae). *Zool. Middle East* **2020**, *66*, 107–117.

Mahmoudi Shamsabad, M.; Assadi, M.; Parducci, L. Impact of Climate Change Implies the Northward Shift in Distribution of the Irano-Turanian Subalpine Species Complex *Acanthophyllum squarrosum*. *J. Asia-Pac. Biodivers.* **2018**, *11*, 566–572.

Marsh, A.; Rico-Ramirez, M.; Matt Jolly, W.; Han, D.; Daham, A. Predicting Vegetation Phenology in Response to Climate Change Using Bioclimatic Indices in Iraq. *J. Water Clim. Change* **2019**, *10*, 835–851.

Melkonyan, A. Climate Change Impact on Water Resources and Crop Production in Armenia. *Agric. Water Manage.* **2015**, *161*, 86–101.

Middleton, N. J.; Sternberg, T. Climate Hazards in Drylands: A Review. *Earth-Sci. Rev.* **2013**, *126*, 48–57.

Mohammadi, A.; Almasieh, K.; Nayeri, D.; Adibi, M. A.; Wan, H. Y. Comparison of Habitat Suitability and Connectivity Modelling for Three Carnivores of Conservation Concern in an Iranian Montane Landscape. *Landscape Ecol.* **2022**, *37*, 411–430.

Mohammadi, A.; Almasieh, K.; Nayeri, D.; Ataei, F.; Khani, A.; Lopez-Bao, J. V.; Penteriani, V.; Cushman, S. A. Identifying Priority Core Habitats and Corridors for Effective Conservation of Brown Bears in Iran. *Sci. Rep.* **2021**, *11*, 1044.

Molotoks, A.; Stehfest, E.; Doelman, J.; Albanito, F.; Fitton, N.; Dawson, T. P.; Smith, P. Global Projections of Future Cropland Expansion to 2050 and Direct Impacts on Biodiversity and Carbon Storage. *Glob. Change Biol.* **2018**, *24*, 5895–5908.

Mountain Research Initiative EDW Working Group. Elevation-Dependent Warming in Mountain Regions of the World. *Nat. Clim. Change* **2015**, *5*, 424–430.

Naboureh, A.; Li, A.; Ebrahimy, H.; Bian, J.; Azadbakht, M.; Amani, M.; Lei, G.; Nan, X. Assessing the Effects of Irrigated Agricultural Expansions on Lake Urmia Using Multi-Decadal Landsat Imagery and a Sample Migration Technique within Google Earth Engine. *Int. J. Appl. Earth Obs. Geoinf.* **2021**, *105*.

Noroozi, J.; Minaei, M.; Khalvati, S.; Kaveh, A.; Nafisi, H.; Nazari, B.; Zare, G.; Vitek, E.; Hülber, K.; Schneeweiss, G. M. Hotspots of (Sub)alpine Plants in the Irano-Anatolian Global Biodiversity Hotspot Are Insufficiently Protected. *Divers. Distrib.* **2022**, *29*, 244–253.

Noroozi, J.; Naqinezhad, A.; Talebi, A.; Doostmohammadi, M.; Plutzar, C.; Rumpf, S. B.; Asgarpour, Z.; Schneeweiss, G. M. Hotspots of Vascular Plant Endemism in a Global Biodiversity Hotspot in Southwest Asia Suffer from Significant Conservation Gaps. *Biol. Conserv.* **2019**, *237*, 299–307.

Nouri, M. Drought Assessment Using Gridded Data Sources in Data-Poor Areas with Different Aridity Conditions. *Water Resour. Manage.* **2023**, *37*, 4327–4343.

Nouri, M. Unveiling Precipitation Trend Characteristics in Changing Poorly Gauged Regions: Leveraging Alternative Raster Sources. *Water Resour. Manage.* **2025**, *39*, 1129–1147.

Nouri, M.; Ebrahimipak, N. A.; Hosseini, S. N. Estimating Reference Evapotranspiration for Water-Limited Windy Areas under Data Scarcity. *Theor. Appl. Climatol.* **2022**, *150*, 593–611.

Nouri, M.; Homaee, M. Contribution of Soil Moisture Variations to High Temperatures over Different Climatic Regimes. *Soil Tillage Res.* **2021a**, *213*, 105115.

Nouri, M.; Homaee, M. Spatiotemporal Changes of Snow Metrics in Mountainous Data-Scarce Areas Using Reanalyses. *J. Hydrol.* **2021b.**

Nouri, M.; Veysi, S. CMIP6 Multi-Model Ensemble Projection of Reference Evapotranspiration Using Machine Learning Algorithms. *Agric. Water Manage.* **2024**, *306*.

Oliver, T. H.; Marshall, H. H.; Morecroft, M. D.; Brereton, T.; Prudhomme, C.; Huntingford, C. Interacting Effects of Climate Change and Habitat Fragmentation on Drought-Sensitive Butterflies. *Nat. Clim. Change* **2015**, *5*, 941–945.

Oliver, T. H.; Morecroft, M. D. Interactions between Climate Change and Land Use Change on Biodiversity: Attribution Problems, Risks, and Opportunities. *WIREs Clim. Change* **2014**, *5*, 317–335.

Peel, M. C.; Finlayson, B. L.; McMahon, T. A. Updated World Map of the Köppen–Geiger Climate Classification. *Hydrol. Earth Syst. Sci. Discuss.* **2007**, *4*, 439–473.

Pepin, N. C.; Arnone, E.; Gobiet, A.; Haslinger, K.; Kotlarski, S.; Notarnicola, C.; Palazzi, E.; Seibert, P.; Serafin, S.; Schöner, W.; Terzago, S.; Thornton, J. M.; Vuille, M.; Adler, C. Climate Changes and Their Elevational Patterns in the Mountains of the World. *Rev. Geophys.* **2022**, *60*.

Pepin, N. C.; Lundquist, J. D. Temperature Trends at High Elevations: Patterns across the Globe. *Geophys. Res. Lett.* **2008**, *35*.

Qin, Y.; Abatzoglou, J. T.; Siebert, S.; Huning, L. S.; AghaKouchak, A.; Mankin, J. S.; Hong, C.; Tong, D.; Davis, S. J.; Mueller, N. D. Agricultural Risks from Changing Snowmelt. *Nat. Clim. Change* **2020**, *10*, 459–465.

Rahimi, J.; Ebrahimpour, M.; Khalili, A. Spatial Changes of Extended De Martonne Climatic Zones Affected by Climate Change in Iran. *Theor. Appl. Climatol.* **2013**, *112*, 409–418.

Ramanathan, V.; Carmichael, G. Global and Regional Climate Changes Due to Black Carbon. *Nat. Geosci.* **2008**, *1*, 221–227.

Rangwala, I.; Miller, J. R. Climate Change in Mountains: A Review of Elevation-Dependent Warming and Its Possible Causes. *Clim. Change* **2012**, *114*, 527–547.

Renton, M.; Shackelford, N.; Standish, R. J. Habitat Restoration Will Help Some Functional Plant Types Persist under Climate Change in Fragmented Landscapes. *Glob. Change Biol.* **2012**, *18*, 2057–2070.

Rincón, V.; Velázquez, J.; Gülçin, D.; López-Sánchez, A.; Jiménez, C.; Özcan, A. U.; López-Almansa, J. C.; Santamaría, T.; Sánchez-Mata, D.; Çiçek, K. Mapping Priority Areas for Connectivity of Yellow-Winged Darter (*Sympetrum flaveolum*, Linnaeus 1758) under Climate Change. *Land* **2023**, *12*.

Saemian, P.; Elmi, O.; Vishwakarma, B. D.; Tourian, M. J.; Sneeuw, N. Analyzing the Lake Urmia Restoration Progress Using Ground-Based and Spaceborne Observations. *Sci. Total Environ.* **2020**, *739*, 139857.

Schleussner, C. F.; Rogelj, J.; Schaeffer, M.; Lissner, T.; Licker, R.; Fischer, E. M.; Knutti, R.; Levermann, A.; Frieler, K.; Hare, W. Science and Policy Characteristics of the Paris Agreement Temperature Goal. *Nat. Clim. Change* **2016**, *6*, 827–835.

Şekercioğlu, Ç. H.; Anderson, S.; Akçay, E.; Bilgin, R.; Can, Ö. E.; Semiz, G.; Tavşanoğlu, Ç.; Yokeş, M. B.; Soyumert, A.; İpekdal, K.; Sağlam, İ. K.; Yücel, M.; Nüzhet Dalfes, H. Turkey's Globally Important Biodiversity in Crisis. *Biol. Conserv.* **2011**, *144*, 2752–2769.

Seneviratne, S. I.; Corti, T.; Davin, E. L.; Hirschi, M.; Jaeger, E. B.; Lehner, I.; Orlowsky, B.; Teuling, A. J. Investigating Soil Moisture–Climate Interactions in a Changing Climate: A Review. *Earth-Sci. Rev.* **2010**, *99*, 125–161.

Seneviratne, S. I.; Lüthi, D.; Litschi, M.; Schär, C. Land–Atmosphere Coupling and Climate Change in Europe. *Nature* **2006**, *443*, 205–209.

Shahsavarzadeh, R.; Hemami, M. R.; Farhadinia, M. S.; Fakheran, S.; Ahmadi, M. Spatially Heterogeneous Habitat Use across Distinct Biogeographic Regions in a Wide-Ranging Predator, the Persian Leopard. *Biodivers. Conserv.* **2023**.

Tabari, H.; Hosseinzadeh Talaee, P.; Ezani, A.; Shifteh Some'e, B. Shift Changes and Monotonic Trends in Autocorrelated Temperature Series over Iran. *Theor. Appl. Climatol.* **2011**, *109*, 95–108.

Tabor, K.; Williams, J. W. Globally Downscaled Climate Projections for Assessing the Conservation Impacts of Climate Change. *Ecol. Appl.* **2010**, *20*, 554–565.

Toros, H. Spatio-Temporal Variation of Daily Extreme Temperatures over Turkey. *Int. J. Climatol.* **2012**, *32*, 1047–1055.

Türkeş, M.; Türp, M. T.; An, N.; Ozturk, T.; Kurnaz, M. L. Impacts of Climate Change on Precipitation Climatology and Variability in Turkey. In *Water Resources of Turkey*; Harmancioglu, N. B.; Altinbilek, D., Eds.; Springer International Publishing: Cham, 2020; pp 467–491.

Vaghefi, S. A.; Keykhai, M.; Jahanbakhshi, F.; Sheikholeslami, J.; Ahmadi, A.; Yang, H.; Abbaspour, K. C. The Future of Extreme Climate in Iran. *Sci. Rep.* **2019**, *9*, 1464.

Vaissi, S. Response of Iranian Lizards to Future Climate Change by Poleward Expansion, Southern Contraction, and Elevation Shifts. *Sci. Rep.* **2022**, *12*, 2348.

Waha, K.; Krummenauer, L.; Adams, S.; Aich, V.; Baarsch, F.; Coumou, D.; Fader, M.; Hoff, H.; Jobbins, G.; Marcus, R.; Mengel, M.; Otto, I. M.; Perrette, M.; Rocha, M.; Robinson, A.; Schleussner, C. F. Climate Change Impacts in the Middle East and Northern Africa (MENA) Region and Their Implications for Vulnerable Population Groups. *Reg. Environ. Change* **2017**, *17*, 1623–1638.

Walther, G. R.; Post, E.; Convey, P.; Menzel, A.; Parmesan, C.; Beebee, T. J. C.; Fromentin, J. M.; Hoegh-Guldberg, O.; Bairlein, F. Ecological Responses to Recent Climate Change. *Nature* **2002**, *416*, 389–395.

Willett, W.; Rockström, J.; Loken, B.; Springmann, M.; Lang, T.; Vermeulen, S.; Garnett, T.; Tilman, D.; DeClerck, F.; Wood, A.; Jonell, M.; Clark, M.; Gordon, L. J.; Fanzo, J.; Hawkes, C.; Zurayk, R.; Rivera, J. A.; De Vries, W.; Majele Sibanda, L.; Afshin, A.; Chaudhary, A.; Herrero, M.; Agustina, R.; Branca, F.; Lartey, A.; Fan, S.; Crona, B.; Fox, E.; Bignet, V.; Troell, M.; Lindahl, T.; Singh, S.; Cornell, S. E.; Srinath Reddy, K.; Narain, S.; Nishtar, S.; Murray, C. J. L. Food in the Anthropocene: The EAT–Lancet Commission on Healthy Diets from Sustainable Food Systems. *Lancet* **2019**, *393*, 447–492.

You, Q.; Chen, D.; Wu, F.; Pepin, N.; Cai, Z.; Ahrens, B.; Jiang, Z.; Wu, Z.; Kang, S.; AghaKouchak, A. Elevation-Dependent Warming over the Tibetan Plateau: Patterns, Mechanisms and Perspectives. *Earth-Sci. Rev.* **2020**, *210*.

Zeinoddini, M.; Bakhtiari, A.; Ehteshami, M. Long-Term Impacts from Damming and Water Level Manipulation on Flow and Salinity Regimes in Lake Urmia, Iran. *Water Environ. J.* **2015**, *29*, 71–87.

INDEX

Q

R

For Product Safety Concerns and Information please contact our
EU representative GPSR@taylorandfrancis.com Taylor & Francis
Verlag GmbH, Kaufingerstraße 24, 80331 München, Germany